FRANKLIN BOOK COMPANY, INC.
7804 MONTGOMERY AVENUE
ELKINS PARK, PA 19027 U.S.A.

215-635-5252
TELEFAX 215-635-6155

Printed in 1995

Handbook
of
Marine Science

Handbook
of
Marine Science

Volume II

EDITORS

F. G. Walton Smith, Ph.D., A.R.C.S.

Frederick A. Kalber, Ph.D.

COORDINATING EDITOR

Norman A. Alldridge, Ph.D.

Published by

CRC PRESS, Inc.
18901 Cranwood Parkway · Cleveland, Ohio 44128

HANDBOOK OF MARINE SCIENCE

VOLUME II

International Standard Book Number (ISBN)
Complete Set 0-87891-388-X
Volume II 0-87819-390-1

Library of Congress Catalog Card Number 73-88624

ADVISORY BOARD AND CONTRIBUTORS

Norman A. Alldridge, Ph.D.*
Associate Professor
Biology Department
Case Western Reserve University
2040 Adelbert Road
Cleveland, Ohio 44106

George D. Grice, Ph.D.
Zoologist, Senior Scientist
Woods Hole Oceanographic Institution
Woods Hole, Massachusetts 02543

Frederick A. Kalber, Ph.D.
President
Hydrobiological Services, Inc.
P.O. Box 2327
Naranja, Florida 33030

F. G. Walton Smith, Ph.D., A.R.C.S.
Dean Emeritus, School of Marine and
    Atmospheric Science
University of Miami
President, International Oceanographic
    Foundation
10 Rickenbacker Causeway
Miami, Florida 33149

*CRC Press gratefully acknowledges the services of Dr. Alldridge as coordinating editor for this volume.

Narragansett Marine Laboratory Occasional Publications

Nature

Oceanography and Marine Biology An Annual Review (United Kingdom)

Oceanology

Pacific Science

Pergamon Press

Prentice-Hall, Inc.

President's Science Advisory Council, U.S. Government Printing Office

Proceedings of the Pacific Science Congress 1966, Tokyo Scientific Council

Progress in Oceanography (United Kingdom)

Publications of the California State Water Quality Control Board

Publications of the Carnegie Institute of Washington

Publications of the Institute of Marine Science, University of Texas

Publications of Michigan State University Museum, Biological Series

Rapport(s) et Procès – Verbaux des Reunions Conseil Permanent Internationale pour l'Exploration de la Mer (Denmark)

Research Memoirs, The Sears Foundation for Marine Research

Revista Sociedad Mexicana de Historia Natural (Mexico)

Rit Fiskideildar (Reykjavik)

Smithsonian Miscellaneous Collections

Systematic Zoology

Statistical Digest, U.S. Department of the Interior

U.S. Department of Commerce, National Marine Fisheries Service Special Scientific Report

U.S. National Marine Fisheries, Fishery Bulletin

University of California, Publications in Zoology

Virginia Journal of Science

Zoologica

Deep-Sea Research and Oceanographic Abstracts (United Kingdom)

Discovery Reports (United Kingdom)

Ecological Monographs

Elsevier Publishing Company

Fauna Caribaea

Fisheries Research Board of Canada Bulletin

Fisheries Research Board of Canada Progress Reports

Fishery Investigations (United Kingdom)

Food and Agriculture Organization of the United Nations, Department of Fisheries

Gulf Research Reports

Hydrobiologia (Netherlands)

Interscience

Journal of the American Fisheries Society

Journal du Conseil, Conseil International pour l'Exploration de la Mer (Denmark)

Journal of the Elisha Mitchell Scientific Society

Journal of the Fisheries Research Board of Canada

Journal of the Marine Biological Association of India

Journal of the Marine Biological Association of the United Kingdom

Journal of the British Museum of Natural History, Economic Series

Journal of Marine Research

Journal of Zoology (United Kingdom)

Limnology and Oceanography

Marine Biology

Marine Research (Scotland)

Memoirs of the Faculty of Fisheries, Kagoshima University (Japan)

Memoirs of the Geological Society of America

Memoirs of the Kobe Marine Observatory

PREFACE

Oceanography is essentially interdisciplinary in its nature. Rather than the elucidation of basic principles of chemistry, physics, biology, or other scientific disciplines, it is the application of these principles towards an understanding of the nature of the ocean environment. The ocean is a complicated system of energy pathways, both within it and across its boundaries, air-sea, ocean-land, seafloor, and internal boundaries of biological membranes of living organisms in the sea. Associated with the energy pathways are the chemical transformations which take place.

In view of the foregoing, the oceanographer is more likely than most scientists to be involved in more than one discipline. This is the justification for preparing a special handbook of constants, tables, and properties covering all of the major disciplines as they relate to the ocean within one or two volumes.

Quite clearly, since there exist a number of handbooks for the basic disciplines and the CRC *Handbook of Chemistry and Physics* alone is a very bulky volume today in its 55th edition, the task of assembling a handbook for oceanographers becomes largely a question of elimination rather than compilation. Here the guideline has been to identify those tables that are most frequently used by the oceanographer. There will be many occasions that he will need to consult handbooks of chemistry and physics and the like, but it is hoped the present compilation will be of especial value.

Volume I covers chemical oceanography, physical oceanography, atmospheric science (relation between ocean and atmosphere), marine geology, ocean engineering, and a number of tables of conversions and constants most used by oceanographers.

The second and third volumes cover various facets of marine productivity. The data are divided into sections on the basis of trophic level, habitat and economic or biological productivity.

Volume II contains sections on primary productivity, phytoplankton, zooplankton, and marine fisheries. Data within the sections are grouped on the basis of geography, ocean depth, chemical parameters, or taxonomic catagory.

The intent is to provide investigators who have specific questions about marine ecosystems with an easy access to closely grouped tables and figures pertinent to their investigations. While no single organization of all the materials will best fit the needs of all users it is hoped that this one will prove to be of value to many.

ACKNOWLEDGEMENTS

Acknowledgement is made to the publishers, publications, and associations listed below, whose material has been used in this Handbook. We appreciate their assistance and cooperation.

Allan Hancock Monographs in Marine Biology

American Association for the Advancement of Science

American Geographical Society, Atlas of the Marine Environment

American Society of Geologists

American Naturalist

Annual Review of Microbiology

Annual Reviews in Oceanography and Marine Biology

Antarctic Research

Arctic Institute of North America (Canada)

Australian Journal of Marine and Freshwater Research

Biological Bulletin, Wood's Hole

Biological Reviews (United Kingdom)

Boletin del Instituto Espanol de Oceanografia (Spain)

Bulletin of the Bernice P. Bishop Museum

Bulletin of the Bingham Oceanographic Collection, Yale University

Bulletin of the Inter-American Tropical Tuna Commission

Bulletin of Marine Science of the Gulf and Caribbean

Bulletin of the Raffles Museum

Bulletin of the Scripps Institute of Oceanography of the University of California, San Diego

Bulletin of the U.S. National Museum

California Cooperative Oceanic Fisheries Investigations Atlas

Canadian Journal of Botany

Ciencia (Mexico)

Cushman Foundation for Foraminiferal Research Special Publication

TABLE OF CONTENTS

Section 1

Primary Productivity

1.1 Annual, By Regions

Table 1.1−1
POSITIONS, DATES, AND VALUES FOR DAILY PRIMARY ORGANIC PRODUCTION

Grams of Carbon Fixed Beneath a Square Meter of Sea Surface in the Pacific

Latitude	Longitude	Date	gC m^{-2} day^{-1}	Latitude	Longitude	Date	gC m^{-2} day^{-1}
		1963				1963	
17°37′ N	70°27′ E	23 May	0.09	11°56′ N	60°53′ E	13 Aug	0.52
15°40′ N	70°07′ E	24 May	0.09	10°12′ N	60°04′ E	14 Aug	0.49
13°50′ N	70°07′ E	25 May	0.03	07°13′ N	59°57′ E	16 Aug	0.52
11°59′ N	69°55′ E	26 May	0.06	04°05′ N	59°58′ E	17 Aug	0.88
09°46′ N	70°06′ E	27 May	0.07	01°16′ N	60°08′ E	19 Aug	0.16
08°09′ N	70°02′ E	28 May	0.12	02°00′ S	59°59′ E	20 Aug	0.31
05°48′ N	70°03′ E	29 May	0.04	05°04′ S	60°03′ E	22 Aug	0.28
03°33′ N	69°54′ E	30 May	0.09	07°21′ S	59°44′ E	24 Aug	0.15
01°30′ N	70°01′ E	31 May	0.04	11°39′ S	58°02′ E	26 Aug	0.46
				22°58′ S	59°45′ E	4 Sept	0.05
01°07′ S	71°00′ E	1 June	0.06	25°55′ S	60°01′ E	5 Sept	0.09
02°23′ S	70°24′ E	2 June	0.10	29°24′ S	60°05′ E	7 Sept	0.27
04°22′ S	69°24′ E	3 June	0.02	31°58′ S	59°51′ E	8 Sept	0.41
06°48′ S	70°07′ E	4 June	0.08	34°57′ S	60°05′ E	9 Sept	0.43
08°35′ S	69°55′ E	6 June	0.09	38°22′ S	59°51′ E	11 Sept	0.25
11°10′ S	70°02′ E	7 June	0.28	40°54′ S	60°01′ E	12 Sept	0.09
13°15′ S	69°51′ E	8 June	0.10				
15°25′ S	69°58′ E	9 June	0.27			1963	
17°18′ S	70°05′ E	10 June	0.06				
19°30′ S	69°51′ E	11 June	0.04	19°14′ S	56°33′ E	25 Sept	0.04
21°40′ S	67°06′ E	25 June	0.04	17°38′ S	54°58′ E	26 Sept	0.13
23°47′ S	69°05′ E	26 June	0.04	14°53′ S	55°02′ E	27 Sept	0.85
26°34′ S	70°12′ E	27 June	0.02	11°34′ S	54°56′ E	29 Sept	0.49
30°34′ S	69°55′ E	29 June	0.06	08°14′ S	55°00′ E	30 Sept	0.69
32°52′ S	69°52′ E	30 June	0.06				
				00°24′ S	54°33′ E	5 Oct	1.17
35°09′ S	69°59′ E	1 July	0.14	02°45′ N	53°51′ E	6 Oct	0.69
37°12′ S	70°10′ E	2 July	0.09	05°52′ N	52°56′ E	7 Oct	0.55
30°11′ S	79°42′ E	5 July	0.31	08°57′ N	52°17′ E	8 Oct	0.28
27°31′ S	80°08′ E	6 July	0.03	12°04′ N	51°31′ E	9 Oct	1.98
20°02′ S	79°50′ E	8 July	0.06	13°11′ N	51°28′ E	15 Oct	0.81
14°15′ S	79°52′ E	9 July	0.29	14°44′ N	51°02′ E	15 Oct	0.77
14°44′ S	79°44′ E	10 July	0.15	15°27′ N	52°50′ E	16 Oct	1.64
11°28′ S	80°00′ E	11 July	0.28	16°27′ N	54°39′ E	17 Oct	1.03
08°37′ S	79°34′ E	12 July	0.12	17°26′ N	56°29′ E	17 Oct	2.38
05°53′ S	79°57′ E	13 July	0.10	16°29′ N	57°09′ E	18 Oct	1.23
03°13′ S	80°02′ E	14 July	0.11	15°18′ N	57°43′ E	18 Oct	2.14
00°33′ S	80°08′ E	15 July	0.12	14°21′ N	58°18′ E	19 Oct	0.95
01°54′ N	79°52′ E	16 July	0.09	13°12′ N	58°58′ E	19 Oct	1.40
04°18′ N	80°08′ E	17 July	0.27	12°15′ N	59°42′ E	20 Oct	0.45

Table 1.1–1 (*Continued*)
POSITIONS, DATES, AND VALUES FOR DAILY PRIMARY ORGANIC PRODUCTION

Latitude	Longitude	Date	gC m^{-2} day^{-1}	Latitude	Longitude	Date	gC m^{-2} day^{-1}
		1963				**1964**	
14°09′ N	61°07′ E	21 Oct	0.68	37°01′ S	75°19′ E	9 Apr	0.20
15°58′ N	62°33′ E	22 Oct	0.59	34°31′ S	74°47′ E	10 Apr	0.09
23°43′ N	66°21′ E	28 Oct	2.20	31°26′ S	74°57′ E	11 Apr	0.04
22°33′ N	65°56′ E	29 Oct	0.84	29°00′ S	74°51′ E	12 Apr	0.02
20°39′ N	64°41′ E	30 Oct	1.00	26°27′ S	75°02′ E	13 Apr	0.04
21°31′ N	64°06′ E	30 Oct	2.01	24°20′ S	74°52′ E	14 Apr	0.17
22°23′ N	63°32′ E	31 Oct	1.37	21°58′ S	74°55′ E	15 Apr	0.05
23°19′ N	62°50′ E	31 Oct	3.30	19°44′ S	75°20′ E	16 Apr	0.03
				16°43′ S	74°53′ E	17 Apr	0.15
24°00′ N	62°04′ E	1 Nov	1.99	14°10′ S	74°55′ E	18 Apr	0.06
24°48′ N	61°37′ E	1 Nov	1.61	11°47′ S	74°42′ E	19 Apr	0.09
23°57′ N	60°58′ E	1 Nov	2.17	09°21′ S	75°08′ E	20 Apr	0.08
23°08′ N	60°32′ E	2 Nov	1.62	06°50′ S	75°02′ E	21 Apr	0.18
22°48′ N	59°34′ E	2 Nov	1.76	04°11′ S	75°00′ E	22 Apr	0.14
22°22′ N	60°05′ E	3 Nov	6.58	02°14′ S	75°14′ E	23 Apr	0.06
21°31′ N	60°41′ E	3 Nov	5.79	01°04′ N	75°07′ E	28 Apr	0.08
20°44′ N	61°15′ E	4 Nov	2.09	03°52′ N	74°57′ E	29 Apr	0.14
20°02′ N	62°00′ E	4 Nov	1.80	06°51′ N	72°02′ E	30 Apr	0.14
19°17′ N	62°29′ E	4 Nov	1.74				
18°31′ N	63°08′ E	5 Nov	1.10			**1964**	
18°32′ N	64°39′ E	6 Nov	1.21				
				18°02′ N	65°08′ E	17 May	0.17
				15°36′ N	64°59′ E	19 May	0.17
				13°36′ N	65°03′ E	20 May	0.08
		1964		11°28′ N	65°04′ E	21 May	0.09
				10°04′ N	64°59′ E	22 May	0.14
16°13′ N	63°29′ E	29 Jan	0.54	07°55′ N	64°56′ E	23 May	0.09
15°42′ N	60°52′ E	30 Jan	0.75	06°01′ N	64°59′ E	24 May	0.15
15°22′ N	58°12′ E	31 Jan	1.09	03°59′ N	65°02′ E	25 May	0.22
				02°01′ N	65°03′ E	26 May	0.10
14°22′ N	54°18′ E	1 Feb	0.72	00°30′ S	65°07′ E	28 May	0.30
13°50′ N	52°59′ E	2 Feb	0.72	02°38′ S	65°01′ E	29 May	0.27
13°11′ N	50°22′ E	3 Feb	0.84	04°40′ S	65°02′ E	30 May	0.24
09°28′ N	54°52′ E	5 Feb	0.31	06°00′ S	65°10′ E	31 May	0.36
07°10′ N	55°05′ E	6 Feb	0.30				
05°02′ N	55°01′ E	7 Feb	0.33	08°00′ S	64°59′ E	1 June	0.41
02°31′ N	55°04′ E	8 Feb	0.44	09°58′ S	64°55′ E	2 June	0.28
00°31′ S	54°56′ E	10 Feb	0.14	12°12′ S	65°29′ E	4 June	0.14
02°51′ S	54°58′ E	11 Feb	0.47	14°11′ S	65°17′ E	5 June	0.18
06°28′ S	55°12′ E	16 Feb	0.13	16°10′ S	64°50′ E	6 June	0.05
08°42′ S	55°07′ E	17 Feb	0.15	19°23′ S	65°30′ E	8 June	0.12
10°47′ S	55°15′ E	18 Feb	0.21	22°06′ S	64°55′ E	23 June	0.07
12°33′ S	54°33′ E	19 Feb	0.10	24°01′ S	65°00′ E	24 June	0.07
14°57′ S	54°43′ E	20 Feb	0.07	26°06′ S	64°58′ E	25 June	0.05
17°14′ S	54°38′ E	21 Feb	0.09	28°28′ S	65°03′ E	27 June	0.08
				30°06′ S	64°58′ E	28 June	0.03
19°57′ S	54°58′ E	3 Mar	0.06	34°34′ S	64°55′ E	30 June	0.00
23°07′ S	54°50′ E	4 Mar	0.11				
26°00′ S	54°52′ E	5 Mar	0.11	37°58′ S	64°59′ E	2 July	0.14
28°22′ S	55°02′ E	6 Mar	0.07	40°57′ S	64°27′ E	4 July	0.09
30°50′ S	55°02′ E	7 Mar	0.11	29°38′ S	49°23′ E	11 July	0.14
33°13′ S	55°10′ E	8 Mar	0.19	29°11′ S	31°37′ E	29 July	0.40
35°42′ S	55°15′ E	9 Mar	0.28	29°12′ S	32°06′ E	30 July	0.38
				28°35′ S	32°40′ E	31 July	0.37
40°04′ S	75°00′ E	4 Apr	0.11	27°38′ S	33°24′ E	31 July	0.41
42°23′ S	74°54′ E	5 Apr	0.24				

Table 1.1–1 (*Continued*)
POSITIONS, DATES, AND VALUES FOR DAILY PRIMARY ORGANIC PRODUCTION

Latitude	Longitude	Date	gC m^{-2} day^{-1}	Latitude	Longitude	Date	gC m^{-2} day^{-1}
		1964				1964	
26° 35′ S	35° 57′ E	1 Aug	0.11	29° 29′ S	31° 44′ E	9 Sept	1.37
24° 53′ S	39° 18′ E	3 Aug	0.21	29° 18′ S	31° 33′ E	9 Sept	3.14
23° 46′ S	43° 07′ E	4 Aug	0.47	29° 32′ S	31° 18′ E	25 Sept	1.85
23° 20′ S	43° 36′ E	12 Aug	1.44	29° 26′ S	31° 33′ E	25 Sept	0.01
23° 19.6′ S	43° 33.3′ E	12 Aug	0.37	29° 29′ S	32° 04′ E	26 Sept	0.27
23° 13′ S	43° 13′ E	12 Aug	0.73	25° 34′ S	33° 19′ E	28 Sept	0.66
22° 34′ S	41° 16′ E	13 Aug	0.20	25° 10′ S	33° 15′ E	28 Sept	2.17
23° 04′ S	38° 35′ E	16 Aug	0.43	25° 17′ S	34° 04′ E	29 Sept	1.57
23° 48′ S	36° 48′ E	17 Aug	1.00				
24° 19′ S	35° 46′ E	17 Aug	1.21	22° 30′ S	36° 07′ E	1 Oct	0.28
24° 42′ S	35° 23′ E	18 Aug	0.69	21° 11′ S	36° 23′ E	2 Oct	0.42
24° 48′ S	34° 59′ E	19 Aug	1.44	20° 42′ S	35° 50′ E	3 Oct	3.18
26° 01′ S	33° 04′ E	22 Aug	3.05	20° 14′ S	35° 16′ E	4 Oct	0.89
26° 57′ S	33° 53′ E	23 Aug	0.39	19° 10′ S	36° 19′ E	9 Oct	0.50
27° 58′ S	35° 16′ E	24 Aug	0.00	18° 52′ S	37° 41′ E	9 Oct	0.43
29° 22′ S	37° 31′ E	26 Aug	0.24	18° 33′ S	39° 48′ E	10 Oct	0.23
30° 09′ S	38° 39′ E	27 Aug	0.22	18° 04′ S	41° 52′ E	12 Oct	0.62
30° 51′ S	40° 11′ E	28 Aug	0.36	17° 41′ S	42° 31′ E	13 Oct	0.43
32° 22′ S	42° 55′ E	29 Aug	0.43	16° 46′ S	43° 45′ E	15 Oct	0.89
32° 58′ S	43° 37′ E	30 Aug	0.45	16° 13′ S	43° 41′ E	17 Oct	1.38
33° 13′ S	42° 53′ E	30 Aug	0.22	15° 25′ S	44° 22′ E	20 Oct	0.34
34° 08′ S	41° 15′ E	31 Aug	0.51	14° 24′ S	46° 08′ E	21 Oct	0.27
				12° 56′ S	46° 43′ E	29 Oct	1.25
34° 57′ S	38° 49′ E	1 Sept	0.40	12° 36′ S	45° 55′ E	29 Oct	0.77
35° 44′ S	36° 47′ E	2 Sept	0.87	10° 34′ S	44° 23′ E	31 Oct	0.65
34° 15′ S	36° 04′ E	3 Sept	0.31				
32° 55′ S	35° 21′ E	4 Sept	0.71	08° 45′ S	43° 39′ E	1 Nov	0.67
31° 57′ S	34° 18′ E	5 Sept	0.35	07° 03′ S	42° 34′ E	2 Nov	0.79
30° 45′ S	32° 58′ E	6 Sept	0.99	05° 10′ S	41° 40′ E	3 Nov	0.40
30° 10′ S	32° 09′ E	6 Sept	0.57	04° 17′ S	41° 10′ E	5 Nov	0.55
29° 45′ S	31° 42′ E	8 Sept	1.64	03° 07′ S	40° 39′ E	5 Nov	2.38

(From Ryther, J. H., Hall, J. R., Pease, A. K., Bakun, A., and Jones, M. M., Primary organic production in relation to the chemistry and hydrography of the western Indian Ocean, *Limnol. Oceanogr.*, 11(3), 372, 1966. With permission.)

Table 1.1–2
ANNUAL RATE OF CARBON FIXATION
g/sq. m of sea surface

	Total depth (m)	gC/m^2/yr
Long Island Sound	25	380
Continental Shelf	25–50	160
	50–1,000	135
	1,000–2,000	100
North Central Sargasso Sea	>5,000	78

(From Ryther, J. H. and Yentsch, C. S., Primary production of continental shelf waters off New York, *Limnol. Oceanogr.*, 3(3), 334, 1958. With permission.)

Figure 1.1—1

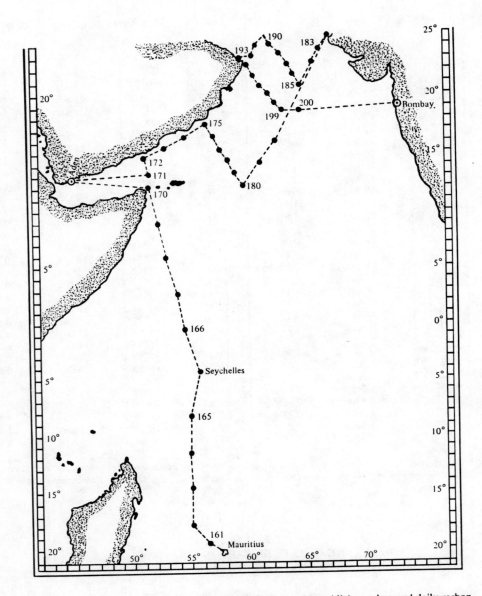

Figure 1.1—1. Cruise course in western Indian Ocean with total living carbon and daily carbon production (A) and vertical distribution of temperature, phosphate, and integrated primary productivity (B) (C) along numbered transects.

(From Ryther, J. H. and Menzel, D. W., on the produciton, composition, and distribution of organic matter in the western Arabian Sea, *Deep-Sea Res.*, 12, 200, © 1965, Pergamon Press. With permission.)

Figure 1.1–1 (*Continued*)

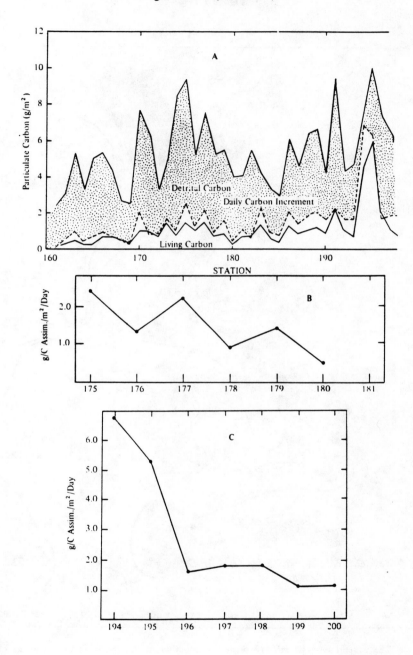

Figure 1.1–2

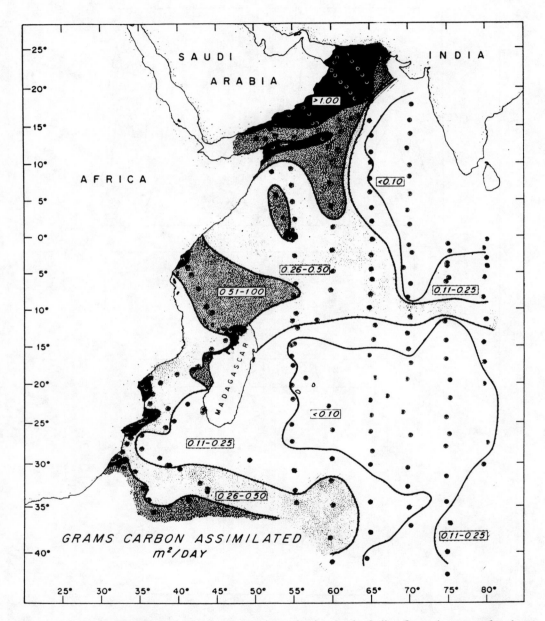

Figure 1.1–2. The general level of primary organic production in the Indian Ocean in grams of carbon assimilated per square meter per day.

(From Ryther, J. H., Hall, J. R., Pease, A. K., Bakun, A., and Jones, M. M., Primary organic production in relation to the chemistry and hydrography of the western Indian Ocean, *Limnol. Oceanogr.*, 11(3), 375, 1966. With permission.)

Figure 1.1–3

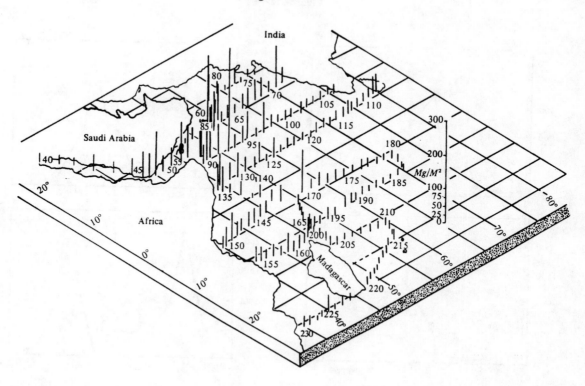

Figure 1.1–3. The distribution of total chloroplastic pigment (mg/m²) in the western Indian Ocean during the southwest Monsoon Period, July 30 to November 12, 1963.

Figure 1.1—4

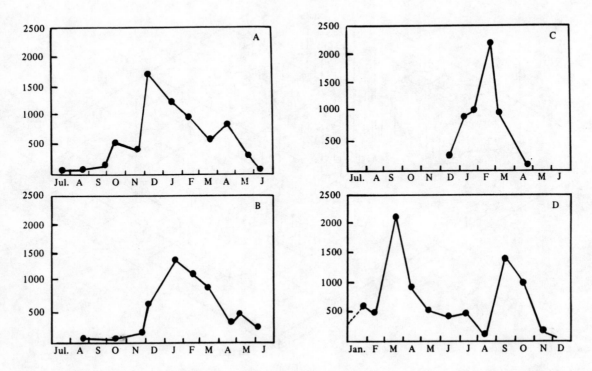

Figure 1.1—4.   Plant pigments (Harvey units/m³) in the northern (A), intermediate (B), and southern (C) regions of the Antarctic Ocean, and in the English Channel (D).

(From Hart, T. J., Phytoplankton periodicity in Antarctic surface waters, Discovery Rep., 21, 261—356, 1942. With permission.)

Table 1.1–3
CHLOROPHYLL–α AND NUTRIENT SALT CONCENTRATION OF PLANKTON COLORED LAYER AT THE BOTTOM OF SEA ICE OFF WAINWRIGHT, ALASKA

(70°55′ N., 159°12′ E.). 27–29 July, 1964

Chlorophyll-a µg/l.	Chloride °/oo	Phosphate-P P mg./l.	Silicate-Si SiO$_2$ mg./l.
180.0	2.80	0.075	1.5
221.0	1.60	0.07	2.1
91.5	0.62	0.025	0.5
93.6	1.40	0.07	0.8
251.0	2.60	0.01	2.1
427.0	–	–	–
41.0	1.10	0.01	0.7
36.4	0.25	0.01	0.4
33.0	0.59	0.01	0.6
72.0	0.59	0.02	1.0
62.7	0.64	0.01	0.8
118.0	0.12	0.02	0.8
47.0	1.80	0.01	0.4
10.2	0.55	0.01	0.4
120.3	1.13	0.027	0.93

Table 1.1–4
MEAN % NITRATE UPTAKE FOR ATLANTIC AND PACIFIC OCEANS TO 1965

Area	Dates	Mean % NO$_3$ uptake
NE Pacific coast, Seattle–Juneau	Sept. 1964	20.3 (5)*
NE Pacific coast, Juneau–Cape Spencer	Nov. 1964	27.6 (3)
N Pacific, Vancouver–Honolulu	Feb. 1965	8.7 (8)
NW Atlantic, Georges Bank–Caribbean	Mar. 1962	28.8 (8)
Bermuda	Sept. 1962 (to Jan. 1963)	8.3 (7)
NW Atlantic, Gulf of Maine	Apr. 1963	39.5 (2)

*(x) = number of samples.

(From Dugdale, R. C. and Goering, J. J., Uptake of new and regenerated forms of nitrogen in primary productivity, *Limnol. Oceanogr.*, 12, 204, 1967. With permission.)

Table 1.1—5
DIVISION OF THE OCEAN INTO PROVINCES
ACCORDING TO THEIR LEVEL OF PRIMARY ORGANIC
PRODUCTION

Province	% Ocean	Area (km²)	Mean productivity (g C/m²/yr)	Total productivity (10³ tons of C/yr)
Open ocean	90.0	326.0 × 10⁶	50	16.3
Coastal zone*	9.9	36.0 × 10⁶	100	3.6
Upwelling areas	0.1	3.6 × 10⁵	300	0.1
Total				20.0

*Indicates offshore areas of high productivity.

Table 1.1—6
ESTIMATES OF ANNUAL PRIMARY PRODUCTION
IN AREAS OFF THE WASHINGTON AND OREGON COASTS
1961

Area	Annual production (g C m⁻² yr⁻¹)	Range (g C m⁻² yr⁻¹)	Mean daily production (g C m⁻² day⁻¹)
Oceanic	61	43—78	0.17
Plume	60	46—73	0.16
River mouth	88		0.24
Upwelling	152		0.42

(From Anderson, G. C., The seasonal and geographic distribution of primary productivity off the Washington and Oregon coasts, *Limnol. Oceanogr.*, 9, 298, 1964. With permission.)

Figure 1.1–5

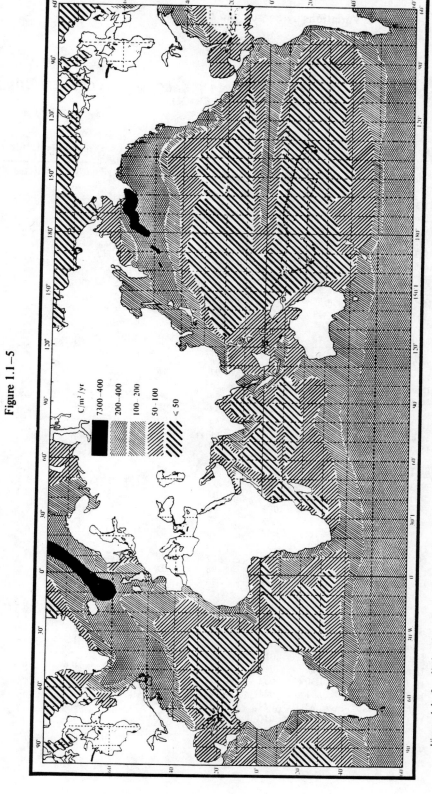

C/m²/yr	
≥300–400	
200–400	
100–200	
50–100	
≤ 50	

Figure 1.1–5. Estimation of organic production in the oceans. Contours are in grams of carbon biologically fixed under each square meter of sea surface per year.

1.2 Variations by Season and Depth

Table 1.2—1
PHYSICAL, CHEMICAL AND BIOLOGICAL DATA 0—300 M, IN THE SARGASSO SEA OFF BERMUDA AT SIX TIMES DURING A 24-HR PERIOD

April 27—28, 1960

	Depth (m)	T (°C)	Sal. (°/oo)	σ_t	NO$_2$-N μg A/L	NO$_3$-N μg A/L	PO$_4$-P μg A/L	Chlα μg/L	Carbon/assim/m³/day C¹⁴	Chl-Rad.
Time: 1330	0	21.30	36.69	25.70	.010	.09	.08	.04	4.1	0.9
	15	20.36	36.62	25.92	.008	.29	.07	.05	3.7	1.6
	25	20.28	36.61	25.94	.008	.14	.05	.06	2.1	1.7
	50	19.58	36.60	26.11	.016	.07		.09	0.6	1.3
	75	19.25	36.59	26.19	.021	.17	.05	.31	—	—
	100	18.82	36.58	26.29	.126	.10	.05	.54	0.4	0.6
	125	18.76	36.58	26.31	.024	.34	.09	.16	—	—
	150	18.72	36.58	26.32	.010	.40	.11	—	—	—
	200	18.54	36.57	26.36	.040	.31	.10	—	—	—
	300	18.03	36.49	.43	.016	.80	.18	—	—	—
Time: 1600	0	21.06	36.64	25.74	.003	.27	.10	.13	—	3.0
	15	20.31	36.62	25.93	.010	.02	.07	—	—	—
	25	20.23	36.62	25.95	.021	.08	.06	—	—	—
	50	19.52	36.60	26.13	.008	.07	.07	.12	—	1.8
	75	19.37	36.58	26.15	.008	.16	—	.22	—	—
	100	18.89	36.58	26.28	.024	.04	.06	.81	—	0.9
	125	18.76	36.60	26.32	.042	.34	.06	.10	—	—
	150	18.70	36.59	26.33	.018	.45	.09	.07	—	—
	200	18.60	36.58	26.35	.029	.60	.12	—	—	—
	300	18.07	36.49	26.42	.005	1.73	.21	—	—	—
Time: 2100	0	21.45	36.65	25.64	.021	.22	.07	.05	5.8	1.2
	15	20.39	36.61	25.90	.008	.09	.07	.05	4.2	1.6
	25	20.31	36.60	25.92	.008	.08	.06	.08	2.8	2.3
	50	19.74	36.60	26.06	.005	.08	.09	.08	0.6	1.2
	75	19.51	36.60	26.12	.003	.17	.07	.09	—	—
	100	19.15	36.57	26.20	.008	.16	.07	.40	0.5	0.5
	125	18.76	36.57	26.30	.137	.20	.10	.65	—	—
	150	18.69	36.57	26.32	.024	.54	.13	.12	—	—
	200	18.53	36.56	26.36	.029	.98	.10	—	—	—
	300	18.00	36.46	26.42	.021	.84	.25	—	—	—
Time: 0030	0	21.54	36.62	25.59	.008	.08	.11	.04	4.3	0.9
	15	20.39	36.60	25.89	.003	.12	.07	.05	3.4	1.6
	25	19.87	36.59	26.02	.016	.25	.07	.05	2.8	1.4
	50	19.40	36.59	26.15	.003	.12	.07	—	1.9	—
	75	19.01	36.58	26.24	.013	.09	.06	.24	—	—
	100	18.84	36.58	26.29	.047	.28	.09	.54	0.6	0.6
	125	18.77	36.57	26.30	.032	.22	.09	.31	—	—
	150	18.72	36.57	26.31	.024	.27	.14	.06	—	—
	200	18.63	36.57	26.33	.024	.94	.09	—	—	—
	300	18.15	36.49	26.40	.010	2.18	.11	—	—	—

Table 1.2–1 (*Continued*)
PHYSICAL, CHEMICAL AND BIOLOGICAL DATA 0–300 M, IN THE SARGASSO SEA OFF BERMUDA AT SIX TIMES DURING A 24-HR PERIOD

	Depth (m)	T (°C)	Sal. (°/oo)	σ_t	NO_2-N µg A/L	NO_3-N µg A/L	PO_4-P µg A/L	Chlα µg/L	Carbon/assim/m³/day C^{14}	Chl-Rad.
Time: 0400	0	21.00	36.62	25.74	.003	.09	.13	.04	–	0.9
	15	20.43	36.61	25.89	.003	.15	.09	.05	–	1.6
	25	20.22	36.63	25.96	.008	.07	.04	.04	–	1.1
	50	19.27	36.58	26.18	.003	.09	.09	.10	–	1.5
	75	19.00	36.58	26.25	.008	.11	.06	.41	–	–
	100	18.84	36.58	26.29	.108	.30	.07	.45	–	0.5
	125	18.75	36.57	26.30	.029	.33	.10	.20	–	–
	150	18.75	36.57	26.30	.023	.84	.14	–	–	–
	200	18.65	36.57	26.34	.010	1.55	.11	–	–	–
	300	18.40	36.53	26.36	.013	1.15	.13	–	–	–
Time: 0800	0	20.98	36.61	25.74	.021	.12	.07	.07	5.0	1.7
	15	20.74	36.61	25.80	.008	.20	.06	.06	4.3	1.9
	25	20.02	36.60	25.99	.018	.30	.05	.06	1.5	1.7
	50	19.34	36.58	26.16	.010	.10	–	.12	0.9	0.8
	75	18.89	36.58	26.28	.042	.24	.06	.19	–	–
	100	18.78	36.58	26.31	.016	.20	–	.27	0.9	0.3
	125	18.72	36.57	26.31	.029	.87	.09	.05	–	–
	150	18.69	36.57	26.32	.008	.94	.10	.08	–	–
	200	18.60?	36.56	26.34	.028	.33	.17	–	–	–
	300	18.35?	36.53	26.37	.018	.53	.11	–	–	–

(From Ryther, J. H., Menzel, D. W., and Vaccaro, R. F., Diurnal variations in some chemical and biological properties of the Sargasso Sea, *Limnol. Oceanogr.*, 6(2), 150, 1961. With permission.)

Table 1.2–2

DISTRIBUTION OF THE PLANKTONIC BIOMASS AT A 24-HR STATION IN THE SOUTHWESTERN PART OF THE BERING SEA*

Sampled layer, m	$0^{00}-1^{05}$	$3^{55}-4^{30}$	$7^{55}-9^{07}$	$12^{01}-13^{05}$	$16^{00}-17^{00}$	$20^{00}-21^{05}$
0–10	23.9	18.2	10.2	8.4	17.9	14.8
10–25	36.1	23.2	28.8	13.5	11.9	41.2
25–50	22.5	32.6	22.9	31.5	37.8	24.2
50–100	6.1	15.8	26.8	37.5	19.2	5.0
100–200	3.2	3.1	2.9	2.5	4.1	4.7
200–500	8.2	7.1	8.4	6.6	9.5	10.1

* % mg/m³.

Table 1.2–3

DISTRIBUTION OF THE PLANKTONIC BIOMASS AT A DIURNAL STATION IN THE KURILE-KAMCHATKA REGION[a]

Sampled layer, m	$19^{40}-0^{50}$	$2^{00}-5^{45}$	$9^{00}-13^{40}$ [b]
0–10	8.1	4.8	0.4
10–25	29.3	25.0	11.6
25–50	23.0	17.2	50.2
50–125	12.8	15.8	9.5
125–200	11.3	17.1	9.6
200–400	9.9	11.7	10.3
400–750	2.7	4.8	5.8
750–1250	2.0	2.0	2.1
1250–1500	1.0	1.6	1.3

[a] % mg/m³.
[b] Sampling started in the lower layers and terminated in the upper layers. Therefore, the data on the amount of plankton in the upper 200 m layer apply to the last 20 min of the indicated period of time.

Table 1.2–4
INITIAL CONTROL (IPC) AND PARTICULATE CARBON (μgm C/L) REMAINING AFTER 90-DAY DECOMPOSITION EXPERIMENTS (FPC) AND ESTIMATES OF LIVING (Δ) AND DETRITAL CARBON (FPC)

IPC	FPC	Δ	Initial chlorophyll (μg/l)	Δ chlorophyll	Detrital C (%)
63	30	33	0.378	87	48
37	31	6	0.083	72	84
44	27	17	0.206	82	61
34	22	12	0.130	92	65
55	33	23	0.270	85	60
86	56	30	0.340	88	65
74	52	22	0.253	87	70
67	29	38	0.349	109	43
79	38	41	0.405	101	48
34	21	13	0.146	90	62
45	28	17	0.226	75	62
40	21	19	–	–	49
74	53	21	0.208	101	72
56	42	14	–	–	75

* All samples from 1 m 5–20°N lat., 50–60°W long.

Table 1.2–5
VERTICAL DISTRIBUTION OF CHLOROPLAST FLUORESCENCE AND EXTRACTED CHLOROPHYLL

Long. 64°41'W, Lat. 13°28'N
December 3, 1964

Depth (m)	In vivo chlorophyll micro fluorescence*	Extract chlorophyll (mg/m³)
1	15	0.10
25	25	0.15
50	38	0.21
100	16	0.20
600	1	0.01
1,000	2	0.002

* Total count of micro fluorescence per one sweep across slide.

Figure 1.2—1

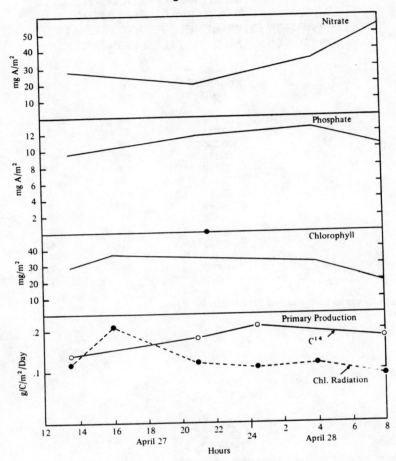

Figure 1.2.—1. Diurnal variations in nitrate, phosphate, chlorophyll and primary production in the middle of the Sargasso Sea.

(From Ryther, J. H., Menzel, D. W., and Vaccaro, R. F., Diurnal variations in some chemical and biological properties of the Sargasso Sea, *Limnol. Oceanogr.*, 6(2), 152, 1961. With permission.)

Figure 1.2–2

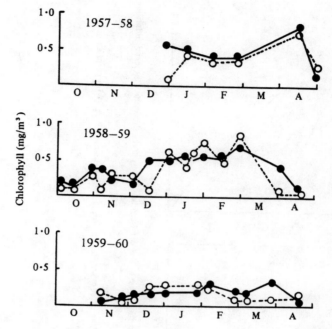

Figure 1.2–2. Comparison of observed (–○–) and theoretical (–●–) chlorophyll–*a* concentrations during three winters in the Sargasso Sea off Bermuda.

(From Steele, J. H. and Menzel, D. W., Conditions for maximum primary production in the mixed layer, *Deep-Sea Res.*, 9, 48, © 1962 Pergamon Press. With permission.)

Figure 1.2–3

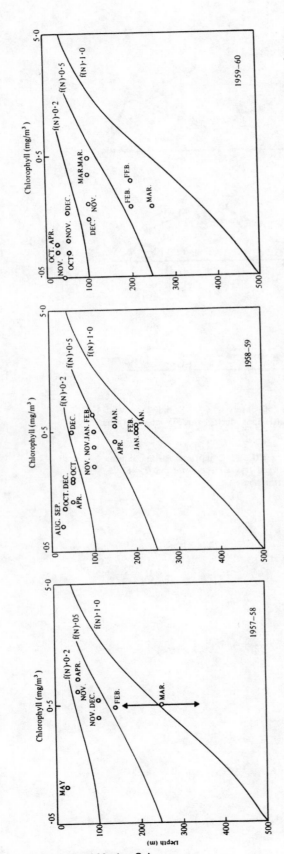

Figure 1.2–3. Relations between mixed layer depth and chlorophyll-*a* during three winters in the Sargasso Sea off Bermuda. The curves show the theoretical relations for different degrees of nutrient limitation.

(From Steele, J. H. and Menzel, D. W., Conditions for maximum primary production in the mixed layer, *Deep-Sea Res.*, 9, 46, © 1962 Pergamon Press. With permission.)

Figure 1.2—4

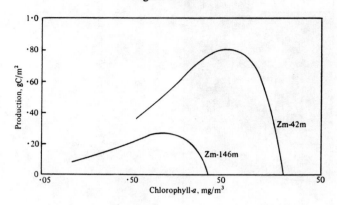

Figure 1.2—4. The relationship between net production and chlorophyll-*a* concentration for two depths of mixed layer (Zm) off Bermuda.

(From Steele, J. H. and Menzel, D. W., Conditions for maximum primary production in the mixed layer, *Deep-Sea Res.,* 9, 45, © 1962 Pergamon Press. With permission.)

Table 1.2—6

PHOTOSYNTHESIS (C[14] UPTAKE)

For 24 Hr at 1,500 ft. candles of Water
from Different Depths
December 18, 1958, off Bermuda

Depth (m)	Mg c assim m^3/24 hr
0	3.05
17	2.10
34	2.20
60	2.90
120	0.47
150	0.00
200	0.00
250	0.00
300	0.00
350	0.00
400	0.00

(From Menzel, D. W. and Ryther, J. H., The annual cycle of primary production in the Sargasso Sea off Bermuda, *Deep-Sea Res.,* 6, 362, © 1960 Pergamon Press. With permission.)

Figure 1.2–5

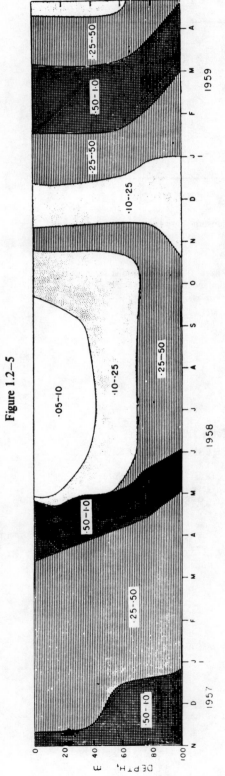

Figure 1.2–5.   A three-year seasonal depth profile of chlorophyll-*a* (mg/m³) off Bermuda.

(From Ryther, J. H. and Menzel, D. W., The annual cycle of primary production in the Sargasso Sea off Bermuda, *Deep-Sea Res.*, 6, 360, © 1960 Pergamon Press. With permission.)

Figure 1.2–6

Chlorophyll, g/mg/m³

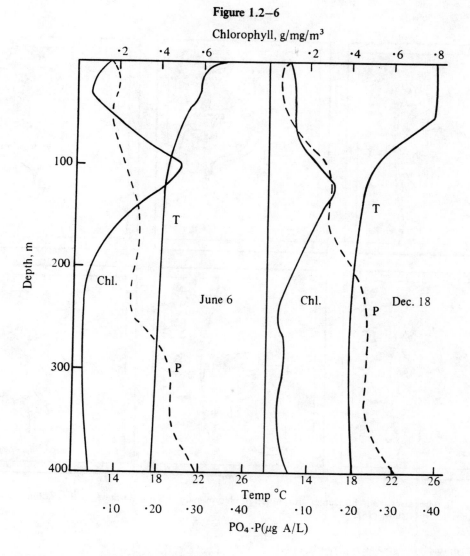

Figure 1.2–6. The vertical distribution of chlorophyll-*a* (Chl), temperature (T), and phosphate-phosphorus (P) off Bermuda, 1958.

(From Menzel, D. W., and Ryther, J. H., The annual cycle of primary production in the Sargasso Sea off Bermuda, *Deep-Sea Res.*, 6, 361, © 1960 Pergamon Press. With permission.)

Figure 1.2–7

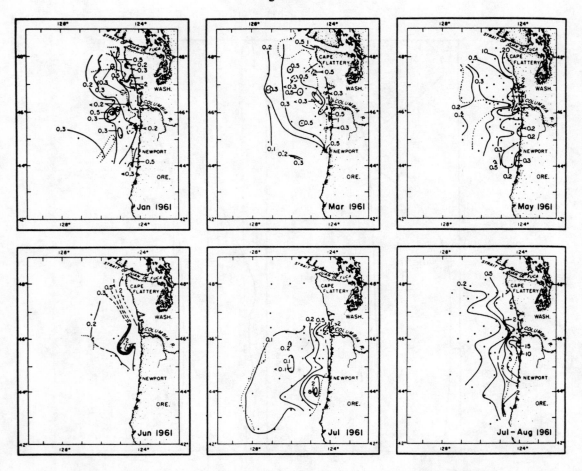

Figure 1.2–7. The seasonal and horizontal distribution of chlorophyll-*a* in surface waters off the Washington and Oregon coasts. January, 1961–June, 1962.

Note: The dotted line represents the boundary of the Columbia River Plume.

Figure 1.2–7 (*Continued*)

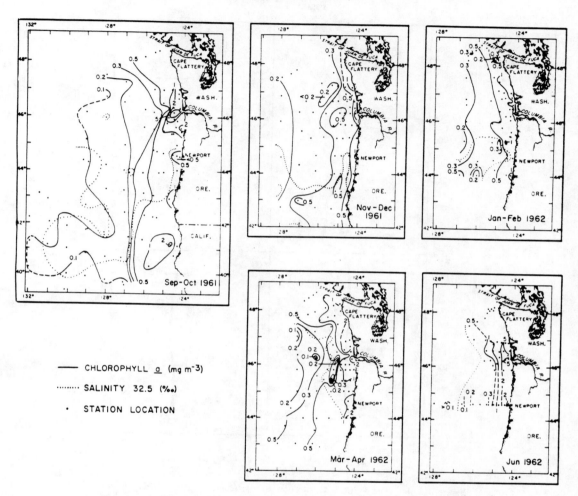

(From Anderson, G. C., The seasonal and geographic distribution of primary productivity off the Washington and Oregon coasts, *Limnol. Oceanogr.*, 9, 288, 1964. With permission.)

Figure 1.2−8

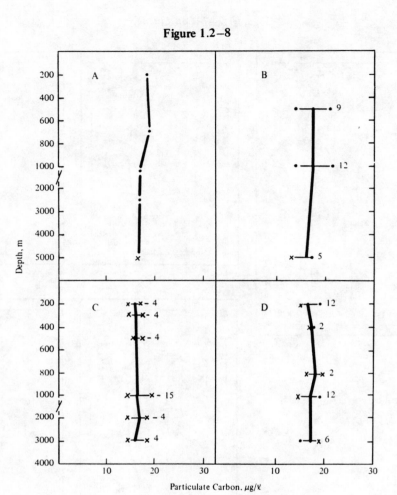

Figure 1.2−8. The vertical distribution of particulate organic carbon in the central North Atlantic Ocean. Numbers at each depth indicate the numbers of samples the range (x) and mean (•) are based upon. A. Caribbean Sea. 13°35'N lat., 67° 04'W long. November 1964. B. Tropical Atlantic Ocean. 5−20°N lat., 50−60° W long. October−November 1964. C. Subtropical Atlantic Ocean. 22−36°N lat., 65°W long. April 1962. D. Tropical Atlantic Ocean. 5−20°N lat., 50−60°W long. May−June 1965.

Figure 1.2—9

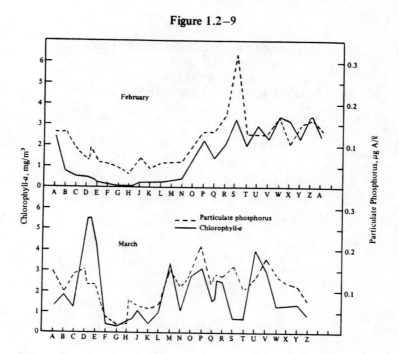

Figure 1.2—9. Quantities of particulate phosphorus and of chlorophyll-*a* in the euphotic zone in the western Atlantic off Massachusetts, February and March, 1957.

Table 1.2—7
PRIMARY PRODUCTION ON THE FLADEN GROUND (GM CARBON/M²)

Year	Spring	Summer	Autumn	Total
1951	28.0	8.9	28.0	64.9
1952	30.4	22.4	29.5	82.3
1953	26.1	17.6	13.7	57.4

(From Steele, J. H., Plant production on the Fladen Ground, *J. Mar. Biol. Assoc. U.K.*, 35, 1—33, 1956. With permission.)

Table 1.2—8
PRIMARY PRODUCTION RATES FOR
WATER COLUMN 0—100 M AT
NOON STATIONS IN
THE EASTERN TROPICAL PACIFIC

Stn.	Prod.
6	5.2
8	7.1
13	3.2
15	1.8
23	11.5
32	5.0
34	9.5
49	28.3
56	11.0
58	11.8
62	31.0
72	8.0
74	5.0
76	74.0
79	10.8
86	3.2
100	14.8
122	23.4
139*	15.0

* In mgC/m^2/hr.; obtained by the C_{14} method, samples incubated at 1000 ft-candles.

Figure 1.2−10

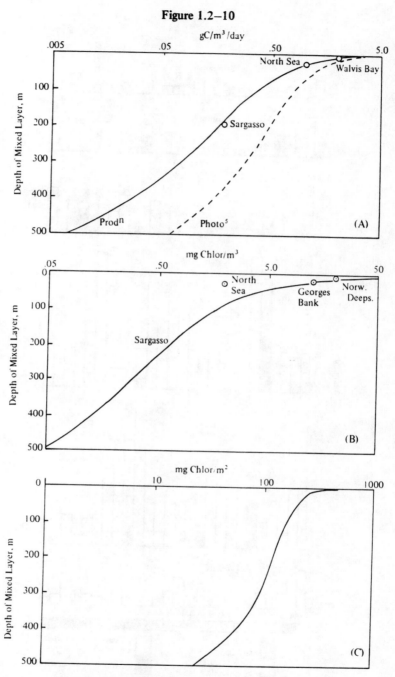

Figure 1.2−10. Relationships between the depth of the mixed layer (A), net production and photosynthetic rates below each square meter of sea surface (B), and chlorophyll-*a* concentration per cubic meter (C), at indicated points in the Atlantic Ocean.

(From Steele, J. H. and Menzel, D. W., Conditions for maximum primary production in the mixed layer, *Deep-Sea Res.*, 9, 44, 1962. With permission.)

Figure 1.2–11

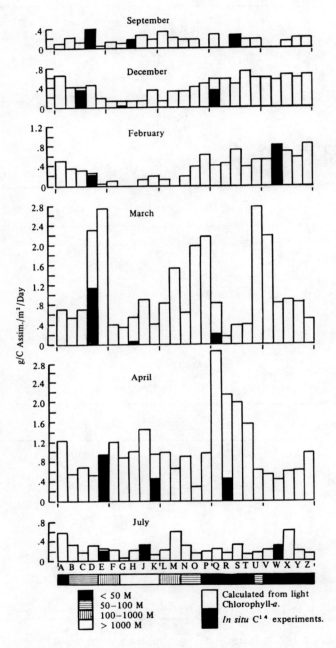

Figure 1.2–11. *In situ* primary production measured by ^{14}C method and/or calculated from chlorophyll, radiation, and light penetration off New Jersey.

(From Ryther, J. H. and Yentsch, C. S., Primary production of continental shelf waters off New York, *Limnol. Oceanogr.*, 3(3), 331, 1958. With permission.)

Figure 1.2—12

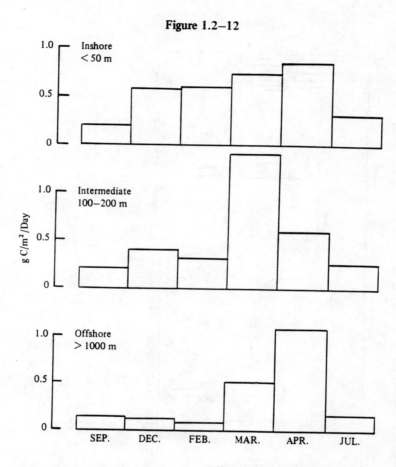

Figure 1.2—12.   Mean daily primary production beneath a square meter of sea surface at five shallow, five intermediate, and five deep stations on the continental shelf off New York.

(From Ryther, J. H. and Yentsch, C. S., Primary production of continental shelf waters off New York, *Limnol. Oceanogr.*, 3(3), 333, 1958. With permission.)

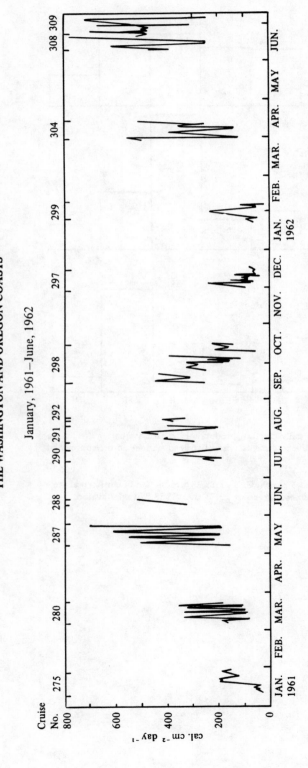

Table 1.2–9

SEASONAL VARIATIONS IN THE AMOUNT OF SOLAR RADIATION OFF THE WASHINGTON AND OREGON COASTS

January, 1961—June, 1962

Figure 1.2–13

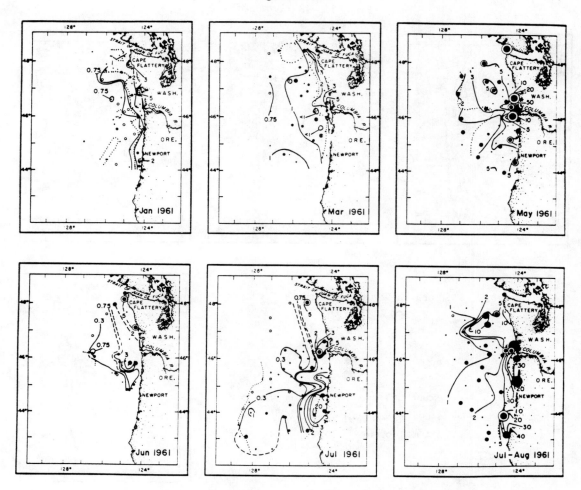

Figure 1.2–13. The seasonal and horizontal distribution of primary productivity in surface waters off the Washington and Oregon coasts. January, 1961–June, 1962.

Figure 1.2–13 (*Continued*)

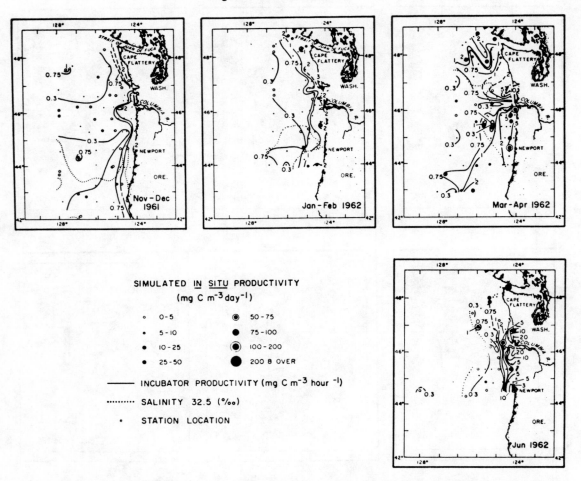

(From Anderson, G. C., The seasonal and geographic distribution of primary productivity off the Washington and Oregon coasts, *Limnol. Oceanogr.*, 9, 290, 1964. With permission.)

Figure 1.2–14

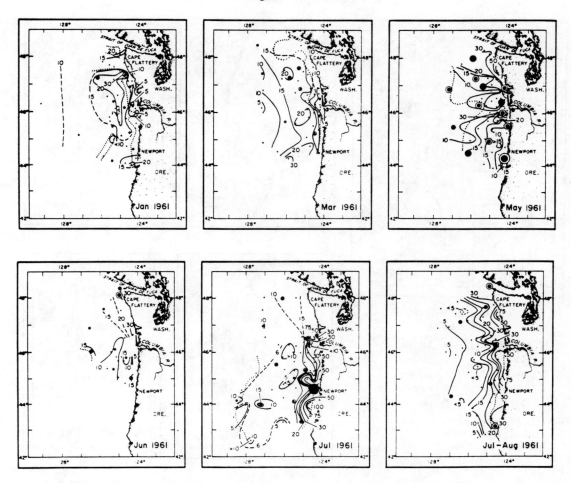

Figure 1.2–14.   The seasonal and horizontal distributions of chlorophyll-*a* and primary productivity in the euphotic zone off the Washington and Oregon coasts, January, 1961–June, 1962.

Note: The dotted line represents the boundary of the Columbia River Plume.

Figure 1.2–14 (Continued)

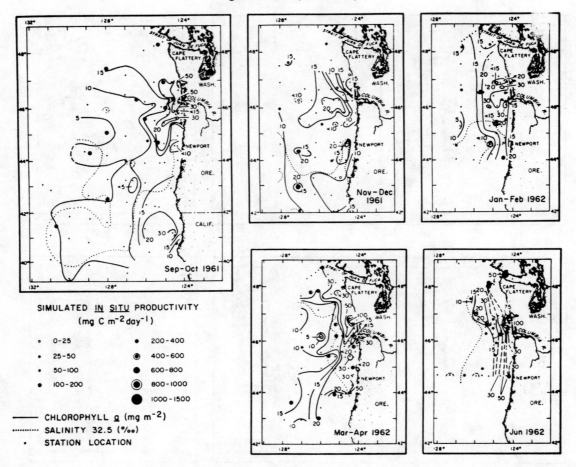

SIMULATED IN SITU PRODUCTIVITY
(mg C m⁻² day⁻¹)

· 0–25	• 200–400
· 25–50	◉ 400–600
· 50–100	● 600–800
· 100–200	⊙ 800–1000
	● 1000–1500

—— CHLOROPHYLL a (mg m⁻²)
········· SALINITY 32.5 (‰)
· STATION LOCATION

(From Anderson, G. C., The seasonal and geographic distribution of primary productivity off the Washington and Oregon coasts, *Limnol. Oceanogr.*, 9, 292, 1964. With permission.)

Figure 1.2–15

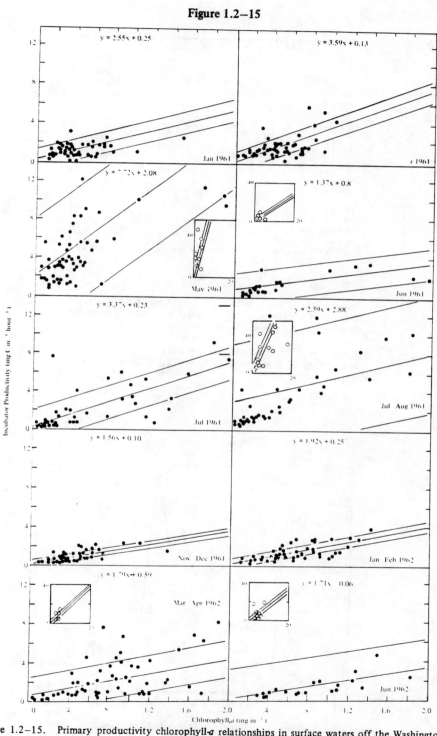

Figure 1.2–15.   Primary productivity chlorophyll-*a* relationships in surface waters off the Washington-Oregon coasts*. January, 1961–June, 1962.

*The insets contain off-scale values.

(From Anderson, G. C., The seasonal and geographic distribution of primary productivity off the Washington and Oregon coasts, *Limnol. Oceanogr.,* 9, 299, 1964. With permission.)

Figure 1.2—16

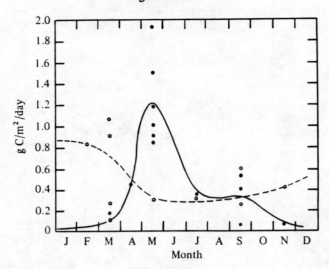

Figure 1.2—16. Primary production as determined by *in situ* [14]C measurements at offshore stations (solid line, filled circles) and shallow stations (broken line, open circles) off New York.

(From Ryther, J. H., Geographic variations in productivity, *The Sea*, Vol. 2, Hill, M. N., Ed., Interscience, New York, 1963, 362. With permission.)

Figure 1.2—17

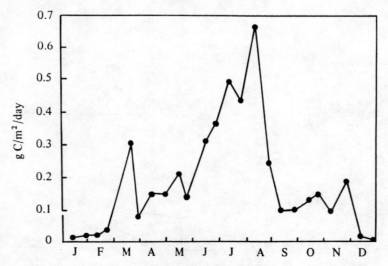

Figure 1.2—17. Primary production in the Great Belt off Denmark, 1955.

(From Steemann Nielsen, E., A survey of recent Danish measurements of the organic productivity in the sea, *Rapp. Cons. Explor. Mer.*, 144, 92—95, 1958. With permission.)

Figure 1.2—18

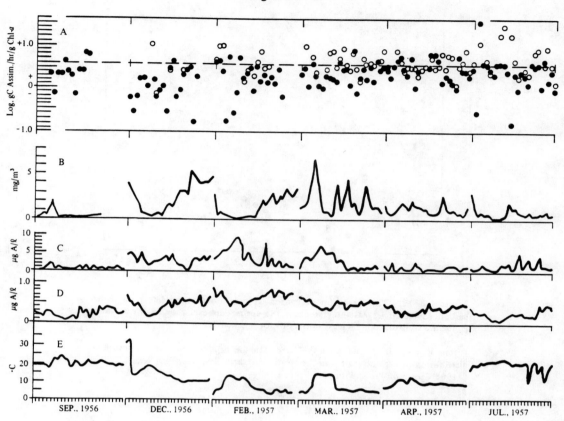

Figure 1.2—18.   Biological, physical and chemical properties of water from 10 meters depth on the continental shelf off New York. (A) Log ratio of photosynthesis/hr at optimum light: chlorophyll-*a* (Open circles are oxygen method; closed circles ^{14}C method). (B) Chlorophyll-*a* concentration. (C) Available nitrogen ($NH^+ + NO_2 + NO_3^-$). (D) Phosphate. (E) Temperature.

(From Ryther, J. H. and Yentsch, C. S., Primary production of continental shelf waters off New York, *Limnol. Oceanogr.*, 3(3), 330, 1958. With permission.)

Figure 1.2–19

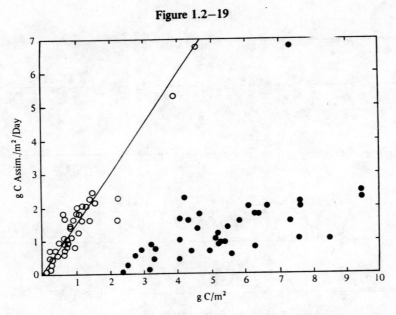

Figure 1.2–19. The regression of carbon assimilation (natural light) on living carbon (open circles) as integrated for the entire euphotic zone in the western Arabian Sea. Broken line is least squares fit for open circles.

(From Ryther, J. H. and Menzel, D. W., On the production, composition, and distribution of organic matter in the western Arabian Sea, *Deep-Sea Res.*, 12, 207, © 1965 Pergamon Press. With permission.)

Figure 1.2–20

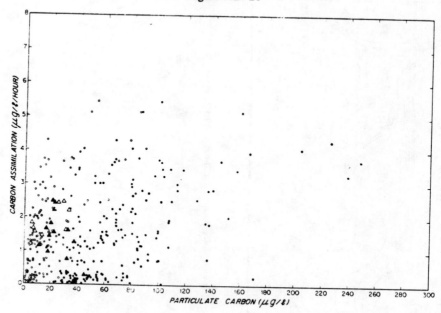

Figure 1.2–20. The regression of carbon assimilation (at 1000 foot candles) on living carbon (open circles and triangles) and on total particulate carbon (filled circles and triangles) in the Indian Ocean.

(From Ryther, J. H. and Menzel, D. W., On the production, composition, and distribution of organic matter in the western Arabian Sea, *Deep-Sea Res.*, 12, 206, © 1965 Pergamon Press. With permission.)

Figure 1.2–21

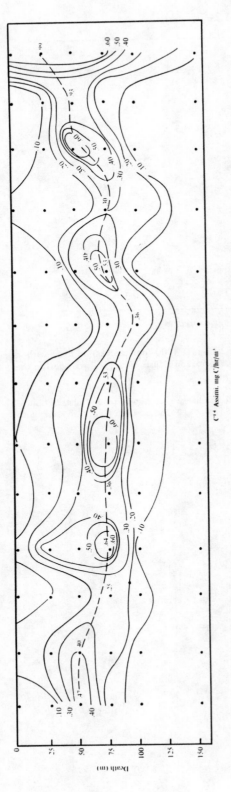

C[14] Assim. mg C/hr/m[3]

Figure 1.2–21.   Profile of ^{14}C assimilation along a section from North Cape, New Zealand, to Cape Howe, Australia, in the East Australian current.

Figure 1.2–22

Figure 1.2–23

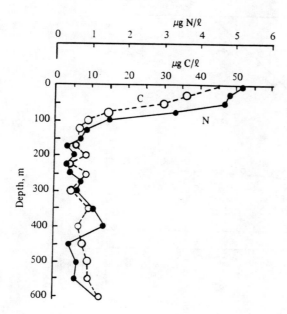

Figure 1.2–22. Profiles of total particulate organic carbon and nitrogen off southern California.

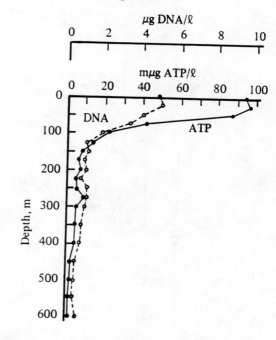

Figure 1.2–23. Profiles of particulate ATP and DNA off southern California.

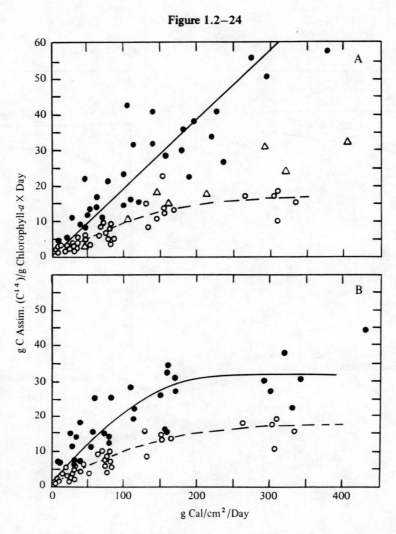

Figure 1.2−24.   (A) The ratio of ^{14}C uptake/chlorophyll for different values of daily incident radiation during summer (filled circles), spring (triangles), and winter (open circles) for 1958−59. (B) The same for the winter of 1958−59 (open circles) and the winter of 1959−60 (filled circles) off Barrow, Alaska.

(From Menzel, D. W. and Ryther, J. H., Annual variations in primary production of the Sargasso Sea off Bermuda, *Deep-Sea Res.*, 7, 287, © 1961 Pergamon Press. With permission.)

Table 1.2–10

AVERAGE CONTENT OF CHLOROPHYLL-*a* AND NUTRIENTS IN VARIOUS SECTIONS OF SEA ICE AND SEA WATER

Collected off Barrow, Alaska, 1964

	Chlorophyll-*a* μg/l	Chloride °/oo	Phosphorous P μg/l		Silicon Si mg/l	
			Phosphate-P	Diatom-P	Dissolved silicate-Si	Diatom-Silicate
Pool water at the surface of sea ice	0.3	0.79	2.0	0.5	0.12	0.006–0.009
Middle section of sea ice	trace	0.92	9.3	–	0.07	–
Plankton-colored layer at the bottom of sea ice	120.3	1.13	27.0	204.0	0.44	2.4–3.7
Sea water around sea ice just after melting	3.1	4.71	10.0	5.3	0.05	0.06–0.09
Sea water in open lead	2.1	13.3	17.0	3.6	0.14	0.04–0.06

1.3 Dissolved Organic Composition Related to Productivity

Table 1.3—1
DISSOLVED ORGANIC MATTER IN SEA-WATER

Locality	mg C/l	mg N/l	Ref.
Atlantic Ocean (Bermuda)	2.35 ± 0.07 (along vertical)	0.244 ± 0.08 (same)	1
Black Sea	2.4	—	2
Black Sea	2.83—3.36 (seasonal variations)	—	3
Sea of Azov	4.63—6.02	—	4
Baltic	2.0—4.6 (maximum in euphotic zone)	—	5
Pacific (3 stations)	0.6—2.7	—	6
North Atlantic	1.04—1.97	—	
Atlantic Ocean	2.40—2.48	0.24—0.26	7
Pacific	0.98—2.68	0.07—0.11	
Greenland Sea	2.0—2.1	0.03—0.38	
North Atlantic	0.2—1.3	0.04—0.40	8
Norwegian Sea	0.45—1.38	0.10—0.21	
North Sea	0.5—1.8	0.08—0.54	
Wadden Sea	1.0—8.0	0.10—0.60	

(From Provasoli, L., Organic regulation of phytoplankton fertility, *The Sea,* Vol. 2, Hill, M. N., Ed., Interscience, New York, 1963, 172. With permission.)

REFERENCES

1. Krogh, A., *Ecol. Monogr.,* 4, 421, 1934.
2. Dazko, V. G., *Dokl. Akad. Nauk S.S.S.R.,* 24, 294, 1939.
3. Dazko, V. G., *Dokl. Akad. Nauk S.S.S.R.,* 77, 1059, 1951.
4. Dazko, V. G., *Akad. Nauk S.S.S.R., Hydrotech. Inst. Novocherkask, Hydrochem.,* Mat., 23, 1, 1955.
5. Kay, H., *Kiel. Meeresforsch.,* 10, 26, 1954.
6. Plunkett, M. A. and Rakestraw, N. W., *Deep-Sea Res.,* 3(suppl.), 12, 1955.
7. Skopintsev, B. A., *Preprints Intern. Oceanogr. Cong. A.A.A.S.,* p. 953, 1959.
8. Duursma, E. K., *Neth. J. Mar. Res.,* 1, 1, 1960.

Table 1.3–2
ORGANIC COMPOUNDS IDENTIFIED IN SEA-WATER

Substances	Quantities	Locality	Method	Ref.
Rhamnoside Dehydroascorbic acid	Up to 0.1 g/l present	Inshore waters, Gulf of Mexico	Activated charcoal absorption, ethanol elution	1
Carbohydrates– arabinose equivalents	0.0–20 mg/l	Estuary, Gulf of Mexico	N-Ethyl carbazole	2
Carbohydrates– sucrose equivalents	0.14–0.45 mg/l	Pacific Coast, U.S.A.	Anthrone and N-ethyl carbazole	3
Carbohydrates– arabinose equivalents	0.0–2.6 mg/l (max. of 12 mg/l at surface, 29° N, 80°, 31′ W)	South Atlantic (30° N–25° N)	N-Ethyl carbazole	4
	0.0–3.0 mg/l (23% = 0.0; 50% = 0.2–1 mg/l)	Continental Shelf, Gulf of Mexico (50 mg/l in red tides of *G. breve*)	N-Ethyl carbazole	5
Citric acid	0.025–0.145 mg/l	Littoral Atlantic French coast		6
Malic acid Acetic and formic acids[a]	0.028–0.277 mg/l <0.1 mg/l	Northeast Pacific, surface and inshore	Chloroform or ether extraction at pH 3; partition chromatography on silica gel column	7
Fatty acids (up to 20 carbons)	0.4–0.5 mg/l (weight of methyl esters)	Gulf of Mexico	Ethyl acetate extraction at pH 2; Gas-liquid chromatography	8
Amino acids – hydrolyzed proteins	Traces to 13 mg/ m³ [b]	Gulf of Mexico, Yucatan Strait Reef (British Honduras), Caribbean	Coprecipitation of organic material with $FeCl_3$ + NaOH; acid hydrolysis; paper and ion-exchange chromatography	9

[a]Acetic, formic, lactic, and glycolic (up to 1.4 mg/l) acids are liberated from breakdown of larger organic molecules during the long extraction procedure (4–5 weeks).
[b]18 amino acids were found in the hydrolysates. The amounts and kind of amino acids vary widely in samples.

Table 1.3–2 (*Continued*)
ORGANIC COMPOUNDS IDENTIFIED IN SEA-WATER

Substances	Quantities	Locality	Method	Ref.
Vitamin B_{12}	Present			10
Plant hormones	Present	North Sea	Chloroform extraction at pH 5; ether extract of residue, measured biologically	

(From Provasoli, L., Organic regulation of phytoplankton fertility, *The Sea,* Vol. 2, Hill, M. N., Ed., Interscience, New York, 1963, 174. With permission.)

REFERENCES

1. Wangersky, P. J., *Science,* 115, 685, 1960.
2. Collier, A., Ray, S. M., Magritsky, A. W., and Bell, J. O., *U.S. Dept. Int. Fish and Wildlife Service, Fish Bull.,* 84, 167, 1953.
3. Lewis, G. J. and Rakestraw, N. W., *J. Mar. Res.,* 14, 253, 1955.
4. Anderson, W. W. and Gehringer, J. W., *Spec. Sci. Rep.* (Fisheries), 265, 1, 1953; 303, 1, 1958.
5. Collier, A., *Limnol. Oceanogr.,* 3, 33, 1958.
6. Creac'h, P., *C.R. Acad. Sci. Paris,* 240, 2551, 1955.
7. Koyama, T., and Thompson, T. G., *Preprints Intern. Oceanogr. Cong. A.A.A.S.,* p. 925, 1959.
8. Slowey, J. F., Jeffrey, L. M., and Hood, D. W., *Geochim. Cosmochim. Acta,* 26, 607, 1962.
9. Tatsumoto, M., Williams, W. T., Prescott, J. M., and Hood, D. W., *J. Mar. Res.,* 19, 89, 1961.
10. Bentley, J. A., *Preprints Intern. Oceanogr. Cong. A.A.A.S.,* p. 910, 1959.

1.4 Photosynthetic Quotient and Gross-Net Differences

Table 1.4—1
SOME MEASUREMENTS OF THE PHOTOSYNTHETIC QUOTIENT
(O_2/-CO_2) IN AQUATIC PLANTS

Organism	Reference	PQ
Chlorella	Manning et al. (1938)	0.98
Chlorella	Sargent (1940)	ca. 1.00
Chlorella	Wassink and Kersten (1944)	1.09—1.12
Diatoms	Wassink and Kersten (1944)	1.12—1.13
Nitzschia closterium	Barker (1935a)	1.05—1.08
Nitzschia palea	Barker (1935a)	1.03—1.05
Peridinium sp.	Barker (1935b)	1.03—1.11
Synechoccus sp.	Fraenkel et al. (1950)	1.08 ± 0.03
Homidium flaccidum	van der Paauw (1932)	0.92—1.07
Gelidium	Tseng and Sweeney (1946)	0.94 ± 0.05
Gigartina	Emerson and Green (1934)	0.99—1.07
Raw seawater	Sargent and Hindman (1943)	1.09 ± 0.23

(From Ryther, J. H., The measurement of primary production, *Limnol. Oceanogr.*, 1(2), 74, 1956. With permission.)

Figure 1.4—1

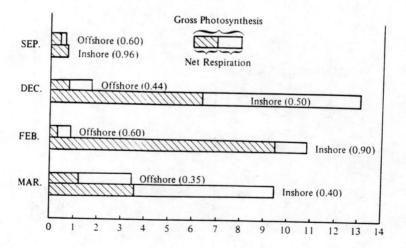

Figure 1.4—1. Comparison of net and gross photosynthesis for inshore and offshore (deeper than 500 fm) locations off Massachusetts. Figures in parentheses are ratios of net: gross.

(From Raymont, J. E. G., Factors affecting primary production — II Light and temperature, *Plankton and Productivity in the Oceans*, Pergamon Press, London. 1963, 227. With permission.)

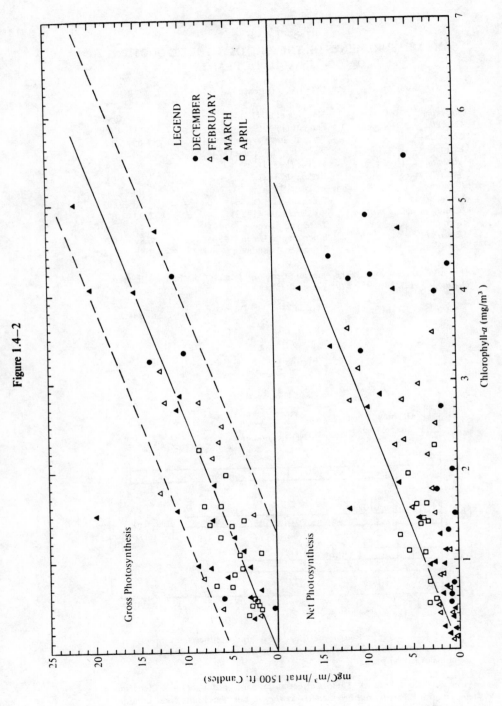

Figure 1.4—2

Figure 1.4—2. Gross and net photosynthesis at 1500 foot candles related to the chlorophyll-*a* concentration. Gross photosynthesis was determined by dark and light oxygen bottle experiments. The envelope corresponds to an oxygen titration difference of ± 0.05 ml/l. Net photosynthesis was determined by C[14] assimilation experiments, uncorrected for respiration. Determinations made off Massachusetts.

Figure 1.4—3

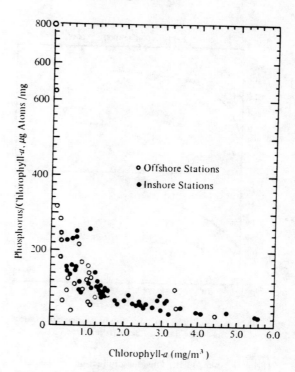

Figure 1.4—3. The variation of the ratio of particulate
phosphates per chlorophyll-*a* as a function of the chloro-
phyll-α content of the water off Massachusetts.

Figure 1.4—4

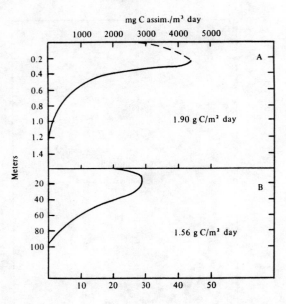

Figure 1.4—4. Primary production in a sewage oxidation pond (A) and in the Sargasso Sea off Bermuda (B).

(From Ryther, J. H., Organic production by plankton algae and its environmental control, in *The Ecology of the Algae*, Spec. Pub. 2, Pymatuning Laboratory of Field Biology, the University of Pittsburgh, 72—83, 1960. With permission.)

Figure 1.4–5

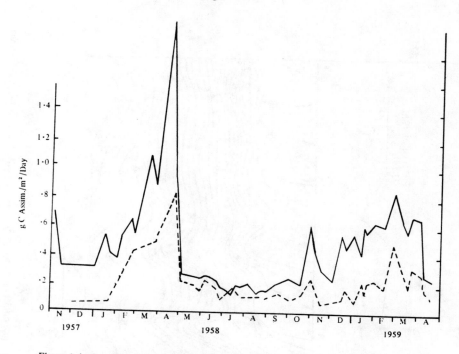

Figure 1.4–5.   The annual cycle of net and gross primary production off Bermuda.

(From Menzel, D. W. and Ryther, J. H., The annual cycle of primary production in the Sargasso Sea off Bermuda, *Deep-Sea Res.*, 6, 363, © 1960 Pergamon Press. With permission.)

Figure 1.4—6

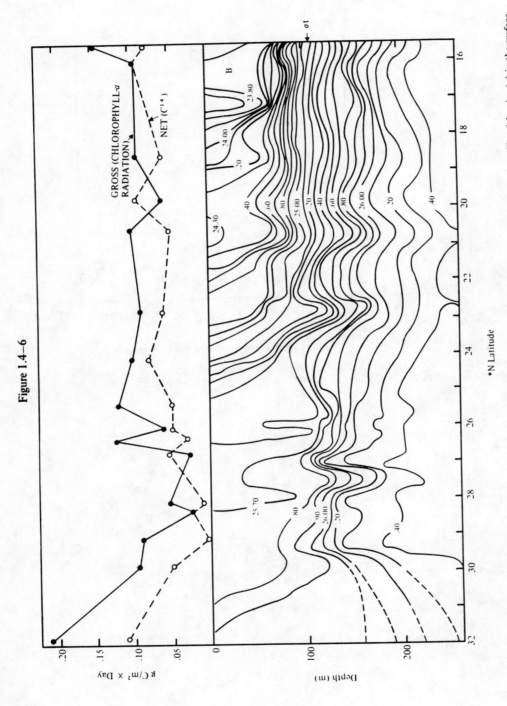

Figure 1.4—6.   A. Gross (closed circles, solid line) and net (open circles, broken line) primary production. B. Vertical profile of density (σ) in the surface water as related to latitude off Bermuda.

(From Ryther, J. H. and Menzel, D. W., Primary production in the southwest Sargasso Sea, January-February, 1960, *Bull. Mar. Sci. Gulf Caribb.*, 11, 383, 1961. With permission.)

Table 1.4–2
EFFECT OF ENRICHMENT ON PHOTOSYNTHESIS OF
NATURAL POPULATIONS*

Location 16° 10′ N, 65°05′ W

Conditions	Type of measurement	Photosynthesis g C/m³/day	Net Gross
No enrichment	^{14}C	0.0018	–
	O_2	0.0125	0.14
P + N enriched[1]	^{14}C	0.11	–
	O_2	0.135	0.82

[1] 100 μg at/l of NO_3-N, 6-6 μg at/l PO_3-P.

* Surface water exposed for three days to natural light in running sea water on deck.

Table 1.4—3
ULTRAPLANKTON RESPIRATION AT STATIONS IN THE WESTERN NORTH ATLANTIC

Mg atoms O m^{-3} day^{-1}

Sample depth (m)[6]	May 23—26, 1966 33°30′N, 72°00′W Sargasso Sea	Aug. 13—17, 1966 33—32°N, 71°20′W Sargasso Sea	Aug. 18—19, 1966 33°N, 75°W Sargasso Sea	Apr. 4, 1966 35°55′N, 73°32′W Gulf Stream	Apr. 2—3, 1966 36°10′N, 74°38′W Slope water
0.1	1.3, 1.9, 1.3	0.9, 0.8	0.8, **0.7**	2.4	3.8
5	–	–	–	–	15.5
10	–	0.5	–	–	–
15	–	–	2.6	–	5.4
20	–	0.2	–	–	–
25	**0.9, 0.2**	–	–	–	–
30	–	–	0.4	–	–
50	0.9, 0.6	0.2	–	3.1	0.5
75	0.6	–	–	–	–
100	0.3	0.3	–	0.7	0.7
150	0.3	–	–	–	–
200	0.3	0.5	–	–	–
250	0.3	0.1	–	0.3	0.9
300	0.5	–	–	–	–
400	0.5	0.3	–	–	–
500	0.2	0.3	0.5	0.5	0.8
550	–	0.3	–	–	–
700	–	0.3	–	–	–
750	–	–	–	0.05	–
800	0.1	0.05	–	–	–

* Bold numbers are in the scattering layer.

Table 1.4—4
ULTRAPLANKTON RESPIRATION IN THE UPPER 500 METERS, AND
P/R RATIOS AT STATIONS IN THE
SOUTHEAST PACIFIC OFF PERU

Total depth (m)	Location	Respiration (mg atoms O m^{-2}/day^{-1})	P/R
41	07°58′S 79°21′W	375	1.6
80	08°20′S 79°33′W	400	–
200	08°17′S 79°52′W	520	0.9
520	07°57′S 80°32′W	568	5.3
568	11°51′S 77°33′W	878	1.0
600	08°34′S 80°00′W	216	1.3
1900	08°16′S 80°52′W	469	–
3100	08°22′S 80°45′W	250	1.0
4500	08°24′S 81°40′W	315	1.3
4500	06°22′S 81°47′W	169	1.1
4900	05°49′S 82°05′W	319	0.7

Table 1.4–5

ULTRAPLANKTON RESPIRATION AT STATIONS IN THE SOUTHEAST PACIFIC

Mg atoms O m^{-3}/day^{-1}

Sample depth (m)	Oct. 4, 5 05°50'S 82°04'W	Oct. 7 06°21'S 82°14'W	Oct. 10, 11 06°21'S 81°46'W	Oct. 14 07°57'S 81°40'W	Oct. 15 08°24'S 81°04'W	Oct. 17 08°16'S 80°52'W	Oct. 17 08°22'S 80°43'W	Oct. 18 08°22'S 80°45'W	Oct. 19 08°17'S 79°52'W	Oct. 19 07°58'S 79°21'W
0.1	0.61		13.9, 0.89, 0.71, 0.63	9.11	2.80	6.3, 8.6	2.35	2.46	4.00	5.46
10							3.12	2.17	7.66	13.32
20	1.31									7.05
25										
30			1.80	6.60	0.80	7.4	2.86	0.91	4.93	
50	1.04		1.96, 0.85	1.04	0.95	3.51	0.28	0.83	3.40	6.50
60										
75										
100	0.45	0.73	0.05	0.67	0.34	0.00	0.00	1.13	2.31	
175										
200	0.21	0.00	0.24	0.22	0.09	0.68			1.33	
250										
300	1.90	0.00	0.00		0.99					
400										
500	0.66	0.00	0.00							

Table 1.4–5 (Continued)

ULTRAPLANKTON RESPIRATION AT STATIONS IN THE SOUTHEAST PACIFIC

Sample depth (m)	Oct. 20 08°20'S 79°33'W	Oct. 20 08°34'S 80°00'W	Oct. 25 11°51'S 77°33'W	Oct. 25 11°53'S 77°49'W	Oct. 26 12°01'S 78°30'W	Oct. 27 11°58'S 78°35'W	Oct. 28 12°00'S 78°46'W	Oct. 31–Nov. 1 08°24'S 80°25'W	Nov. 2 09°03'S 81°29'W	Nov. 5 09°01'S 80°45'W
0.1	2.24	4.16	4.50	0.00	0.32	0.82	1.60, 0.60	0.41	0.93	0.98
10	6.56	—	—	—	—	—	—	—	—	—
20	6.44	3.48	3.65	—	2.80	—	1.65	0.53	0.43	2.53
25	—	—	—	—	—	—	—	—	—	—
30	—	—	—	—	—	0.10	—	—	—	—
50	2.40	1.33, 1.46	2.66	2.06	0.67	1.10	0.22	0.78	0.20	1.42
60	2.40	—	—	—	—	0.70	—	—	—	—
75	—	0.04	3.20	—	—	—	—	—	—	—
100	—	—	4.00	0.80	0.00	—	0.20	0.27	0.39	1.07
175	—	—	—	—	0.13	—	—	—	—	—
200	—	—	—	—	—	—	0.10	0.20	0.29	0.44
250	—	—	—	—	0.22	—	—	—	—	—
300	—	0.27	—	—	0.12	—	—	—	0.24	0.20
400	—	—	—	—	—	—	—	—	0.76	0.62
500	—	—	—	—	—	—	—	—	—	—

Table 1.4–6
STANDARD RELATIONSHIPS BETWEEN CHLOROPHYLL, STANDING CROP OF ORGANIC MATTER, TRANSPARENCY, ORGANIC PRODUCTION, AND NITROGEN AVAILABILITY AND REQUIREMENT

1	2	3	4	5	6	7	8
Chl. a g/l	Standing crop, mg/m^3 dry wt	Extinc. coeff., k	Depth euphotic zone, m	Calc. org. prod., g dry wt/m^2/day	N initially available in euphotic zone, mg	N required to produce existing population, mg	N requirement, mg/day
0	0	0.04	120	0	25,300	0	0
0.1	10	0.07	66	0.20	13,800	66	20
0.5	50	0.13	36	0.50	7600	180	50
1.0	100	0.19	24	0.75	5000	240	75
2.0	200	0.29	15	1.00	3100	300	100
5.0	500	0.50	10	1.40	2100	500	140
10.0	1000	0.79	6	1.80	1260	600	180
20.0	2000	1.30	3.5	2.20	735	700	220

(From Ryther. J. H., Geographic variations in productivity, *The Sea*. Vol. 2, Hill, M. N., Ed., Interscience, New York, 1963, 352. With permission.)

Table 1.4—7
GROSS AND NET ORGANIC PRODUCTION OF
VARIOUS NATURAL AND CULTIVATED SYSTEMS

g/dry wt/sq. m/day

System	Gross	Net
A. Theoretical potential		
Average radiation (200–400 g cal/cm² day)	23–32	8–19
Maximum radiation (750 g cal/cm² day)	38	27
B. Mass outdoor *Chlorella* culture		
Mean		12.4
Maximum		28.0
C. Land (maximum for entire growing seasons)		
Sugar cane		18.4
Rice		9.1
Wheat		4.6
Spartina marsh		9.0
Pine forest (best growing years)		6.0
Tall prairie		3.0
Short prairie		0.5
Desert		0.2
D. Marine (maxima for single days)		
Coral reef	24	(9.6)
Turtle grass flat	20.5	(11.3)
Polluted estuary	11.0	(8.0)
Grand Banks (April)	10.8	(6.5)
Walvis Bay	7.6	
Continental Shelf (May)	6.1	(3.7)
Sargasso Sea (April)	4.0	(2.8)
E. Marine (annual average)		
Long Island Sound	2.1	0.9
Continental Shelf	0.74	(0.40)
Sargasso Sea	0.88	0.40

(From Ryther, J. H., Potential productivity of the sea, *Science*, 130(3376), 606, 1959. With permission.)

Figure 1.4–7

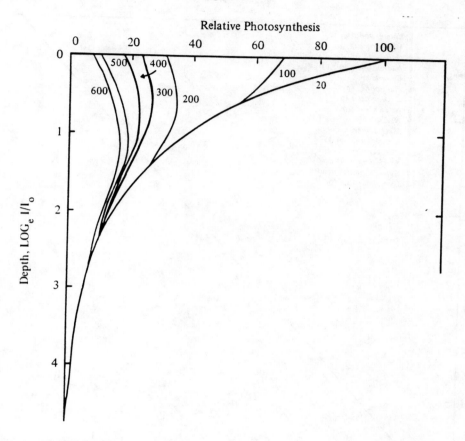

Figure 1.4–7. Relative photosynthesis as a function of water depth for days of different incident radiation. Numbers beside curves show gram calories per square centimeter per day.

(From Ryther, J. H., Potential productivity of the sea, *Science*, 130(3376), 606, 1959. With permission.)

Figure 1.4—8

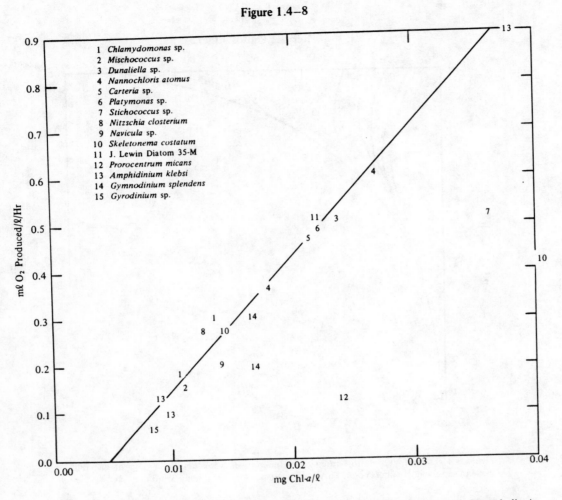

Figure 1.4—8. The relationship between total photosynthesis at constant light intensity and chlorophyll-*a* in some marine plankton algae.

(From Ryther, J. H., The measurement of primary production, *Limnol. Oceanogr.*, 1(2), 80, 1956. With permission.)

Table 1.4–8
OXYGEN METABOLISM (g/m^2) IN THE WATERS OF SOUTHWESTERN PUERTO RICO

Community and location	Depth (m)	Current (m/min)	Temp. (°C)	Date	Time	Method used (key below)	P_{cn}	R	P_g	R_{24}
East La Gata Reef *Porites* reef	0.26	7.6	—	2/27/58	1310	AF	1.75			
La Gata *Thalassia* bed	—	7.3	27.5	2/27/58	1335	AF	2.8			
			27.5	2/27/58	1100	AF	2.8			
La Gata Reef *Porites* and *Thalassia*	0.30	slight	25–28	2/18/58	24 hrs	BC	1.8+	1.8+	8.6+	11.3+
West La Gata Reef *Thalassia*	0.26	6.4	27–28	2/28/58	1100	AF	2.6	—	—	
Margarita Reef *Porites* carpet	0.23	0–9	25–28	2/25/58	24 hrs	A	1.4	0.5–4	44	—
Isla Magueyes, southern shore, *Porites* and *Thalassia*	0.30	0–3	25–29	2/17/58	24 hrs	BC	0.32+	0.75+	12+	18+
La Media Luna, *Millepora* 18m: *Acropora* and *Porites* 65m	0.28–3	5–8	28–30	6/12/59	24 hrs	AC	2.0	1.21	38.4+	29+
Las Palmas, *Thalassia*	0.30	1.8	25–28	2/28/58	1237	AF	1.33			

A. Upstream-downstream change.
B. Diurnal curve—single water mass method.
C. Since there is some admixture of water from other areas, this estimate is minimal.
D. Dark and light bottles.
E. Data over a 24-hr period.
F. With standard diffusion correction.
G. Upstream-downstream measurements were similar and averaged.

P_{cn} : Net daytime community photosynthesis/hr.
R: Nighttime community respiration/hr.
P_g : Net photosynthesis/day with night R added.
R_{24} : 24-hr community respiration based on night R.

Table 1.4—8 (Continued)
OXYGEN METABOLISM (g/m²) IN THE WATERS OF SOUTHWESTERN PUERTO RICO

Community and location	Depth (m)	Current (m/min)	Temp. (°C)	Date	Time	Method used (key below)	P_{cn}	R	P_g	$R_{2,4}$
Channel north of Isla Magueyes, *Thalassia*	1.2—1.5	4.2	26—28	2/13/58	24 hrs	BG	0.50	0.52	10.5	12.5
	1.2—1.5	—	25—28	3/1/58	24 hrs	BG	0.60	0.45	11.0	12.5
	1.2—1.5	—	28—29	5/29/59	24 hrs	BG	0.60	0.60	14.0	14.5
Enrique Reef, *Millepora Thalassia Zoanthus, Dictyota, Porites*	0.1—0.6	6.4	27—29	3/12/58	24 hrs	A	2.0	0.90	20.0	17.3
El Mario Reef	0.39	4—8	—	3/29/59	24 hrs	AF	1.6	1.5	39	36
Bahia Fosforescente										
Total Bay	4	—	—	1/24/57	24 hrs	B	0.08	0.28	5.6	7.7
Plankton only	4	—	—	1/24/57	24 hrs	D	0.06	0.09	1.4	2.3
Thalassia bed	—	—	—	2/12/58	24 hrs	BC	0.85+	0.75+	15+	18+
Thalassia bed	—	—	—	3/13/58	24 hrs	BC	—	—	5+	6+
Total Bay	4	—	—	3/13/58	24 hrs	C	—	—	<48	<48
Total Bay									8.8	7.7
Plankton only	4	—	—	6/1/59	24 hrs	D	-0.18	0.29	1.4	6.9

A. Upstream-downstream change.
B. Diurnal curve—single water mass method.
C. Since there is some admixture of water from other areas, this estimate is minimal.
D. Dark and light bottles.
E. Data over a 24-hr period.
F. With standard diffusion correction.
G. Upstream-downstream measurements were similar and averaged.

P_{cn}: Net daytime community photosynthesis/hr.
R: Nighttime community respiration/hr.
P_g: Net photosynthesis/day with night R added.
$R_{2,4}$: 24-hr community respiration based on night R.

1.5 Relationships to Productivity at Other Trophic Levels

Table 1.5−1

ESTIMATES OF POTENTIAL YIELDS (PER YEAR) AT VARIOUS TROPHIC LEVELS, IN METRIC TONS

	Ecological efficiency factor					
	10%		15%		20%	
Trophic level	Carbon (tons)	Total weight (tons)	Carbon (tons)	Total weight (tons)	Carbon (tons)	Total weight (tons)
0. Phytoplankton (net particulate production)	1.9×10^{10}	–	1.9×10^{10}	–	1.9×10^{10}	–
1. Herbivores	1.9×10^{9}	1.9×10^{10}	2.8×10^{9}	2.8×10^{10}	3.8×10^{9}	3.8×10^{10}
2. 1st stage carnivores	1.9×10^{8}	1.9×10^{9}	4.2×10^{8}	4.2×10^{9}	7.6×10^{8}	7.6×10^{9}
3. 2nd stage carnivores	1.9×10^{7}	1.9×10^{8}	6.4×10^{7}	6.4×10^{8}	15.2×10^{7}	15.2×10^{8}
4. 3rd stage carnivores	1.9×10^{6}	1.9×10^{7}	9.6×10^{6}	9.6×10^{7}	30.4×10^{6}	30.4×10^{7}

Table 1.5−2

DAILY PRIMARY PRODUCTION AND ZOOPLANKTON BIOMASS AT THREE OCEANIC LOCATIONS

Depth		Location		
		5°N-4°S 15°W	1.5°N-1.5°S 35°W	15°S 17-35°W
		mgC/m^2	mgC/m^2	mgC/m^2
0−100 m	Primary production	880	480	252
	Herbivores	397	658	140
	Carnivores	377	572	148
	Total zooplankton	774	1,230	288
0−500 m	Total zooplankton	1,750	1,900	680

Figure 1.5—1

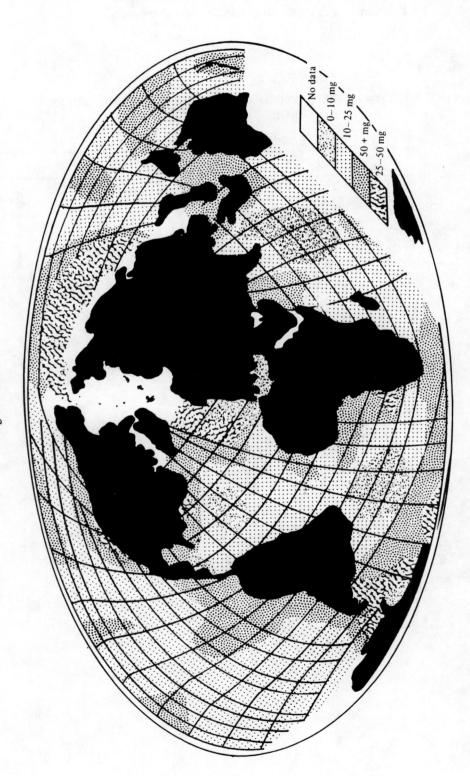

Figure 1.5—1. Biological productivity of the seas. Density of shading is roughly proportional to the degree of biological productivity as measured by the amount of organic matter (in milligrams) produced annually per cubic meter of sea water.

Section 2

Fishery Statistics

2.1 Total World Catch and Catch Potentials

Figure 2.1—1   WORLD CATCH (1968) AND ESTIMATED POTENTIALS BY REGIONS

Letters Refer To Suffix Designations Of Regional Maps In Sections 2.2, 2.3, 2.5 And Corresponding Areas In The Composite Map Of Figure 2.6—1

See special foldout section at back of the book, p. 381.

(Reprinted from FAO Department of Fisheries, *Atlas of the Living Resources of the Seas,* Rome, 1972, by permission of the Food and Agricultural Organization of the United Nations.)

Table 2.1—1
HARVEST OF MARINE FISHERIES
OF THE WORLD TO 1962

In Millions of Metric Tons, Whales Excluded*

Year	Harvest	Annual rate of increase	
1938	18.17		
1948	17.80		
1952	21.92		
1953	22.63		
1954	24.18		4½%
1955	25.34		
1956	26.83		
1957	27.40		
1958	28.39	4%	
1959	31.41	10%	
1960	33.63	7%	8%
1961	36.77	9%	
1962	40.05	9%	

* Three million tons of whales were taken in 1962.

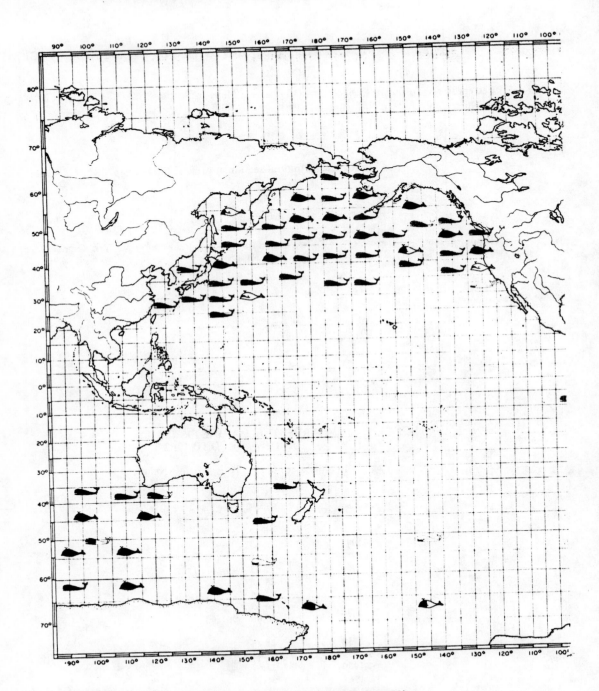

DISTRIBUTION OF WHALE CATCHES (1970)
DISTRIBUTION DES CAPTURES DE BALEINES (1970)
DISTRIBUCION DE LAS CAPTURAS DE CETACEOS (1970)

(From FAO Department of Fisheries, *Atlas of the Living Resources of the Seas*, Food and Agricultural Organization of the
United Nations, Rome, 1972.)

Figure 2.2—1

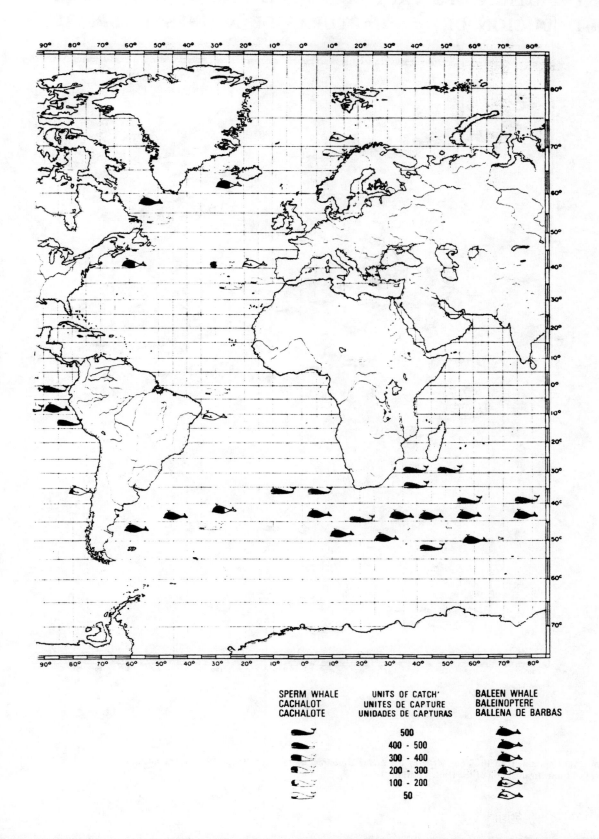

SPERM WHALE CACHALOT CACHALOTE	UNITS OF CATCH' UNITES DE CAPTURE UNIDADES DE CAPTURAS	BALEEN WHALE BALEINOPTERE BALLENA DE BARBAS
	500	
	400 - 500	
	300 - 400	
	200 - 300	
	100 - 200	
	50	

DISTRIBUTION OF CATCHES OF TUNA AND TUNA-LIKE
DISTRIBUTION DES CAPTURES DE THONS ET DES ESPECES
DISTRIBUCION DE LAS CAPTURAS DE ATUNES Y ESPECIES

Figure 2.2–2

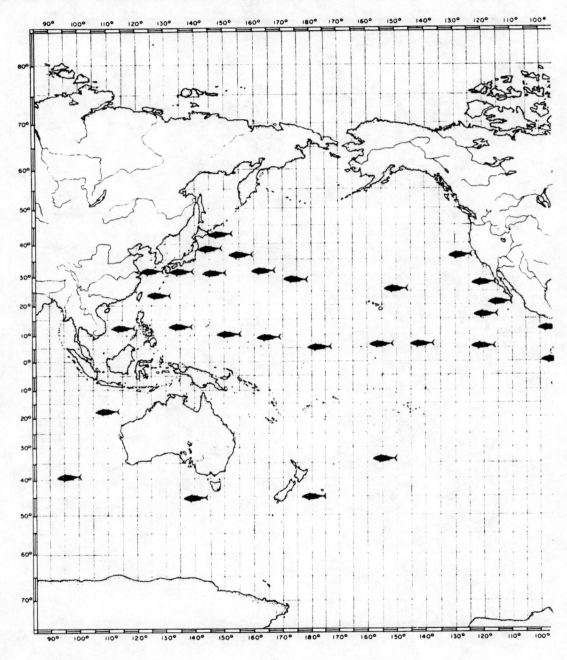

(Reprinted from FAO Department of Fisheries, *Atlas of the Living Resources of the Seas*, Rome, 1972, by permission of the Food and Agricultural Organization of the United Nations.)

FISHES (1969)
VOISINES (1969)
AFINES (1969)

Figure 2.2-2

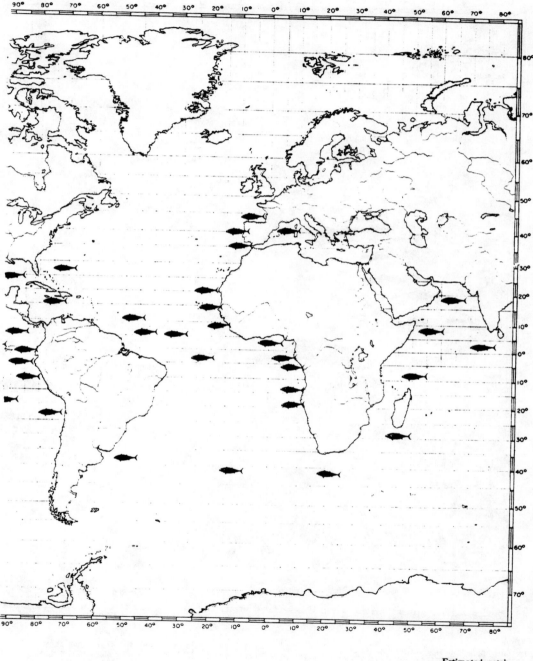

Estimated catch
Captures estimées
Capturas estimadas

= 25 000 t

DISTRIBUTION OF COASTAL PELAGIC CATCHES (1968)
DISTRIBUTION DES CAPTURES PELAGIQUES COTIERES (1968)
DISTRIBUCION DE LAS CAPTURAS PELAGICAS COSTERAS (1968)

Figure 2.2—3

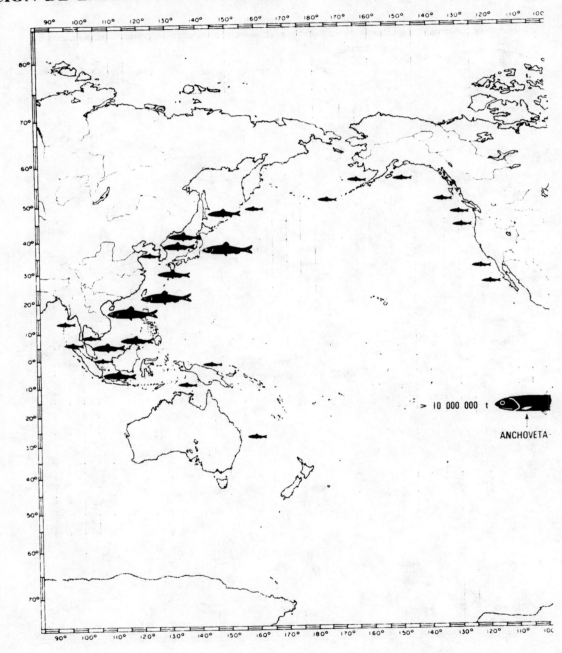

> 10 000 000 t

ANCHOVETA

CATCH BY COUNTRIES BORDERING THE STATISTICAL AREA'
CAPTURES EFFECTUEES DANS L'AIRE STATISTIQUE PAR LES PAYS LIMITROPHES'
CAPTURAS POR PAISES RIBEREÑOS QUE LIMITAN CON EL AREA ESTADISTICA'

CATCH BY COUNTRIES OUTSIDE THE STATISTICAL AREA'
CAPTURES PAR DES PAYS D'AUTRES AIRES STATISTIQUES'
CAPTURAS POR PAISES FUERA DEL AREA ESTADISTICA'

1 000 000 t

250 000 t

25 000 t

(excluding tuna and tuna-like fish)

(à l'exception des thonidés)

(con la excepción de tunidos)

Figure 2.2–3

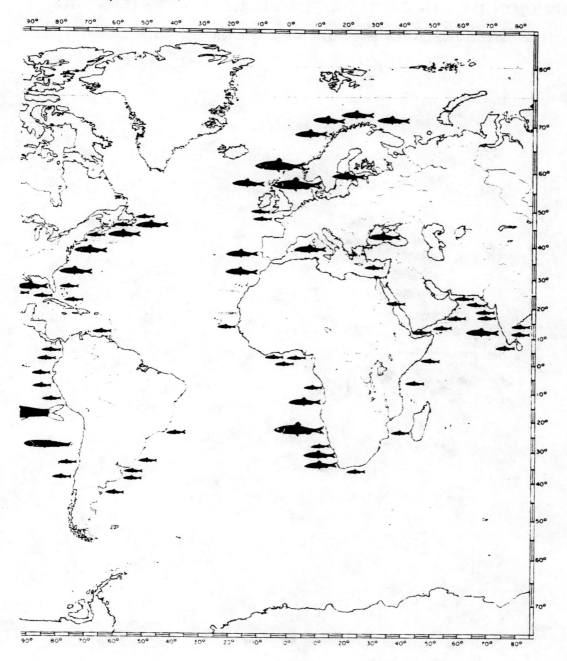

Estimated catch – Captures estimées – Capturas estimadas (1968)
29 000 000 t

Statistical areas are shown on Maps 1.3 and 1.4
Les aires statistiques sont representees sur les cartes 1.3 et 1.4
Las areas estadisticas se indican en los mapas 1.3 y 1.4

(Reprinted from FAO Department of Fisheries, *Atlas of the Living Resources of the Seas*, Rome, 1972, by permission of the Food and Agricultural Organization of the United Nations.)

DISTRIBUTION OF DEMERSAL FISH CATCHES (1968)
DISTRIBUTION DES CAPTURES DEMERSALES (1968)
DISTRIBUCION DE LAS CAPTURAS DE PECES DEMERSALES (1968)

Figure 2.2—4

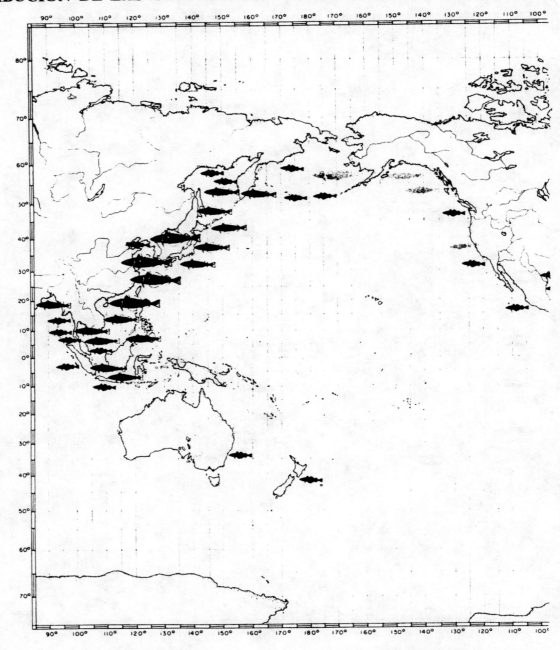

CATCH BY COUNTRIES BORDERING THE STATISTICAL AREA¹
CAPTURES EFFECTUEES DANS L'AIRE STATISTIQUE PAR LES PAYS LIMITROPHES¹
CAPTURAS POR PAISES RIBEREÑOS QUE LIMITAN CON EL AREA ESTADISTICA¹

CATCH BY COUNTRIES OUTSIDE THE STATISTICAL AREA¹
CAPTURES PAR DES PAYS D'AUTRES AIRES STATISTIQUES¹
CAPTURAS POR PAISES FUERA DEL AREA ESTADISTICA¹

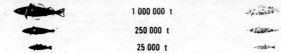

1 000 000 t

250 000 t

25 000 t

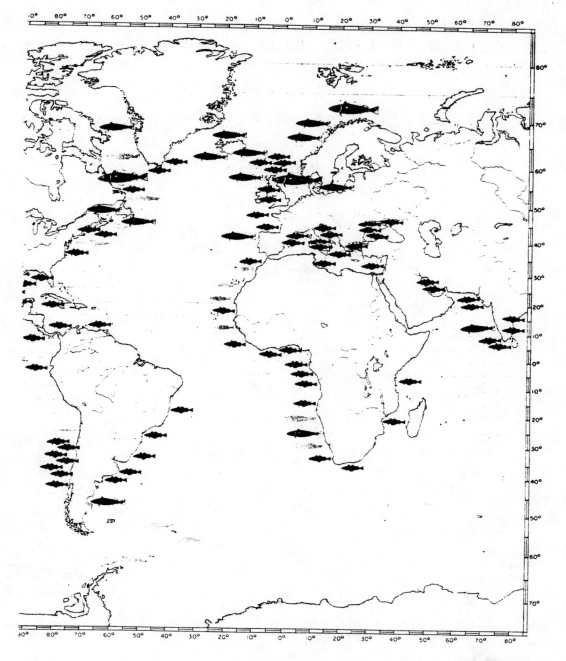

Figure 2.2—4

Estimated catch – Captures estimées – Capturas estimadas (1968)
19 600 000 t

¹ Statistical areas are shown on Maps 1.3 and 1.4
² Les aires statistiques sont représentées sur les cartes 1.3 et 1.4
³ Las áreas estadísticas se indican en los mapas 1.3 y 1.4

(Reprinted from FAO Department of Fisheries, *Atlas of the Living Resources of the Seas*, Rome, 1972, by permission of the Food and Agricultural Organization of the United Nations.)

EAST CENTRAL PACIFIC TUNA DISTRIBUTION
PACIFIQUE CENTRE-EST - DISTRIBUTION DU THON
PACIFICO CENTRO-ORIENTAL - DISTRIBUCION DE ATUNES

Figure 2.2—5

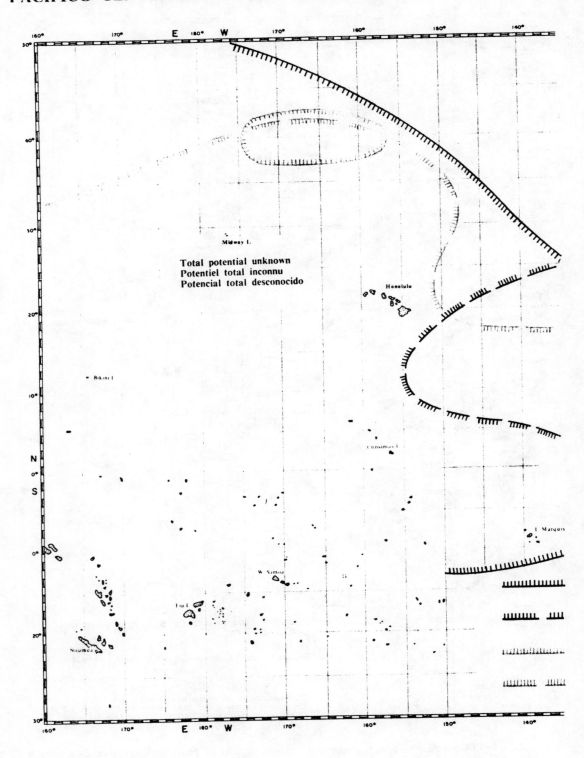

Total potential unknown
Potentiel total inconnu
Potencial total desconocido

Figure 2.2–5

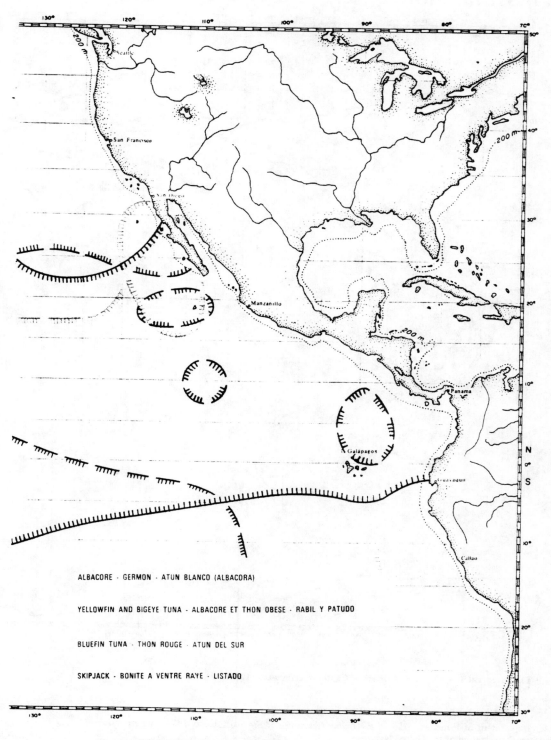

ALBACORE · GERMON · ATUN BLANCO (ALBACORA)

YELLOWFIN AND BIGEYE TUNA · ALBACORE ET THON OBESE · RABIL Y PATUDO

BLUEFIN TUNA · THON ROUGE · ATUN DEL SUR

SKIPJACK · BONITE A VENTRE RAYE · LISTADO

(Reprinted from FAO Department of Fisheries, *Atlas of the Living Resources of the Seas*, Rome, 1972, by permission of the Food and Agricultural Organization of the United Nations.)

DISTRIBUTION OF CRUSTACEAN CATCHES (1970)
DISTRIBUTION DES CAPTURES DE CRUSTACES (1970)
DISTRIBUCION DE LAS CAPTURAS DE CRUSTACEOS (1

Figure 2.2—6

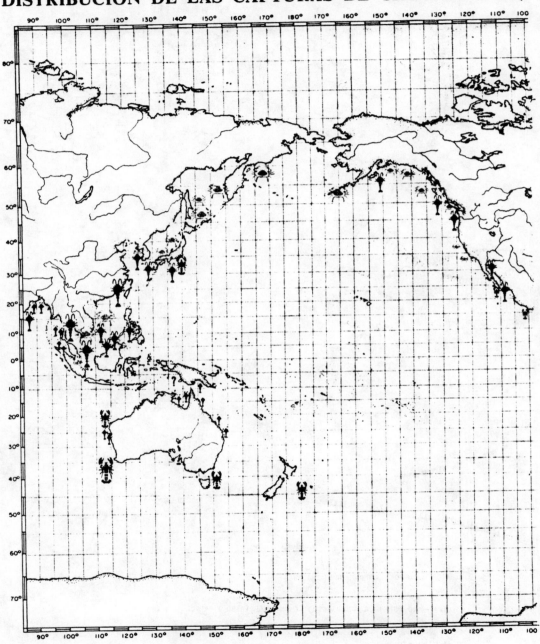

Estimated catch - Captures estimées - Capturas estimadas (1970)

(Reprinted from FAO Department of Fisheries, *Atlas of the Living Resources of the Seas*, Rome, 1972, by permission of the Food and Agricultural Organization of the United Nations.)

Figure 2.2—6

70)

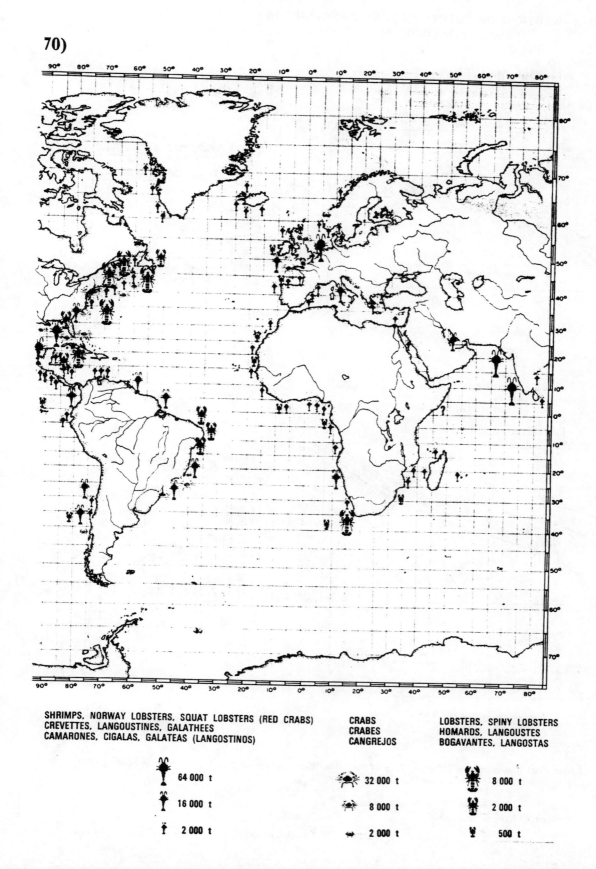

SHRIMPS, NORWAY LOBSTERS, SQUAT LOBSTERS (RED CRABS)
CREVETTES, LANGOUSTINES, GALATHEES
CAMARONES, CIGALAS, GALATEAS (LANGOSTINOS)

CRABS
CRABES
CANGREJOS

LOBSTERS, SPINY LOBSTERS
HOMARDS, LANGOUSTES
BOGAVANTES, LANGOSTAS

64 000 t	32 000 t	8 000 t
16 000 t	8 000 t	2 000 t
2 000 t	2 000 t	500 t

2.3 Vertical Distribution of Common Species by Groups and Regions

Figure 2.3—1   VERTICAL DISTRIBUTIONS OF SPECIES IN NORTHEAST ATLANTIC FISHERIES

Narrow continental shelf
(Norway, Spitzbergen, Iceland, Faroe, East Greenland)
Plateau continental étroit
(Norvège, Spitzbergen, Islande, Iles Féroé, Groënland occidental)
Plataforma continental estrecha
(Noruega, Spitzbergen, Islandia, Feroe, este de Groenlandia)

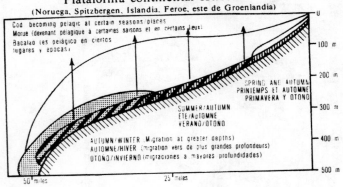

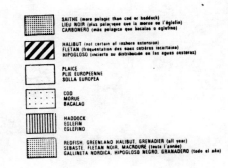

Wide continental shelf
(Baltic, North Sea, West of British Isles)
Plateau continental large
(Baltique, Mer du Nord, ouest des Iles Britanniques)
Plataforma continental ancha
(Báltico, Mar del Norte, oeste de las islas Británicas)

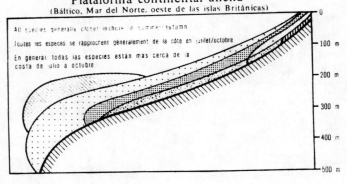

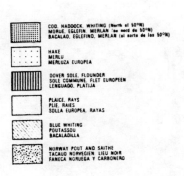

Proyección del mapa base: zona de iguales azimutes, con el centro en 40°N, 35°O.
Preparado por M.J. Pollack y G.G. Pasley, Woods Hole Oceanographic Institute.

Carte de base: projection azimutale équivalente, centrée à 40°N, 35°O.
Etablie par M.J. Pollack et G.G. Pasley, Woods Hole Oceanographic Institute.

Base map projection: azimuthal equal area, centred at 40°N, 35°W.
Prepared by M.J. Pollack and G.G. Pasley, Woods Hole Oceanographic Institute.

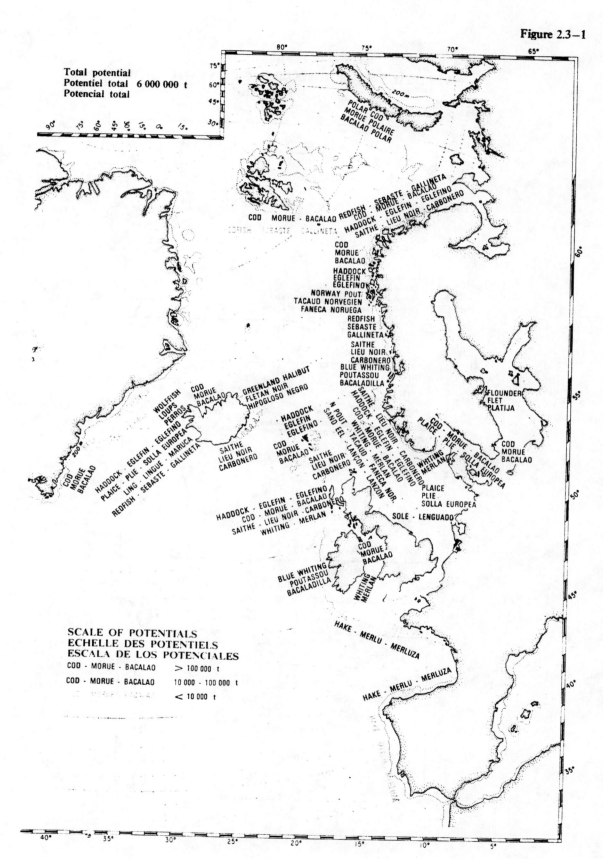

Figure 2.3—1

83

Figure 2.3—2

VERTICAL DISTRIBUTIONS OF SPECIES IN EACH CENTRAL PACIFIC FISHERIES

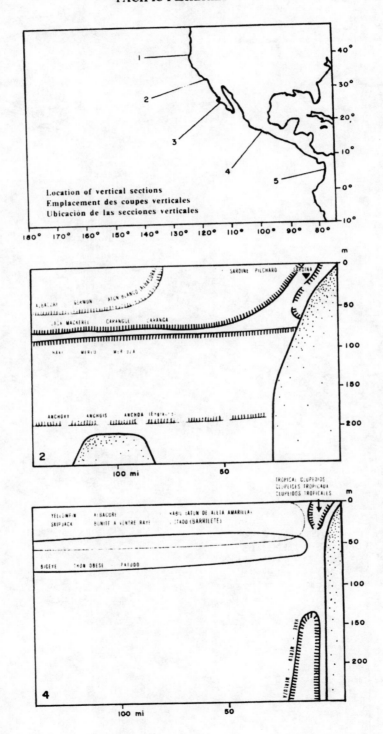

Figure 2.3–2

VERTICAL DISTRIBUTIONS OF SPECIES IN EACH CENTRAL PACIFIC FISHERIES

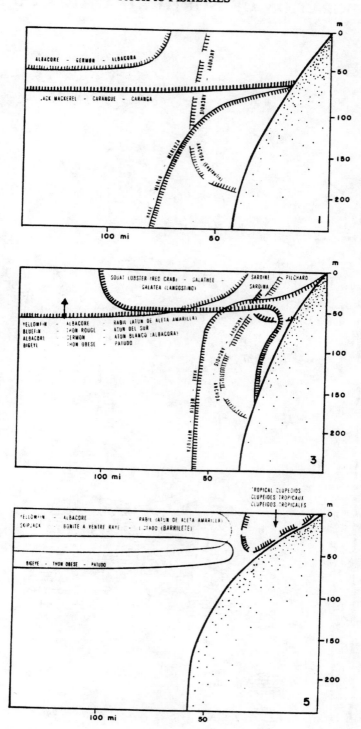

(Reprinted from FAO Department of Fisheries, *Atlas of the Living Resources of the Seas*, Rome, 1972, by permission of the Food and Agricultural Organization of the United Nations.)

MIGRATION OF NORTH ATLANTIC COD STOCKS

Figure 2.4–1

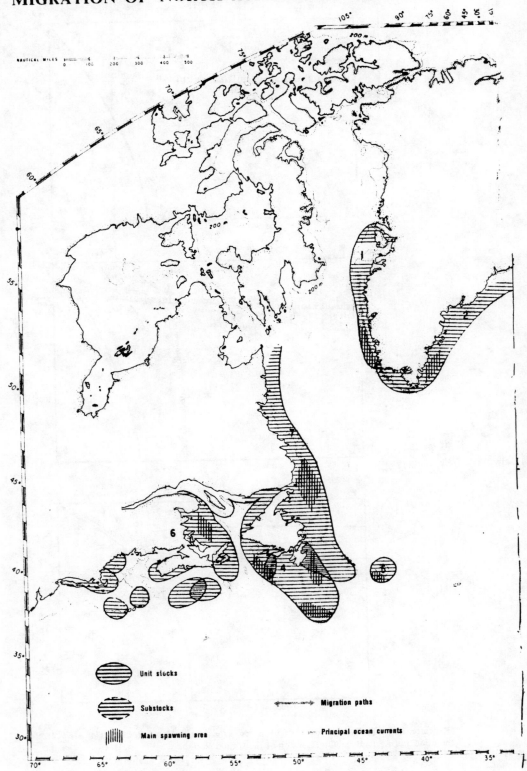

Figure 2.4–1

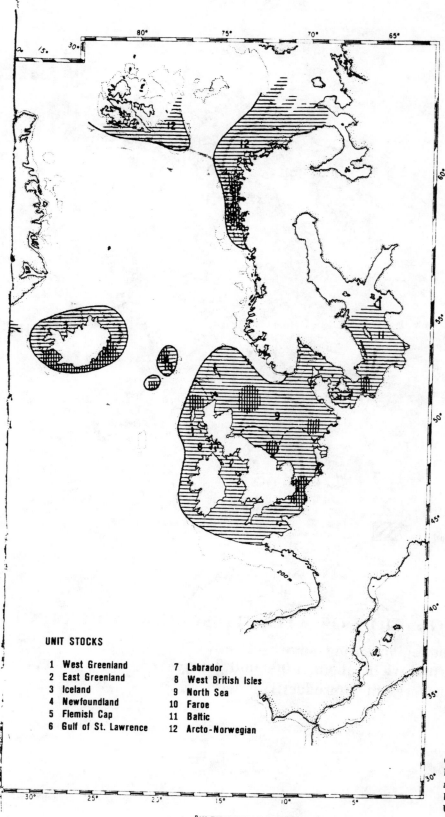

UNIT STOCKS

1	West Greenland	7	Labrador
2	East Greenland	8	West British Isles
3	Iceland	9	North Sea
4	Newfoundland	10	Faroe
5	Flemish Cap	11	Baltic
6	Gulf of St. Lawrence	12	Arcto-Norwegian

Base map projection: azimuthal equal area, centred at 40°N, 15°W
Prepared by M.J. Pollack and G.G. Pasley, Woods Hole Oceanographic Institute.

(Reprinted from FAO Department of Fisheries, *Atlas of the Living Resources of the Seas,* Rome, 1972, by permission of the Food and Agricultural Organization of the United Nations.)

Figure 2.4—2

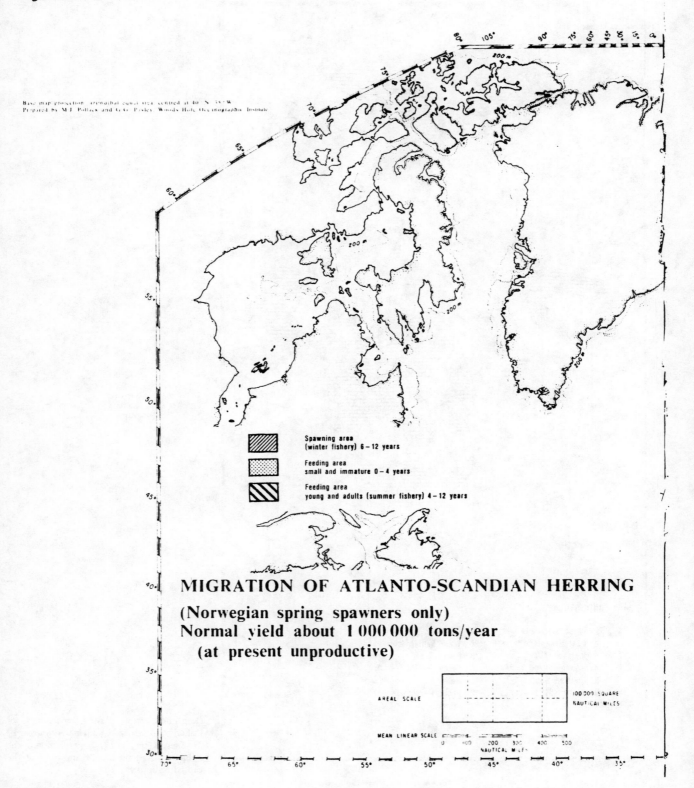

MIGRATION OF ATLANTO-SCANDIAN HERRING

(Norwegian spring spawners only)
Normal yield about 1 000 000 tons/year
(at present unproductive)

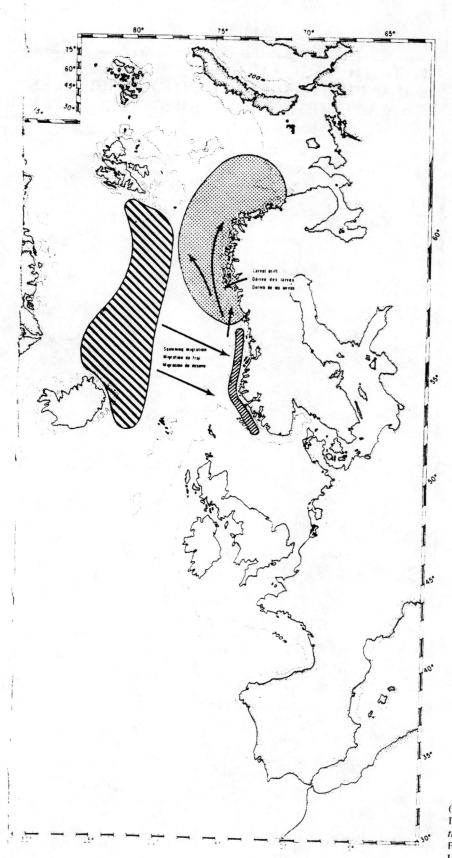

Figure 2.4–2

(Reprinted from FAO Department of Fisheries, *Atlas of the Living Resources of the Seas*, Rome, 1972, by permission of the Food and Agricultural Organization of the United Nations.)

Figure 2.4–3

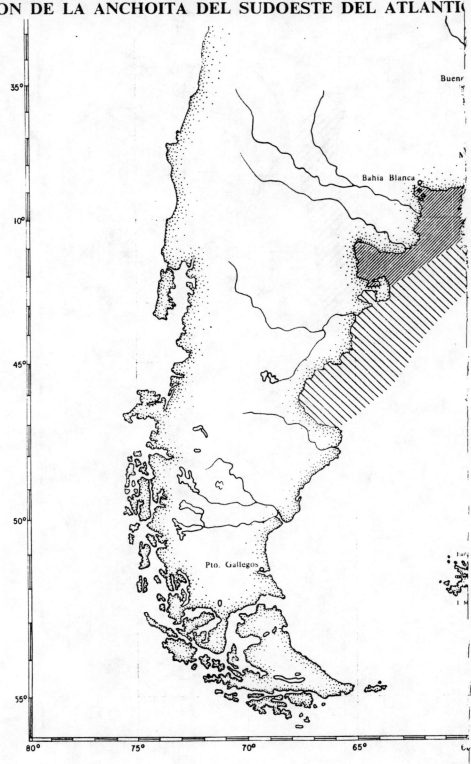

MIGRATION OF SOUTH WEST ATLANTIC ANCHOITA
MIGRATION DE L'ANCHOITA DE L'ATLANTIQUE SUD-OUEST
MIGRACION DE LA ANCHOITA DEL SUDOESTE DEL ATLANTI

Figure 2.4–3

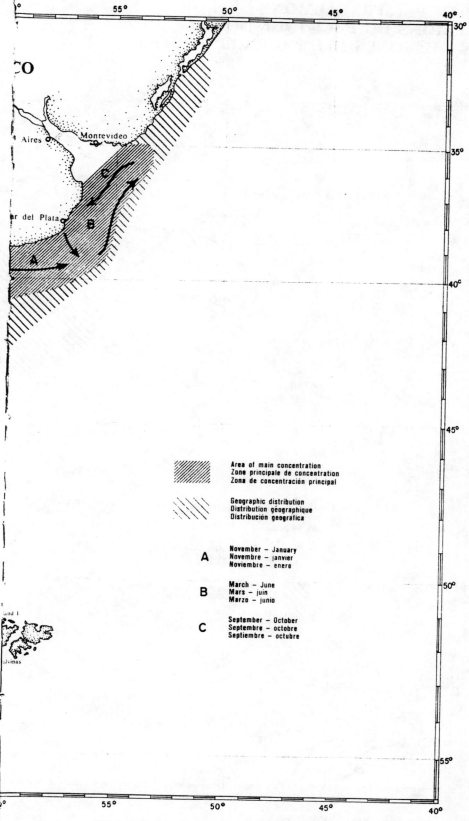

(Reprinted from FAO Department of Fisheries, *Atlas of the Living Resources of the Seas*, Rome, 1972, by permission of the Food and Agricultural Organization of the United Nations.)

MIGRATION OF NORTH PACIFIC SALMONS
MIGRATION DES SAUMONS DU PACIFIQUE NORD
MIGRACION DE LOS SALMONES DEL PACIFICO NORTE

Figure 2.4—4

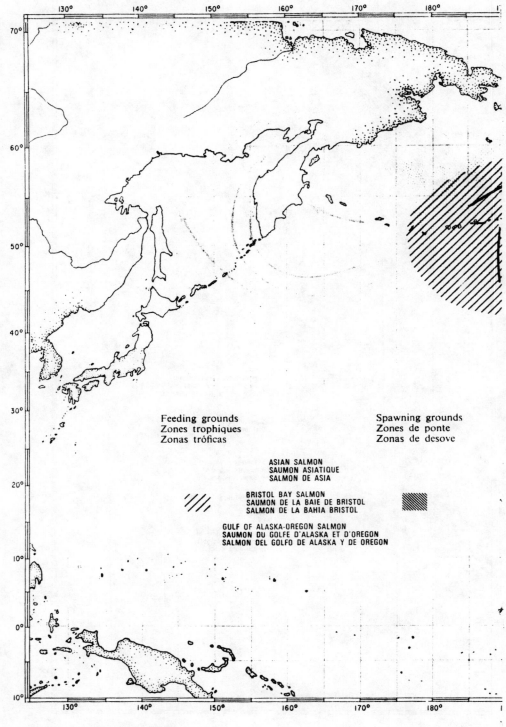

Feeding grounds
Zones trophiques
Zonas tróficas

Spawning grounds
Zones de ponte
Zonas de desove

ASIAN SALMON
SAUMON ASIATIQUE
SALMON DE ASIA

BRISTOL BAY SALMON
SAUMON DE LA BAIE DE BRISTOL
SALMON DE LA BAHIA BRISTOL

GULF OF ALASKA-OREGON SALMON
SAUMON DU GOLFE D'ALASKA ET D'OREGON
SALMON DEL GOLFO DE ALASKA Y DE OREGON

After/d'après/según: INPFC (1965)

Figure 2.4–4

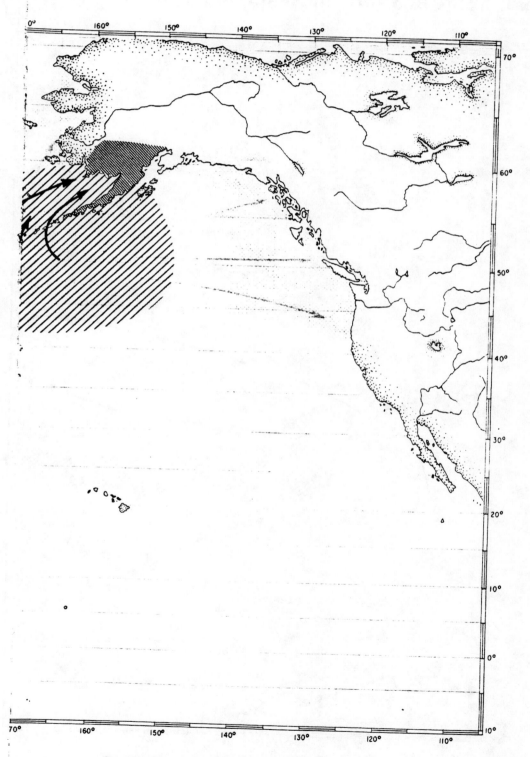

(Reprinted from FAO Department of Fisheries, *Atlas of the Living Resources of the Seas*, Rome, 1972, by permission of the Food and Agricultural Organization of the United Nations.)

MIGRATION OF NORTH PACIFIC ALBACORE
MIGRATION DU GERMON DU PACIFIQUE NORD
MIGRACION DEL ATUN BLANCO DE PACIFICO NORTE

Figure 2.4–5

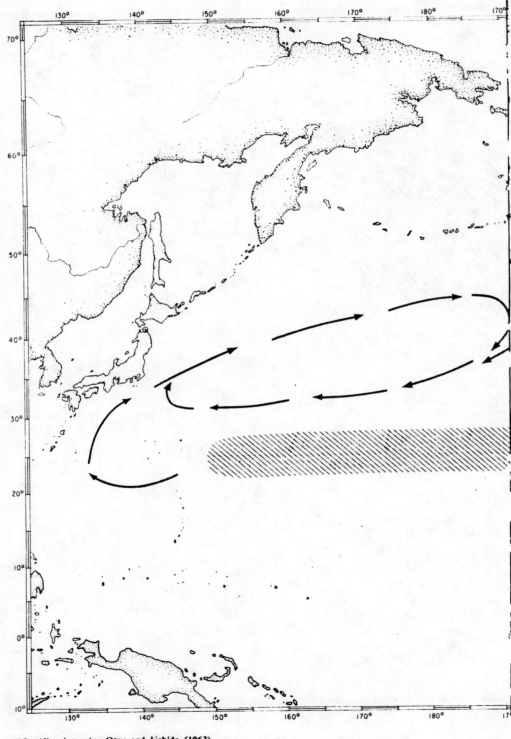

After/d'après/según: Otsu and Uchida (1963)

Figure 2.4—5

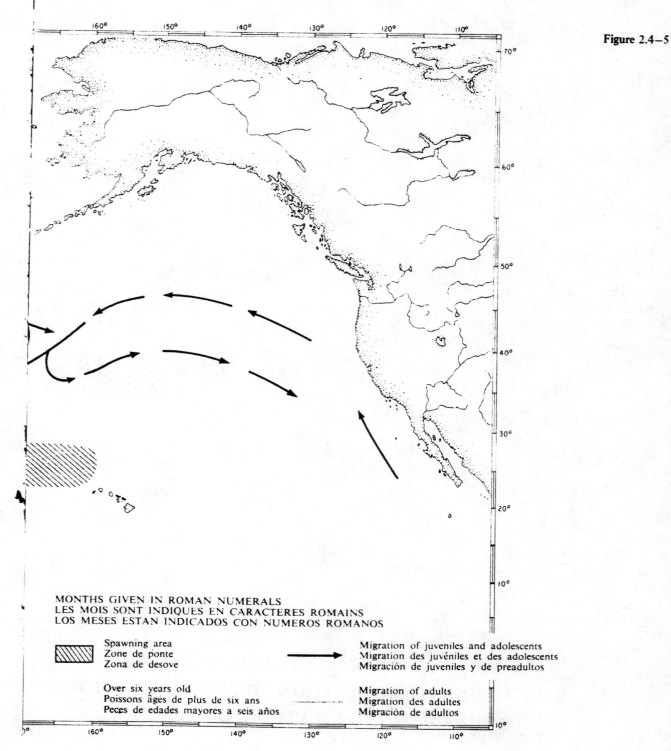

MONTHS GIVEN IN ROMAN NUMERALS
LES MOIS SONT INDIQUES EN CARACTERES ROMAINS
LOS MESES ESTAN INDICADOS CON NUMEROS ROMANOS

Spawning area
Zone de ponte
Zona de desove

Migration of juveniles and adolescents
Migration des juvéniles et des adolescents
Migración de juveniles y de preadultos

Over six years old
Poissons âgés de plus de six ans
Peces de edades mayores a seis años

Migration of adults
Migration des adultes
Migración de adultos

Figure 2.4—6

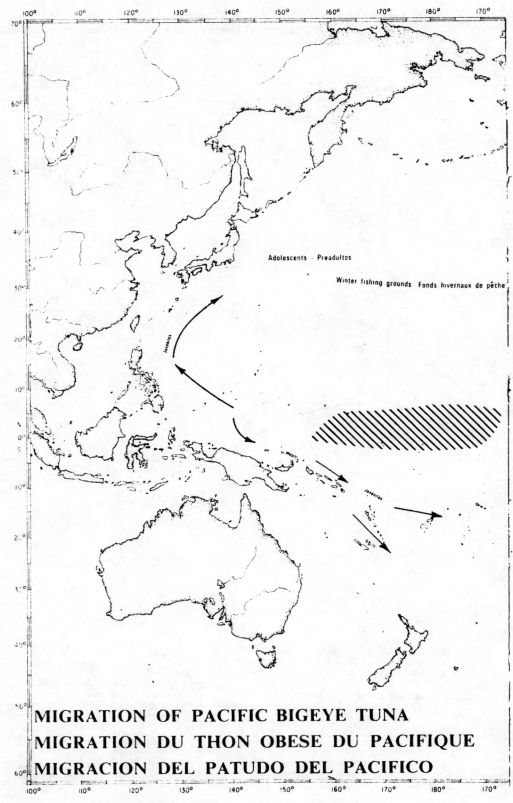

Adolescents · Preadultos

Winter fishing grounds Fonds hivernaux de pêche

MIGRATION OF PACIFIC BIGEYE TUNA
MIGRATION DU THON OBESE DU PACIFIQUE
MIGRACION DEL PATUDO DEL PACIFICO

After/D'après/Según: Nakamura (1969)

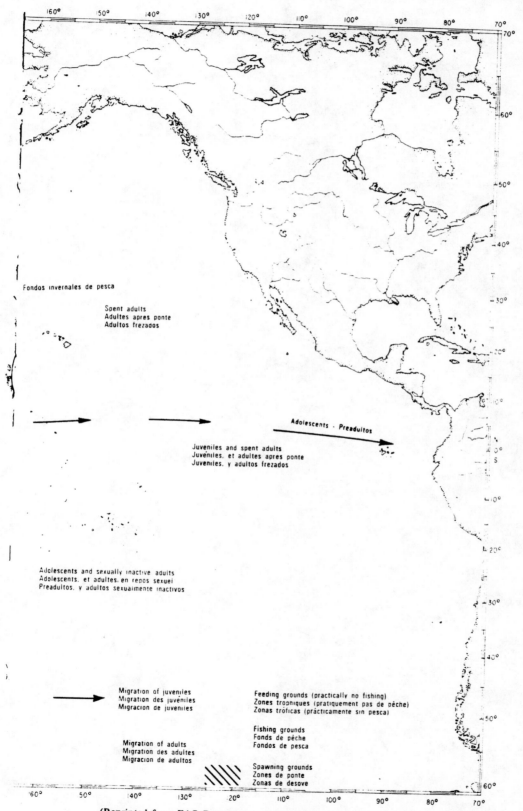

Figure 2.4–6

Fondos invernales de pesca

Spent adults
Adultes apres ponte
Adultos frezados

Adolescents · Preadultos

Juveniles and spent adults
Juvéniles, et adultes apres ponte
Juveniles, y adultos frezados

Adolescents and sexually inactive adults
Adolescents, et adultes, en repos sexuei
Preadultos, y adultos sexualmente inactivos

Migration of juveniles
Migration des juvéniles
Migracion de juveniles

Feeding grounds (practically no fishing)
Zones trophiques (pratiquement pas de pêche)
Zonas tróficas (prácticamente sin pesca)

Fishing grounds
Fonds de pêche
Fondos de pesca

Migration of adults
Migration des adultes
Migracion de adultos

Spawning grounds
Zones de ponte
Zonas de desove

(Reprinted from FAO Department of Fisheries, *Atlas of the Living Resources of the Seas*, Rome, 1972, by permission of the Food and Agricultural Organization of the United Nations.)

97

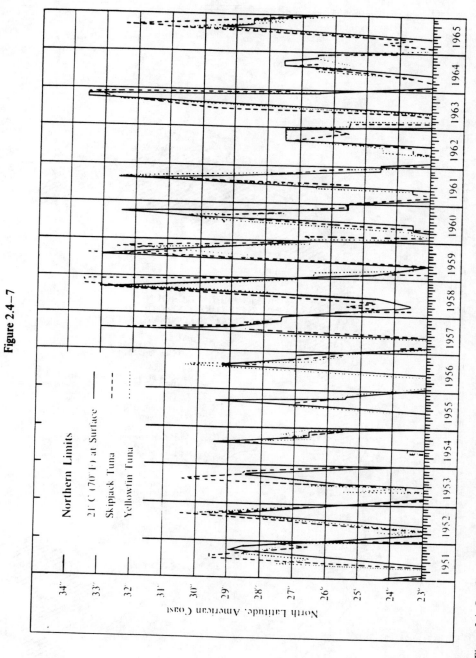

Figure 2.4–7

Figure 2.4–7. Changes in latitudinal position of the 21° C surface isotherm and of the northern limits of commercially caught yellowfin and skipjack tunas off the coast of Baja, California and California, 1957–1965.

MIGRATION OF GULF OF MEXICO PENAEID SHRIMPS
MIGRATION DES CREVETTES PENAEIDES DU GOLFE DU MEXIQUE
MIGRACION DE LOS CAMARONES PENEIDOS DEL GOLFO DE MEXICO

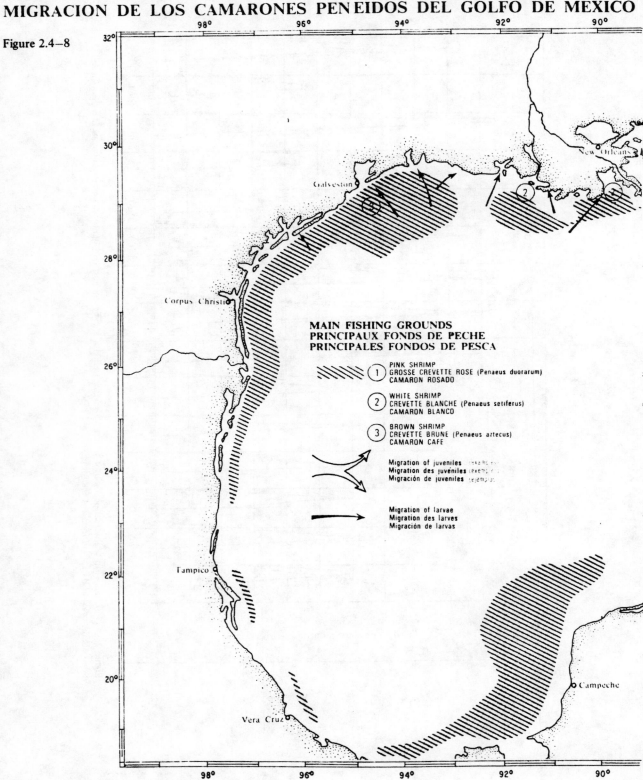

Figure 2.4—8

MAIN FISHING GROUNDS
PRINCIPAUX FONDS DE PECHE
PRINCIPALES FONDOS DE PESCA

1 PINK SHRIMP
GROSSE CREVETTE ROSE (Penaeus duorarum)
CAMARON ROSADO

2 WHITE SHRIMP
CREVETTE BLANCHE (Penaeus setiferus)
CAMARON BLANCO

3 BROWN SHRIMP
CREVETTE BRUNE (Penaeus aztecus)
CAMARON CAFE

Migration of juveniles (example)
Migration des juvéniles (exemple)
Migración de juveniles (ejemplo)

Migration of larvae
Migration des larves
Migración de larvas

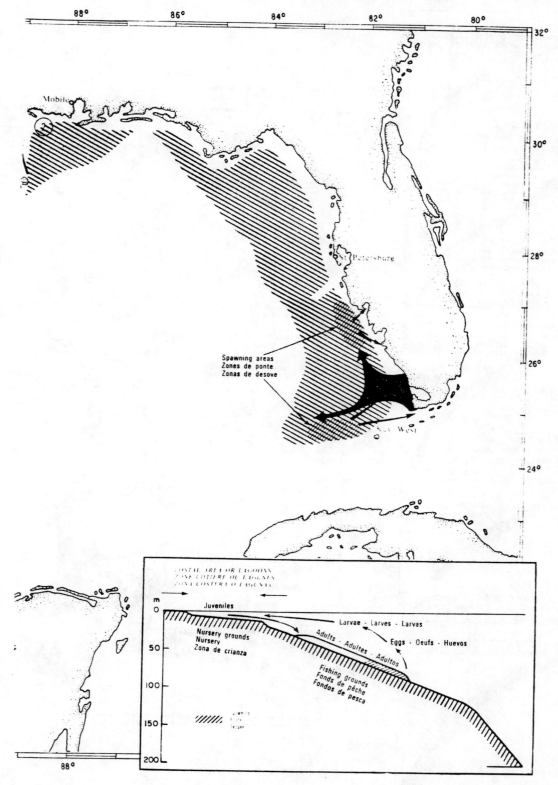

Figure 2.4—8

Figure 2.5–1

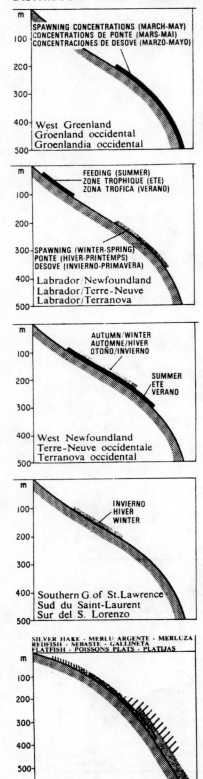

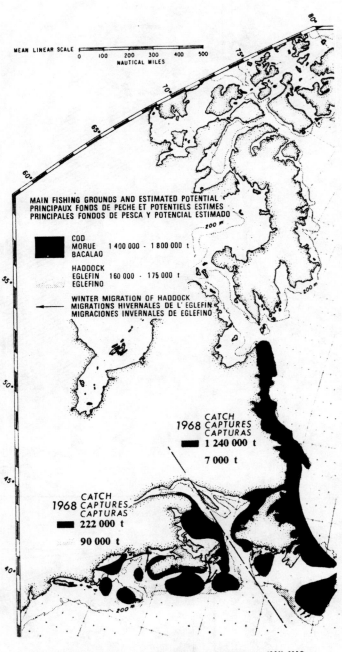

Proyección del mapa base: zona de iguales azimutes, con el centro en 40°N, 35°O.
Preparado por M.J. Pollack y G.G. Pasley, Woods Hole Oceanographic Institute

Figure 2.5—1

NORTH WEST ATLANTIC DEMERSAL RESOURCES
ATLANTIQUE NORD-OUEST - RESSOURCES DEMERSALES
ATLANTICO NORDOCCIDENTAL - RECURSOS DEMERSALES

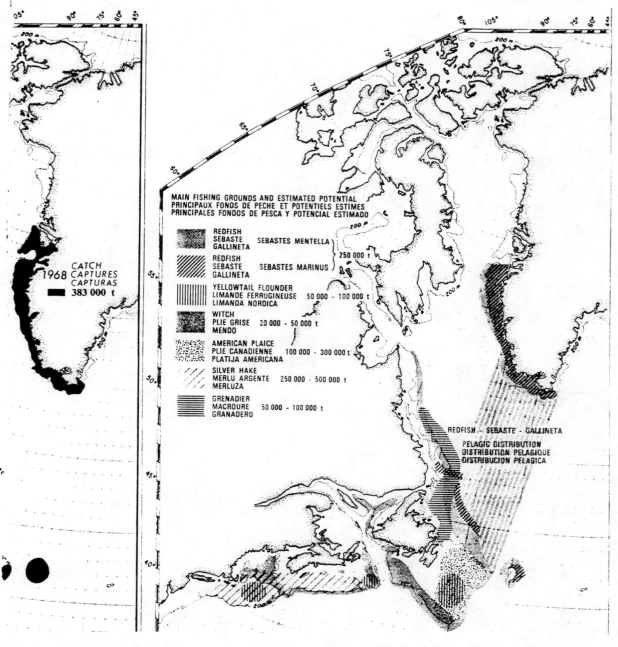

MAIN FISHING GROUNDS AND ESTIMATED POTENTIAL
PRINCIPAUX FONDS DE PECHE ET POTENTIELS ESTIMES
PRINCIPALES FONDOS DE PESCA Y POTENCIAL ESTIMADO

REDFISH
SEBASTE   SEBASTES MENTELLA
GALLINETA   } 250 000 t

REDFISH
SEBASTE   SEBASTES MARINUS
GALLINETA

YELLOWTAIL FLOUNDER
LIMANDE FERRUGINEUSE   50 000 - 100 000 t
LIMANDA NORDICA

WITCH
PLIE GRISE   20 000 - 50 000 t
MENDO

AMERICAN PLAICE
PLIE CANADIENNE   100 000 - 300 000 t
PLATIJA AMERICANA

SILVER HAKE
MERLU ARGENTE   250 000 - 500 000 t
MERLUZA

GRENADIER
MACROURE   50 000 - 100 000 t
GRANADERO

CATCH
CAPTURES
1968 CAPTURAS
383 000 t

REDFISH - SEBASTE - GALLINETA
PELAGIC DISTRIBUTION
DISTRIBUTION PELAGIQUE
DISTRIBUCION PELAGICA

Carte de base: projection azimutale équivalente, centrée à 40°N, 35°O.
Etablie par M.J. Pollack et G.G. Pasley, Woods Hole Oceanographic Institute.

Base map projection: azimuthal equal area, centred at 40°N, 35°W.
Prepared by M.J. Pollack and G.G. Pasley, Woods Hole Oceanographic Institute.

NORTH WEST ATLANTIC PELAGIC RESOURCES
ATLANTIQUE NORD-OUEST - RESSOURCES PELAGIQUES
ATLANTICO NORDOCCIDENTAL - RECURSOS PELAGICOS

Figure 2.5–2

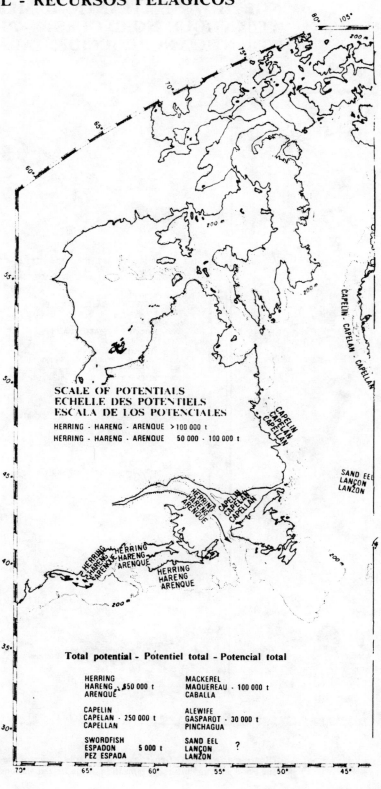

MEAN LINEAR SCALE

0   100   200   300   400   500
NAUTICAL MILES

Proyección del mapa base zona de iguales azimutes, con el centro en 40°N 35°O
Preparado por M J Pollack y G G Pasley, Woods Hole Oceanographic Institute

Carte de base projection azimutale équivalente, centrée a 40°N, 35°O
Etablie par M J Pollack et G G Pasley, Woods Hole Oceanographic Institute

Base map projection azimuthal equal area, centred at 40°N, 35°W
Prepared by M J Pollack and G G Pasley, Woods Hole Oceanographic Institute

SCALE OF POTENTIALS
ECHELLE DES POTENTIELS
ESCALA DE LOS POTENCIALES

HERRING - HARENG - ARENQUE >100 000 t
HERRING - HARENG - ARENQUE 50 000 - 100 000 t

Total potential - Potentiel total - Potencial total

HERRING HARENG ARENQUE	550 000 t	MACKEREL MAQUEREAU CABALLA	100 000 t
CAPELIN CAPELAN CAPELLAN	250 000 t	ALEWIFE GASPAROT PINCHAGUA	30 000 t
SWORDFISH ESPADON PEZ ESPADA	5 000 t	SAND EEL LANCON LANZON	?

(Reprinted from FAO Department of Fisheries, *Atlas of the Living Resources of the Seas*, Rome, 1972, by permission of the Food and Agricultural Organization of the United Nations.)

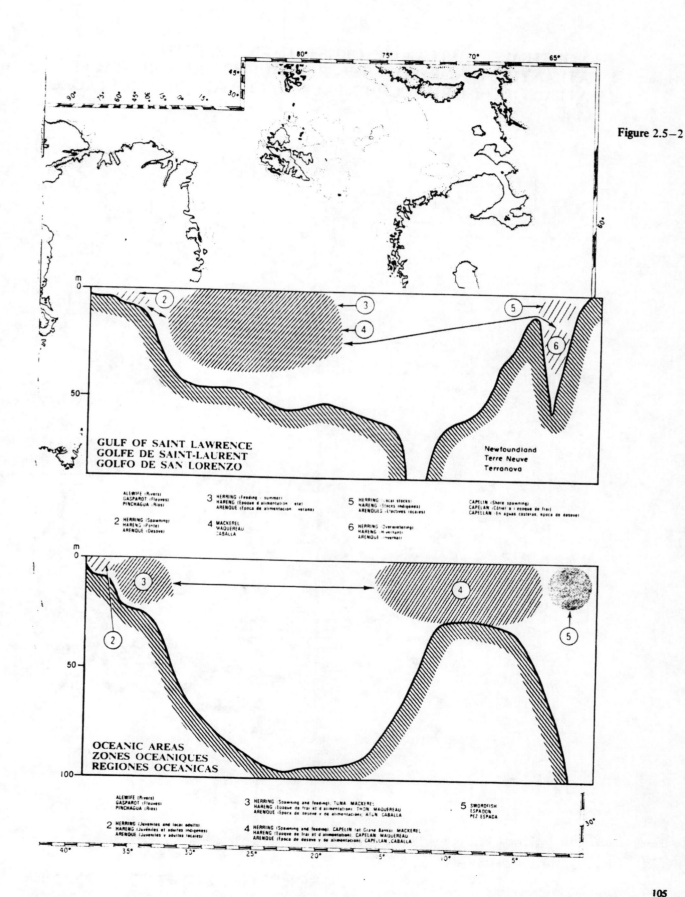

Figure 2.5—2

GULF OF SAINT LAWRENCE
GOLFE DE SAINT-LAURENT
GOLFO DE SAN LORENZO

Newfoundland
Terre Neuve
Terranova

ALEWIFE (Rivers)
GASPAROT (Fleuves)
PINCHAGUA (Rios)

2 HERRING (Spawning)
HARENG (Ponte)
ARENQUE (Desove)

3 HERRING (Feeding summer)
HARENG (Epoque d alimentation ete)
ARENQUE (Epoca de alimentacion verano)

4 MACKEREL
MAQUEREAU
CABALLA

5 HERRING (local stocks)
HARENG (Stocks indigenes)
ARENQUES (L'actives locales)

6 HERRING Overwintering)
HARENG (Hivernant)
ARENQUE (Invernal)

CAPELIN (Shore spawning)
CAPELAN (Côtier a l epoque de frai)
CAPELLAN (En aguas costeras, epoca de desove)

OCEANIC AREAS
ZONES OCEANIQUES
REGIONES OCEANICAS

ALEWIFE (Rivers)
GASPAROT (Fleuves)
PINCHAGUA (Rios)

2 HERRING (Juveniles and local adults)
HARENG (Juvéniles et adultes indigènes)
ARENQUE (Juveniles y adultos locales)

3 HERRING (Spawning and feeding), TUNA, MACKEREL
HARENG (Epoque de frai et d alimentation) THON MAQUEREAU
ARENQUE (Epoca de desove y de alimentacion), ATUN, CABALLA

4 HERRING (Spawning and feeding), CAPELIN (at Grand Banks), MACKEREL
HARENG (Epoque de frai et d alimentation) CAPELAN MAQUEREAU
ARENQUE (Epoca de desove y de alimentacion), CAPELLAN, CABALLA

5 SWORDFISH
ESPADON
PEZ ESPADA

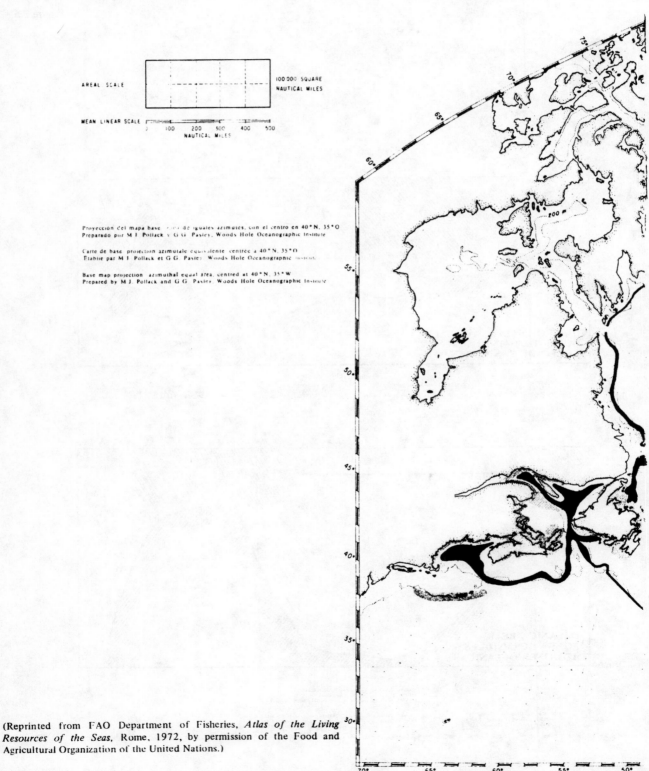

NORTH WEST ATLANTIC CRUSTACEAN RESOURCES
ATLANTIQUE NORD-OUEST - RESSOURCES EN CRUSTACES
ATLANTICO NORDOCCIDENTAL - RECURSOS DE CRUSTACEOS

Figure 2.5–3

AREAL SCALE

100 000 SQUARE
NAUTICAL MILES

MEAN LINEAR SCALE

0   100   200   300   400   500
NAUTICAL MILES

Proyección del mapa base zona de iguales azimutes, con el centro en 40°N, 35°O
Preparado por M J Pollack y G G Pasley, Woods Hole Oceanographic Institute

Carte de base projection azimutale équivalente centrée à 40°N, 35°O
Etablie par M J Pollack et G G Pasley, Woods Hole Oceanographic Institut

Base map projection azimuthal equal area, centred at 40°N, 35°W
Prepared by M J Pollack and G G Pasley, Woods Hole Oceanographic Institute

200 m

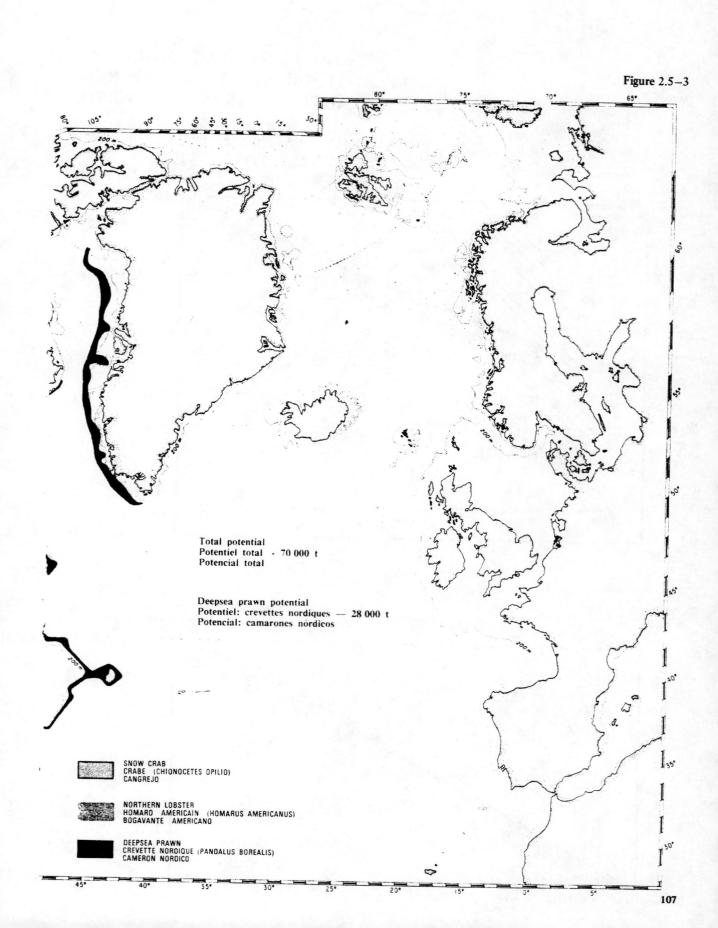

Figure 2.5—3

Total potential
Potentiel total - 70 000 t
Potencial total

Deepsea prawn potential
Potentiel: crevettes nordiques — 28 000 t
Potencial: camarones nórdicos

SNOW CRAB
CRABE (CHIONOCETES OPILIO)
CANGREJO

NORTHERN LOBSTER
HOMARD AMERICAIN (HOMARUS AMERICANUS)
BOGAVANTE AMERICANO

DEEPSEA PRAWN
CREVETTE NORDIQUE (PANDALUS BOREALIS)
CAMERON NORDICO

Figure 2.5—4

NORTH ATLANTIC MOLLUSCS
ATLANTIQUE NORD - RESSOURCES
ATLANTICO NORTE - RECURSOS

INCLUDES RESOURCES OF NORTHWEST AND NORTHEAST ATLANTIC
RESSOURCES DE L'ATLANTIQUE NORD-OUEST ET NORD-EST EGALEMENT
REPRESENTEES SUR CES COUPES
EN ESTAS SECCIONES SE MUESTRAN LOS RECURSOS DEL ATLANTICO
NORDOCCIDENTAL Y NORDORIENTAL

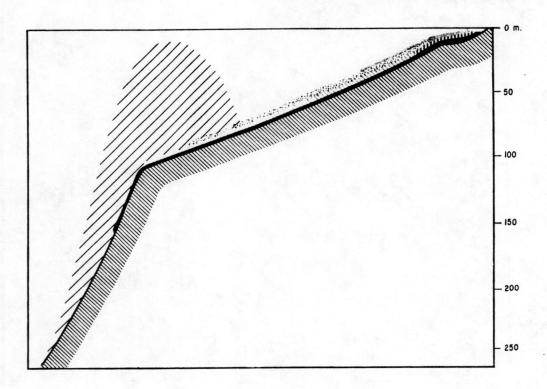

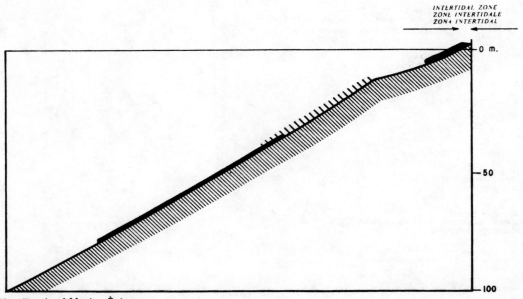

AND CRUSTACEAN RESOURCES
EN MOLLUSQUES ET EN CRUSTACES
DE MOLUSCOS Y CRUSTACEOS

Figure 2.5—4

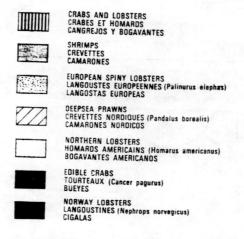

CRABS AND LOBSTERS
CRABES ET HOMARDS
CANGREJOS Y BOGAVANTES

SHRIMPS
CREVETTES
CAMARONES

EUROPEAN SPINY LOBSTERS
LANGOUSTES EUROPEENNES (Palinurus elephas)
LANGOSTAS EUROPEAS

DEEPSEA PRAWNS
CREVETTES NORDIQUES (Pandalus borealis)
CAMARONES NORDICOS

NORTHERN LOBSTERS
HOMARDS AMERICAINS (Homarus americanus)
BOGAVANTES AMERICANOS

EDIBLE CRABS
TOURTEAUX (Cancer pagurus)
BUEYES

NORWAY LOBSTERS
LANGOUSTINES (Nephrops norvegicus)
CIGALAS

DEEPSEA PRAWN : has a large vertical distribution but is caught only by bottom trawl
CREVETTES NORDIQUE : ont une grande dispersion verticale mais ne sont pêchées qu'au chalut de fond
CAMARONES (P. Borealis): tienen una amplia distribución vertical, pero se capturan solamente con redes de arrastre de fondo

NORWAY LOBSTER: in the Irish Sea its extension is limited to a depth of 10 m
LANGOUSTINES: dans la Mer d'Irlande ne descendent pas en dessous de 10 m
CIGALA: en el Mar de Irlanda se extienden hasta la profundidad de 10 m

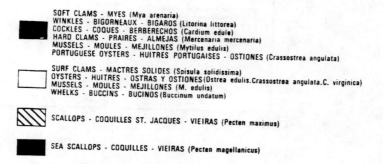

SOFT CLAMS - MYES (Mya arenaria)
WINKLES - BIGORNEAUX - BIGAROS (Litorina littorea)
COCKLES - COQUES - BERBERECHOS (Cardium edule)
HARD CLAMS - PRAIRES - ALMEJAS (Mercenaria mercenaria)
MUSSELS - MOULES - MEJILLONES (Mytilus edulis)
PORTUGUESE OYSTERS - HUITRES PORTUGAISES - OSTIONES (Crassostrea angulata)

SURF CLAMS - MACTRES SOLIDES (Spisula solidissima)
OYSTERS - HUITRES - OSTRAS Y OSTIONES(Ostrea edulis.Crassostrea angulata.C. virginica)
MUSSELS - MOULES - MEJILLONES (M. edulis)
WHELKS - BUCCINS - BUCINOS (Buccinum undatum)

SCALLOPS - COQUILLES ST. JACQUES - VIEIRAS (Pecten maximus)

SEA SCALLOPS - COQUILLES - VIEIRAS (Pecten magellanicus)

NORTH EAST ATLANTIC PELAGIC RESOURCES
ATLANTIQUE NORD-EST - RESSOURCES PELAGIQUES
ATLANTICO NORDORIENTAL - RECURSOS PELAGICOS

Figure 2.5–5

ATLANTO SCANDIAN HERRING   (100 000 t at present unproductive)
HARENG ATLANTICO SCANDINAVE (100 000 t a present improductif)
ARENQUE ATLANTICO ESCANDINAVO (100 000 t actualmente improductivo)

NORTH SEA HERRING
HARENG DE LA MER DU NORD (1 000 000 t)
ARENQUE DEL MAR DEL NORTE

BALTIC HERRING
HARENG DE LA BALTIQUE   (50 000 (?)t)
ARENQUE DEL MAR BALTICO

SHELF HERRING
HARENG DU PLATEAU   200 000 t)
ARENQUE DE LA PLATAFORMA

IRISH SEA AND DUNMORE HERRING
HARENG DE LA MER D'IRLANDE ET DE DUNMORE (25 000 t)
ARENQUE DEL MAR DE IRLANDA Y DE DUNMORE

MACKEREL AND HORSE MACKEREL   potential unknown
MAQUEREA ET CHINCHARD   potentiel inconnu
CABALLA Y JUREL   potencial desconocido

CAPELIN   CAPELAN   CAPELLAN (1 000 000   2 000 000 t)

ALBACORE GERMON   ATUN BLANCO (ALBACORA)   } (70 000 t)

BLUEFIN   THON ROUGE ATUN

PILCHARD SARDINE   SARDINA (300 000 t)

ANCHOVY   ANCHOIS   ANCHOA (300 000 (?)t)

BLUE WHITING   POUTASSOU   BACALADILLA (>1 000 000 t)

MACKEREL   MAQUEREAU   CABALLA (400 000 t)

AREAL SCALE   100 000 SQUARE NAUTICAL MILES

MEAN LINEAR SCALE   0   100   200   300   400   500   NAUTICAL MILES

CAPELIN: NORTH NORWAY – BARENTS SEA
CAPELAN: NORVEGE SEPTENTRIONALE – MER DE BARENTS
CAPELLAN: NORTE DE NORUEGA – MAR DE BARENTS

Summer fishery for capelin 300 000 t (1970)
Pêche d'été du capelan 300 000 t (1970)
Pesca de verano del capelan 300 000 t (1970)

Fishery for spawning capelin about 1 000 000 tons 1970
Pêche du capelan en période de frai environ 1 000 000 t en 1970
Pesca del capellan en periodo de desove aproximadamente 1 000 000 t en 1970

ATLANTO-SCANDIAN HERRING: WEST NORWAY – NORWEGIAN SEA
HARENGS ATLANTO-SCANDINAVES: NORVEGE OCCIDENTALE – MER DE NORVEGE
ARENQUE ATLANTO-ESCANDINAVO: OESTE DE NORUEGA – MAR DE NORUEGA

Summer fishery for young and adult herring
Pêche d'été des harengs jeunes et adultes
Pesca de verano de arenque jóven y adulto

Fishery for spawning, herring and young herring
Pêche des harengs en période de frai et des jeunes
Pesca del arenque en desove y del jóven

Base map projection: azimuthal equal area, centred at 40°N, 35°W
Prepared by M.J. Pollack and G.G. Pasley, Woods Hole Oceanographic Institute

Carte de base projection azimutale équivalente, centrée à 40°N, 35°O
Etablie par M.J. Pollack et G.G. Pasley, Woods Hole Oceanographic Institute

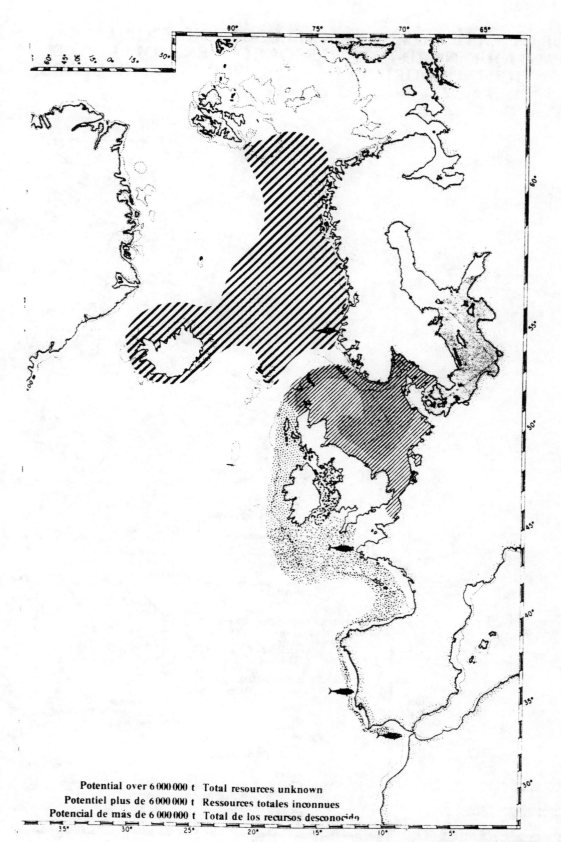

Figure 2.5–5

Potential over 6 000 000 t Total resources unknown
Potentiel plus de 6 000 000 t Ressources totales inconnues
Potencial de más de 6 000 000 t Total de los recursos desconocido

Proyección del mapa base: zona de iguales azimutes, con el centro a 40°N, 35°O
Preparado por M.J. Pollack y G.G. Pasley Woods Hole Oceanographic Institute

(Reprinted from FAO Department of Fisheries, *Atlas of the Living Resources of the Seas*, Rome, 1972, by permission of the Food and Agricultural Organization of the United Nations.)

111

NORTH EAST ATLANTIC MOLLUSCS AND CRUSTACEAN
ATLANTIQUE NORD-EST - RESSOURCES EN MOLLUSQUES
ATLANTICO NORDORIENTAL - RECURSOS DE MOLUSCOS

Figure 2.5—6

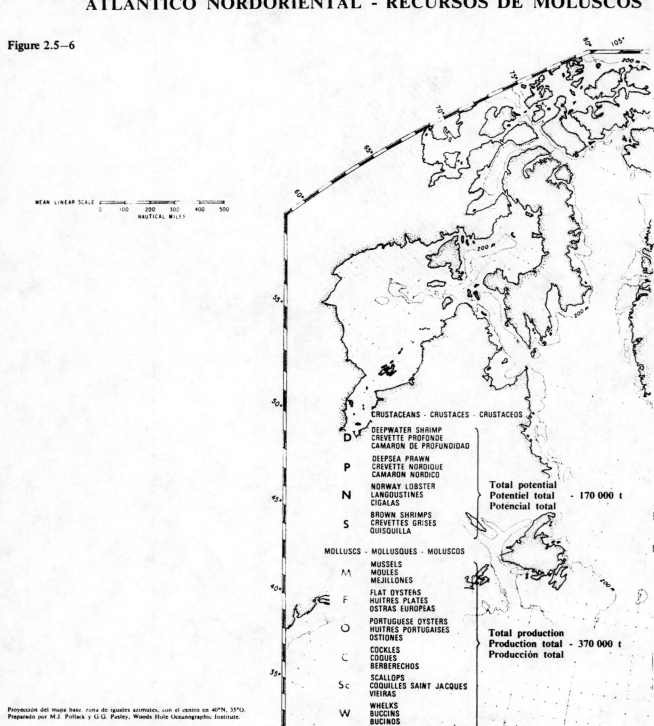

MEAN LINEAR SCALE

0   100   200   300   400   500
NAUTICAL MILES

CRUSTACEANS · CRUSTACES · CRUSTACEOS

D   DEEPWATER SHRIMP
    CREVETTE PROFONDE
    CAMARON DE PROFUNDIDAD

P   DEEPSEA PRAWN
    CREVETTE NORDIQUE
    CAMARON NORDICO

N   NORWAY LOBSTER
    LANGOUSTINES
    CIGALAS

S   BROWN SHRIMPS
    CREVETTES GRISES
    QUISQUILLA

Total potential
Potentiel total   - 170 000 t
Potencial total

MOLLUSCS · MOLLUSQUES · MOLUSCOS

M   MUSSELS
    MOULES
    MEJILLONES

F   FLAT OYSTERS
    HUITRES PLATES
    OSTRAS EUROPEAS

O   PORTUGUESE OYSTERS
    HUITRES PORTUGAISES
    OSTIONES

C   COCKLES
    COQUES
    BERBERECHOS

Sc SCALLOPS
    COQUILLES SAINT JACQUES
    VIEIRAS

W   WHELKS
    BUCCINS
    BUCINOS

E   SQUID
    ENCORNET
    CALAMAR

Total production
Production total - 370 000 t
Producción total

Proyección del mapa base, zona de iguales azimutes, con el centro en 40°N, 35°O.
Preparado por M.J. Pollack y G.G. Pasley, Woods Hole Oceanographic Institute.

Carte de base: projection azimutale équivalente, centrée à 40°N, 35°O.
Etablie par M.J. Pollack et G.G. Pasley, Woods Hole Oceanographic Institute.

Base map projection: azimuthal equal area projection, centred at 40°N, 35°W
Prepared by M.J. Pollack and G.G. Pasley, Woods Hole Oceanographic Institute

RESOURCES
ET EN CRUSTACES
CRUSTACEOS

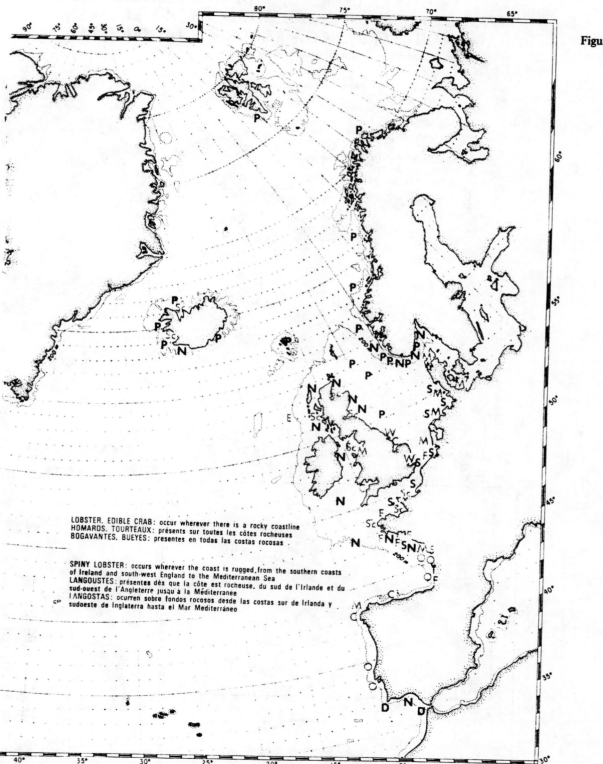

Figure 2.5—6

LOBSTER. EDIBLE CRAB: occur wherever there is a rocky coastline
HOMARDS, TOURTEAUX: présents sur toutes les côtes rocheuses
BOGAVANTES, BUEYES: presentes en todas las costas rocosas .

SPINY LOBSTER: occurs wherever the coast is rugged,from the southern coasts
of Ireland and south-west England to the Mediterranean Sea
LANGOUSTES: présentes dès que la côte est rocheuse, du sud de l'Irlande et du
sud-ouest de l'Angleterre jusqu'à la Méditerranee
LANGOSTAS: ocurren sobre fondos rocosos desde las costas sur de Irlanda y
sudoeste de Inglaterra hasta el Mar Mediterráneo

(Reprinted from FAO Department of Fisheries, *Atlas of the Living Resources of the Seas*, Rome, 1972, by permission of the Food and Agricultural Organization of the United Nations.)

Figure 2.5–7

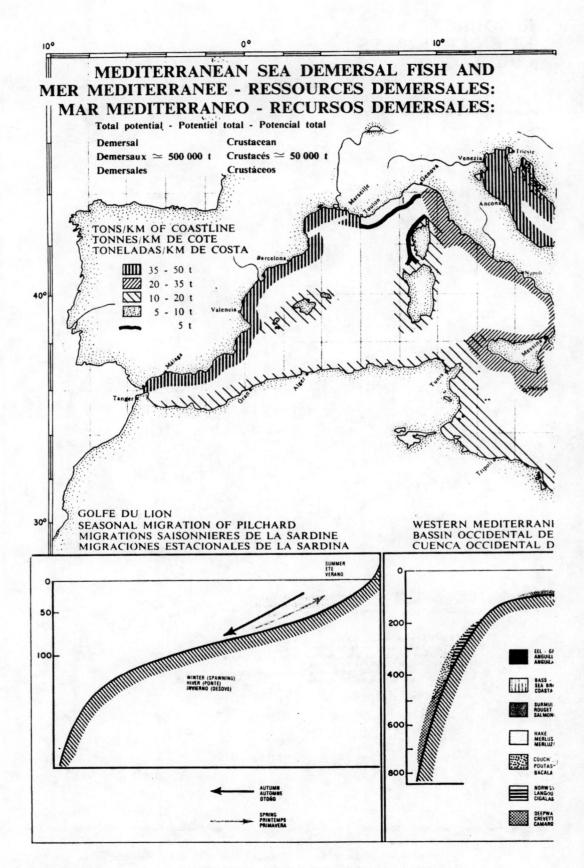

MEDITERRANEAN SEA DEMERSAL FISH AND
MER MEDITERRANEE - RESSOURCES DEMERSALES:
MAR MEDITERRANEO - RECURSOS DEMERSALES:

Total potential - Potentiel total - Potencial total

Demersal	Crustacean
Demersaux ≃ 500 000 t	Crustacés ≃ 50 000 t
Demersales	Crustáceos

TONS/KM OF COASTLINE
TONNES/KM DE COTE
TONELADAS/KM DE COSTA

35 - 50 t
20 - 35 t
10 - 20 t
5 - 10 t
5 t

GOLFE DU LION
SEASONAL MIGRATION OF PILCHARD
MIGRATIONS SAISONNIERES DE LA SARDINE
MIGRACIONES ESTACIONALES DE LA SARDINA

WESTERN MEDITERRANI
BASSIN OCCIDENTAL DE
CUENCA OCCIDENTAL D

SUMMER
ETE
VERANO

WINTER (SPAWNING)
HIVER (PONTE)
INVIERNO (DESOVE)

AUTUMN
AUTOMNE
OTOÑO

SPRING
PRINTEMPS
PRIMAVERA

EEL · GI
ANGUILL
ANGUILA

BASS ·
SEA BR
COASTA

SURMUI
ROUGET
SALMON

HAKE ·
MERLUS
MERLUZ

COUCH ·
POUTAS
BACALA

NORW ·
LANGOU
CIGALAS

DEEPWA
CREVETT
CAMARG

Figure 2.5–7

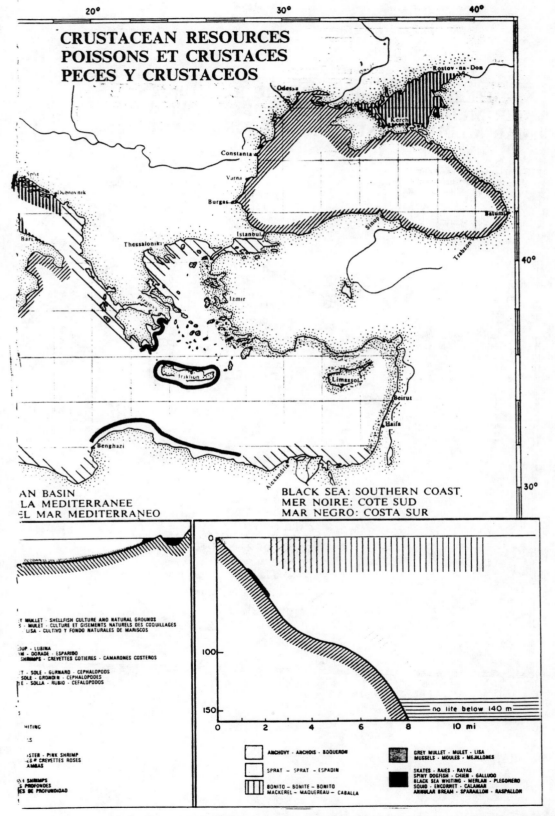

CRUSTACEAN RESOURCES
POISSONS ET CRUSTACES
PECES Y CRUSTACEOS

AN BASIN
LA MEDITERRANEE
EL MAR MEDITERRANEO

BLACK SEA: SOUTHERN COAST
MER NOIRE: COTE SUD
MAR NEGRO: COSTA SUR

no life below 140 m

Y MULLET - SHELLFISH CULTURE AND NATURAL GROUNDS
S - MULET - CULTURE ET GISEMENTS NATURELS DES COQUILLAGES
LISA - CULTIVO Y FONDO NATURALES DE MARISCOS

OUP - LUBINA
M - DORADE - ESPARIBO
SHRIMPS - CREVETTES COTIERES - CAMARONES COSTEROS

T - SOLE - GURNARD - CEPHALOPODS
SOLE - GRONDIN - CEPHALOPODES
E - SOLLA - RUBIO - CEFALOPODOS

HITING

STER - PINK SHRIMP
ES º CREVETTES ROSES
AMBAS

SHRIMPS
S PROFONDES
ES DE PROFUNDIDAD

ANCHOVY - ANCHOIS - BOQUERON

SPRAT - SPRAT - ESPADIN

BONITO - BONITE - BONITO
MACKEREL - MAQUEREAU - CABALLA

GREY MULLET - MULET - LISA
MUSSELS - MOULES - MEJILLONES

SKATES - RAIES - RAYAS
SPINY DOGFISH - CHIEN - GALLUDO
BLACK SEA WHITING - MERLAN - PLEGONERO
SQUID - ENCORNET - CALAMAR
ANNULAR BREAM - SPARAILLON - RASPALLON

(Reprinted from FAO Department of Fisheries, *Atlas of the Living Resources of the Seas,* Rome, 1972, by permission of the Food and Agricultural Organization of the United Nations.)

Figure 2.5–8

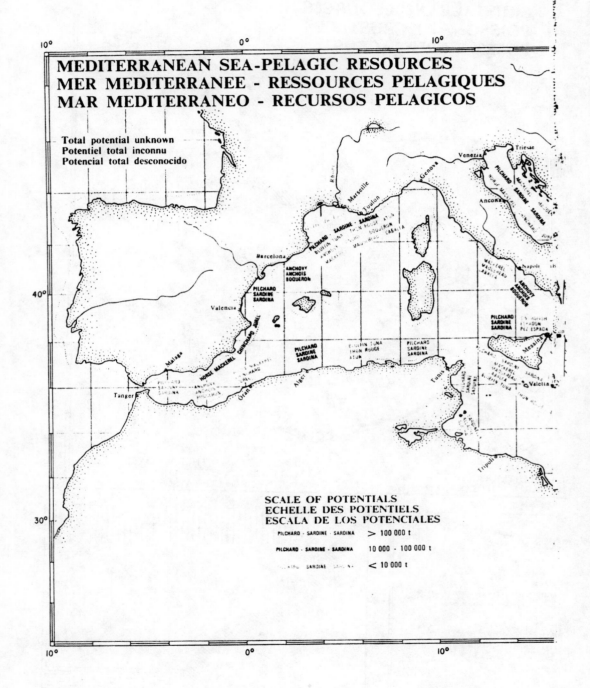

Figure 2.5–8

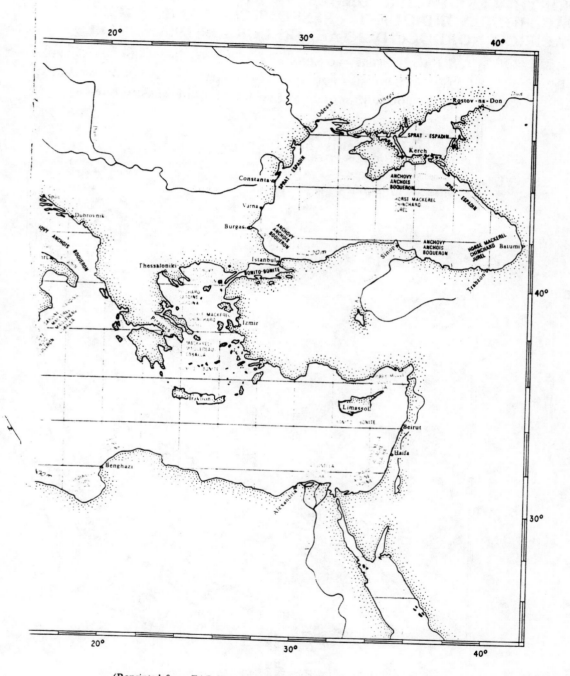

(Reprinted from FAO Department of Fisheries, *Atlas of the Living Resources of the Seas*, Rome, 1972, by permission of the Food and Agricultural Organization of the United Nations.)

Figure 2.5–9

NORTH WEST PACIFIC DEMERSAL RESOURCES
PACIFIQUE NORD-OUEST - RESSOURCES DEMERSALES
PACIFICO NORDOCCIDENTAL - RECURSOS DEMERSALES

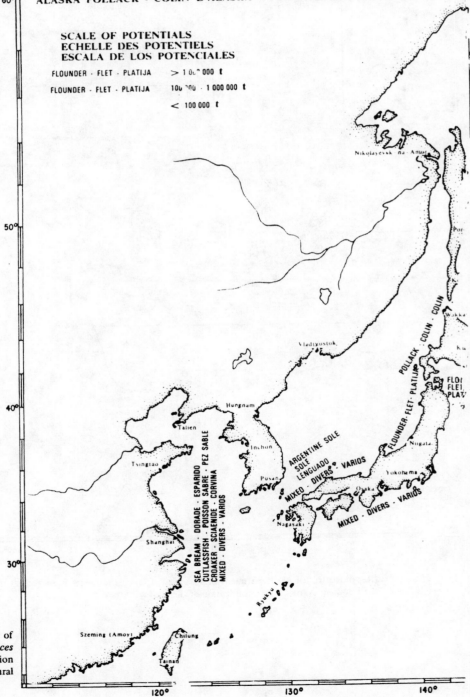

COLD WATER FISH - POSSONS D'EAUX FROIDES - PECES DE AGUAS FRIAS:

WARM WATER FISH - POISSONS D'EAUX CHAUDES - PECES DE AGUAS CAL

ALASKA POLLACK - COLIN D'ALASKA - COLIN DE ALASKA : 1 300 000 t

SCALE OF POTENTIALS
ECHELLE DES POTENTIELS
ESCALA DE LOS POTENCIALES

FLOUNDER · FLET · PLATIJA   > 1 0u˜ 000 t

FLOUNDER · FLET · PLATIJA   10u ˜˜u - 1 000 000 t

< 100 000 t

(Reprinted from FAO Department of Fisheries, *Atlas of the Living Resources of the Seas*, Rome, 1972, by permission of the Food and Agricultural Organization of the United Nations.)

Figure 2.5–9

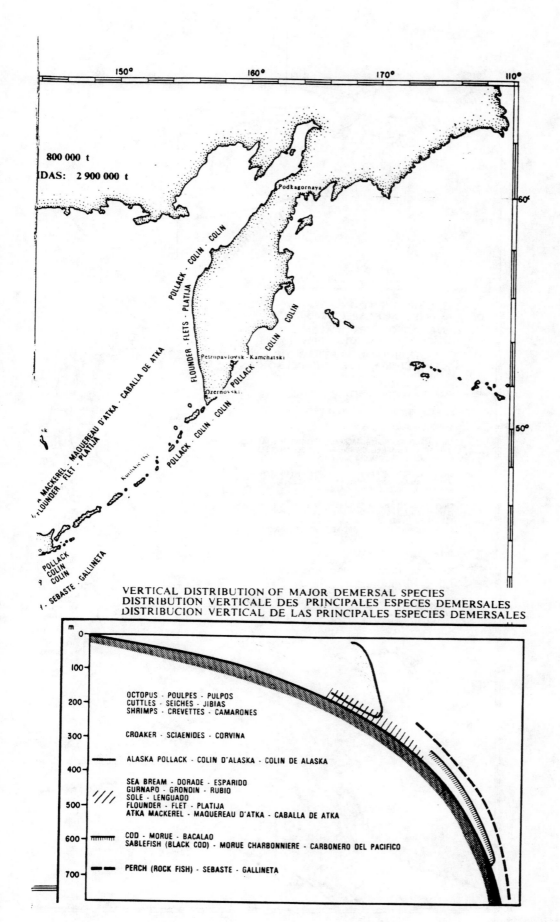

VERTICAL DISTRIBUTION OF MAJOR DEMERSAL SPECIES
DISTRIBUTION VERTICALE DES PRINCIPALES ESPECES DEMERSALES
DISTRIBUCION VERTICAL DE LAS PRINCIPALES ESPECIES DEMERSALES

Figure 2.5—10

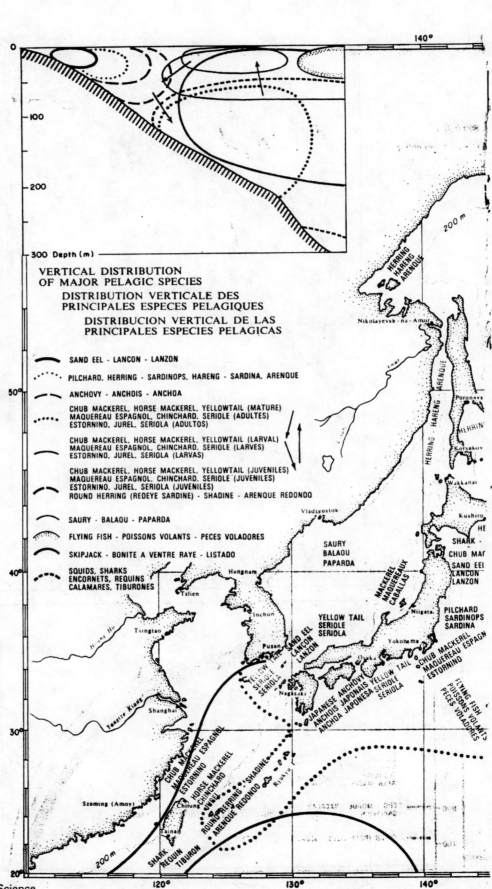

Figure 2.5–10

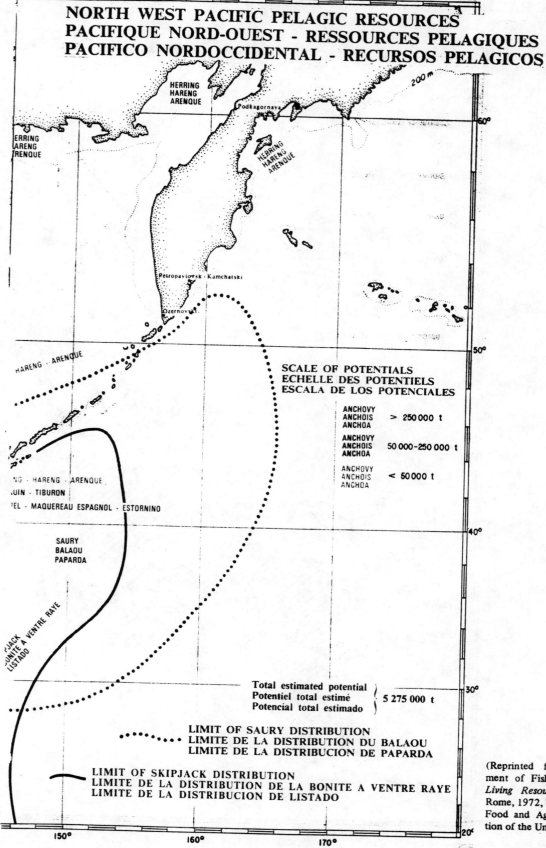

NORTH WEST PACIFIC PELAGIC RESOURCES
PACIFIQUE NORD-OUEST - RESSOURCES PELAGIQUES
PACIFICO NORDOCCIDENTAL - RECURSOS PELAGICOS

SCALE OF POTENTIALS
ECHELLE DES POTENTIELS
ESCALA DE LOS POTENCIALES

ANCHOVY ANCHOIS ANCHOA	> 250 000 t
ANCHOVY ANCHOIS ANCHOA	50 000-250 000 t
ANCHOVY ANCHOIS ANCHOA	< 50 000 t

Total estimated potential
Potentiel total estimé   5 275 000 t
Potencial total estimado

LIMIT OF SAURY DISTRIBUTION
LIMITE DE LA DISTRIBUTION DU BALAOU
LIMITE DE LA DISTRIBUCION DE PAPARDA

LIMIT OF SKIPJACK DISTRIBUTION
LIMITE DE LA DISTRIBUTION DE LA BONITE A VENTRE RAYE
LIMITE DE LA DISTRIBUCION DE LISTADO

(Reprinted from FAO Department of Fisheries, *Atlas of the Living Resources of the Seas*, Rome, 1972, by permission of the Food and Agricultural Organization of the United Nations.)

121

NORTH WEST PACIFIC CRUSTACEAN AND CEPHALOPOD RESOURCES

Figure 2.5—11

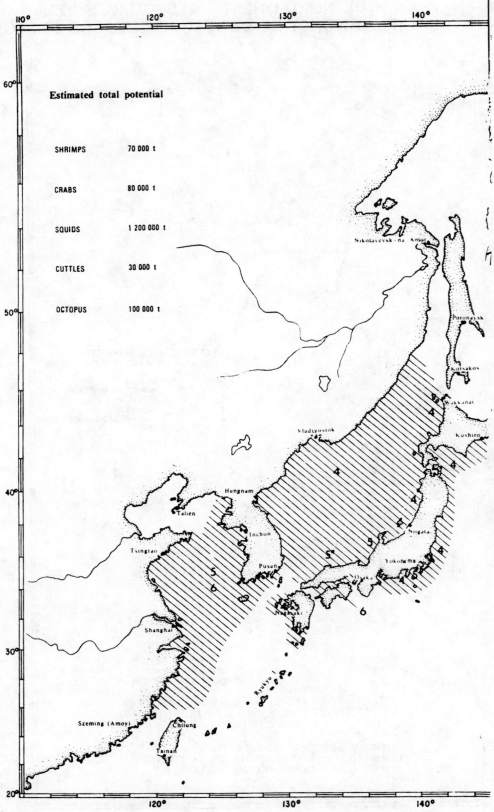

Estimated total potential

SHRIMPS	70 000 t
CRABS	80 000 t
SQUIDS	1 200 000 t
CUTTLES	30 000 t
OCTOPUS	100 000 t

Figure 2.5–11

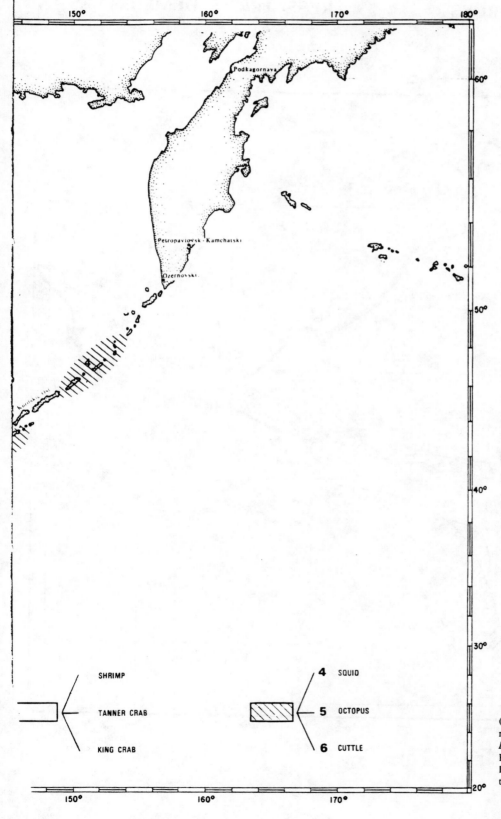

SHRIMP

TANNER CRAB

KING CRAB

4 SQUID

5 OCTOPUS

6 CUTTLE

(Reprinted from FAO Department of Fisheries, *Atlas of the Living Resources of the Seas*, Rome, 1972, by permission of the Food and Agricultural Organization of the United Nations.)

NORTH EAST PACIFIC
DEMERSAL RESOURCES

PACIFIQUE NORD-EST
RESSOURCES DEMERSALES

Figure 2.5—12

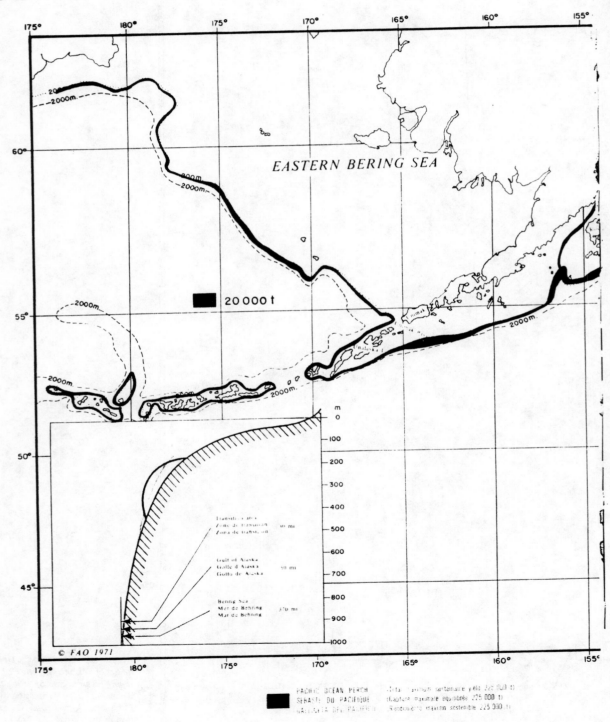

EASTERN BERING SEA

■ 20 000 t

© FAO 1971

■ PACIFIC OCEAN PERCH — (Data — maximum sustainable yield 225 000 t)
SEBASTE DU PACIFIQUE — (Capture maximale équilibrée 225 000 t)
GALLINETA DEL PACIFICO — Rendimiento máximo sostenible 225 000 t)

PACIFICO NORDORIENTAL
RECURSOS DEMERSALES

Figure 2.5–12

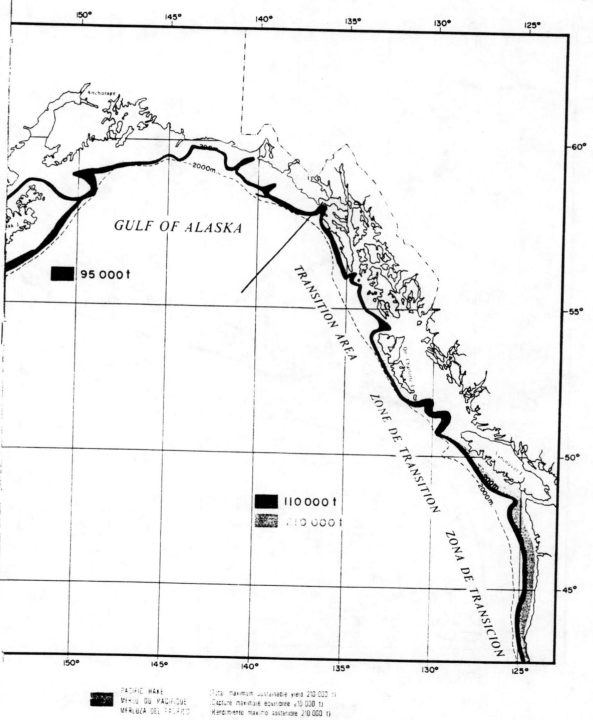

GULF OF ALASKA

■ 95 000 t

TRANSITION AREA

ZONE DE TRANSITION

ZONA DE TRANSICION

■ 110 000 t

░ ≥110 000 t

PACIFIC HAKE   (Total maximum sustainable yield 210 000 t)
MERLU DU PACIFIQUE   (Capture maximale equilibree ≥10 000 t)
MERLUZA DEL PACIFICO (Rendimiento maximo sostenible 210 000 t)

(Reprinted from FAO Department of Fisheries, *Atlas of the Living Resources of the Seas*, Rome, 1972, by permission of the Food and Agricultural Organization of the United Nations.)

NORTH EAST PACIFIC
DEMERSAL RESOURCES

PACIFIQUE NORD-EST
RESSOURCES DEMERSALES

Figure 2.5–13

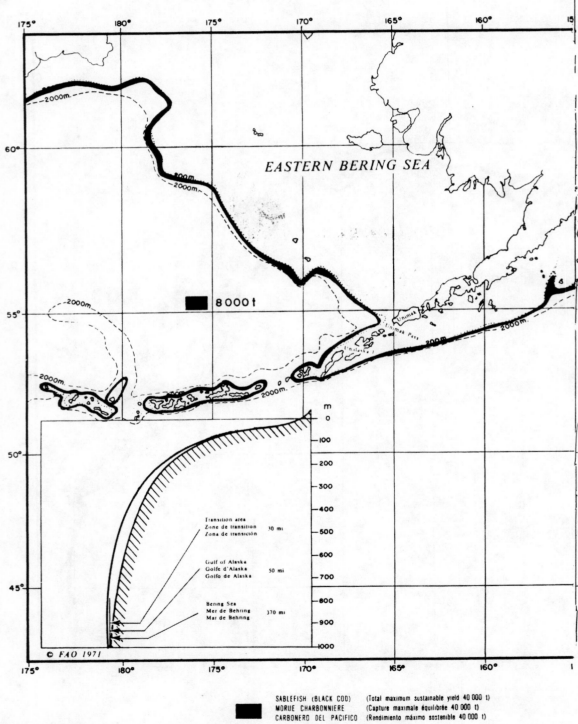

© FAO 1971

SABLEFISH (BLACK COD)   (Total maximum sustainable yield 40 000 t)
MORUE CHARBONNIÈRE   (Capture maximale équilibrée 40 000 t)
CARBONERO DEL PACIFICO   (Rendimiento máximo sostenible 40 000 t)

PACIFICO NORDORIENTAL
RECURSOS DEMERSALES

Figure 2.5–13

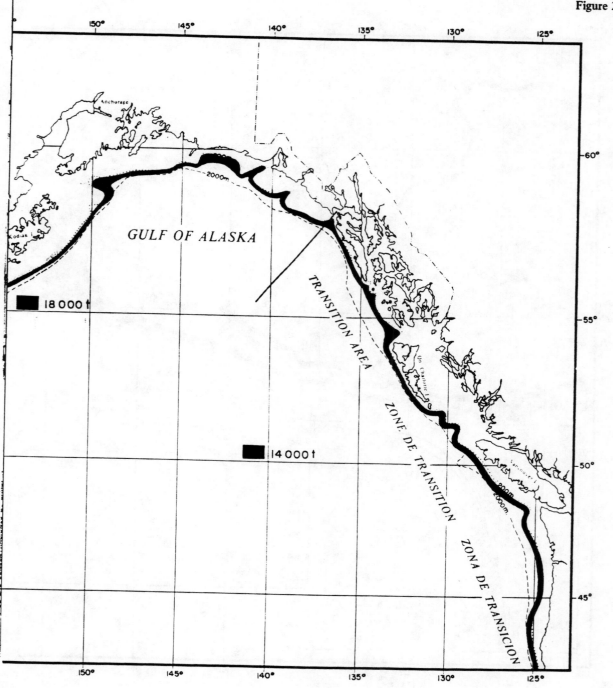

ALASKA POLLACK   (Total maximum sustainable yield 500 000 t)
COLIN D'ALASKA   (Capture maximale equilibree 500 000 t)
COLIN DE ALASKA   (Rendimiento máximo sostenible 500 000 t)

PACIFIC COD   (Total maximum sustainable yield 90 000 t)
MORUE DU PACIFIQUE (Capture maximale equilibrée 90 000 t)
BACALAO DEL PACIFICO (Rendimiento máximo sostenible 90 000 t)

NORTH EAST PACIFIC
DEMERSAL RESOURCES

PACIFIQUE NORD-EST
RESSOURCES DEMERSALES

Figure 2.5–14

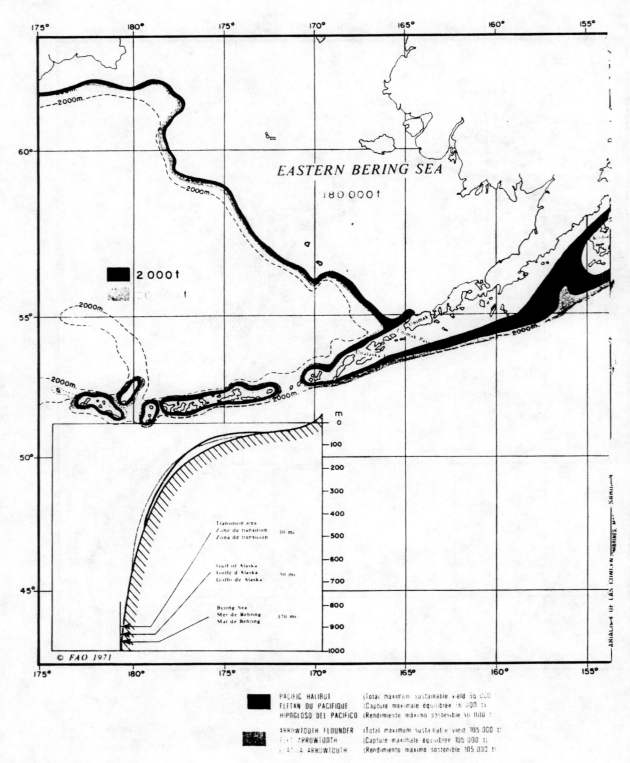

PACIFICO NORDORIENTAL
RECURSOS DEMERSALES

Figure 2.5–14

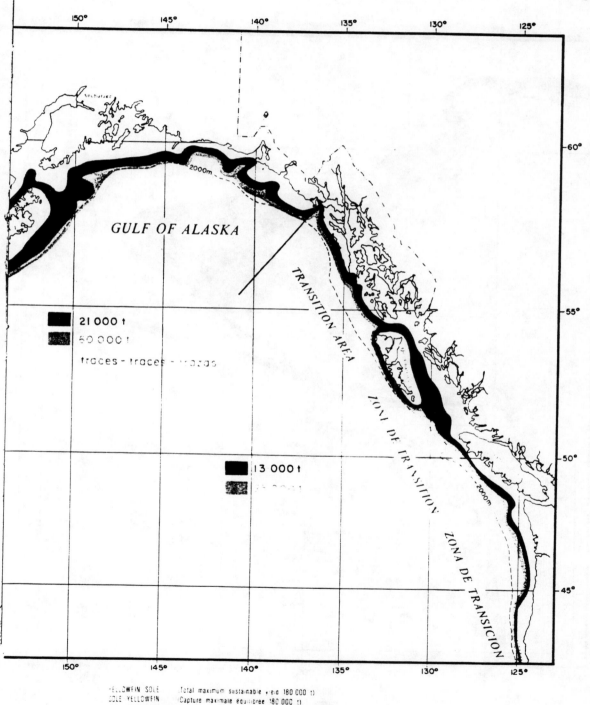

GULF OF ALASKA

21 000 t

60 000 t

traces - traces - trazas

TRANSITION AREA

ZONE DE TRANSITION

ZONA DE TRANSICION

13 000 t

YELLOWFIN SOLE   (Total maximum sustainable yield 180 000 t)
SOLE YELLOWFIN   (Capture maximale équilibrée 180 000 t)
LENGUADO YELLOWFIN (Rendimiento máximo sostenible 180 000 t)

(Reprinted from FAO Department of Fisheries, *Atlas of the Living Resources of the Seas*, Rome, 1972, by permission of the Food and Agricultural Organization of the United Nations.)

NORTH EAST PACIFIC
DEMERSAL RESOURCES

PACIFIQUE NORD-EST
RESSOURCES DEMERSALES

Figure 2.5—15

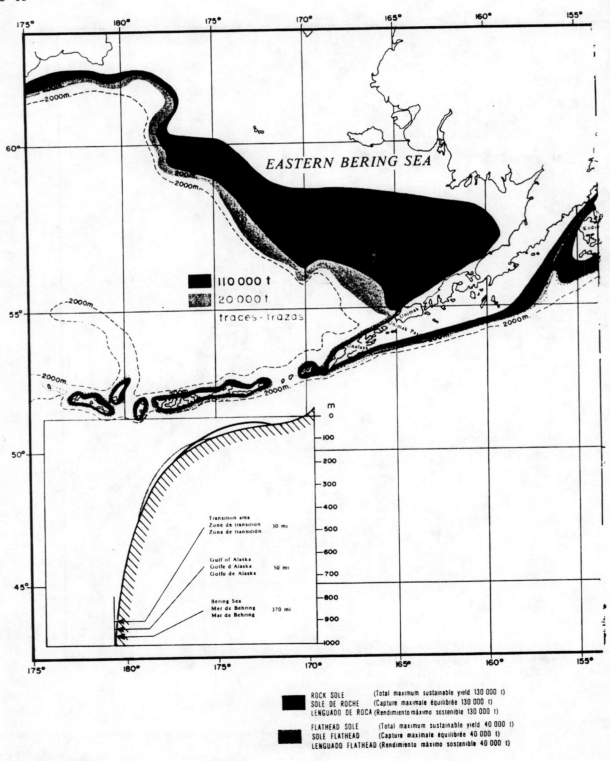

EASTERN BERING SEA

■ 110 000 t

20 000 t

traces - frazas

Transition area
Zone de transition
Zona de transición   30 mi

Gulf of Alaska
Golfe d Alaska
Golfo de Alaska   50 mi

Bering Sea
Mer de Behring
Mar de Behring   370 mi

■ ROCK SOLE   (Total maximum sustainable yield 130 000 t)
SOLE DE ROCHE   (Capture maximale équilibrée 130 000 t)
LENGUADO DE ROCA (Rendimiento máximo sostenible 130 000 t)

■ FLATHEAD SOLE (Total maximum sustainable yield 40 000 t)
SOLE FLATHEAD   (Capture maximale équilibrée 40 000 t)
LENGUADO FLATHEAD (Rendimiento máximo sostenible 40 000 t)

PACIFICO NORDORIENTAL
RECURSOS DEMERSALES

Figure 2.5–15

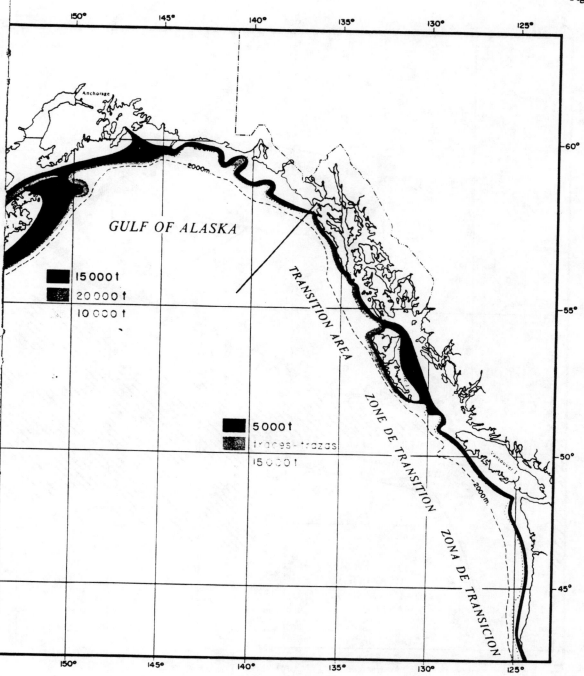

DOVER SOLE   (Total maximum sustainable yield 25 000 t)
SOLE DE DOUVRES   (Capture maximale équilibrée 25 000 t)
LENGUADO DE DOVER   (Rendimiento maximo sostenible 25 000 t)

(Reprinted from FAO Department of Fisheries, *Atlas of the Living Resources of the Seas*, Rome, 1972, by permission of the Food and Agricultural Organization of the United Nations.)

Figure 2.5–16

NORTH EAST PACIFIC PELAGIC RESOURCES
PACIFIQUE NORD-EST - RESSOURCES PELAGIQUES
PACIFICO NORDORIENTAL - RECURSOS PELAGICOS

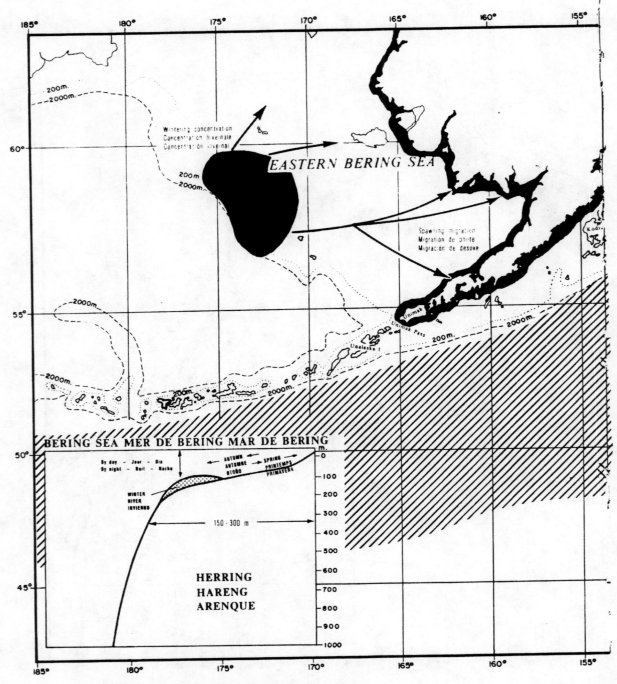

Salmon yield is greater when harvested in coastal waters than
if caught younger on the high seas.
La pêche du saumon donne un rendement plus élevé dans les eaux
côtières qu'en haute mer où elles portent sur des poissons plus jeunes.
Es mayor el rendimiento de salmón pescado en aguas costeras que el
de ejemplares jóvenes capturados en alta mar.

Figure 2.5—16

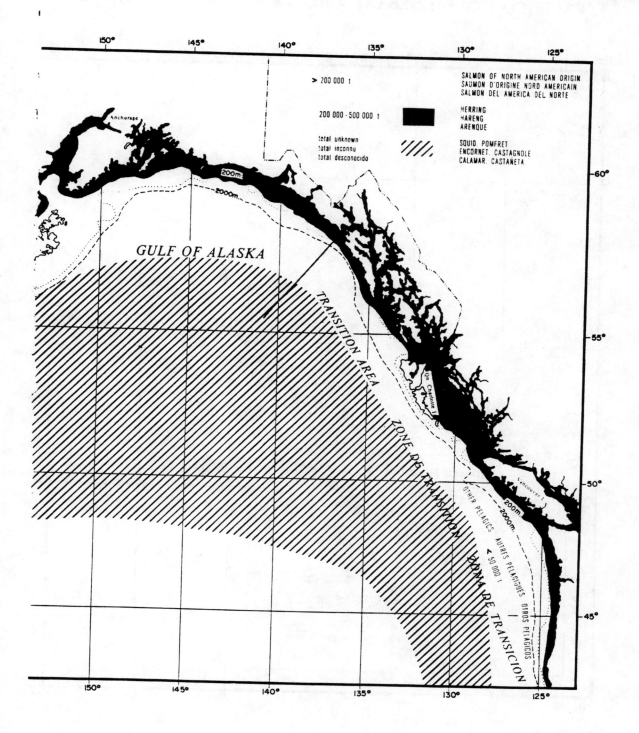

NORTH EAST PACIFIC CRUSTACEAN RESOURCES
PACIFIQUE NORD-EST - RESSOURCES EN CRUSTACES
PACIFICO NORDORIENTAL - RECURSOS DE CRUSTACEOS

Figure 2.5–17

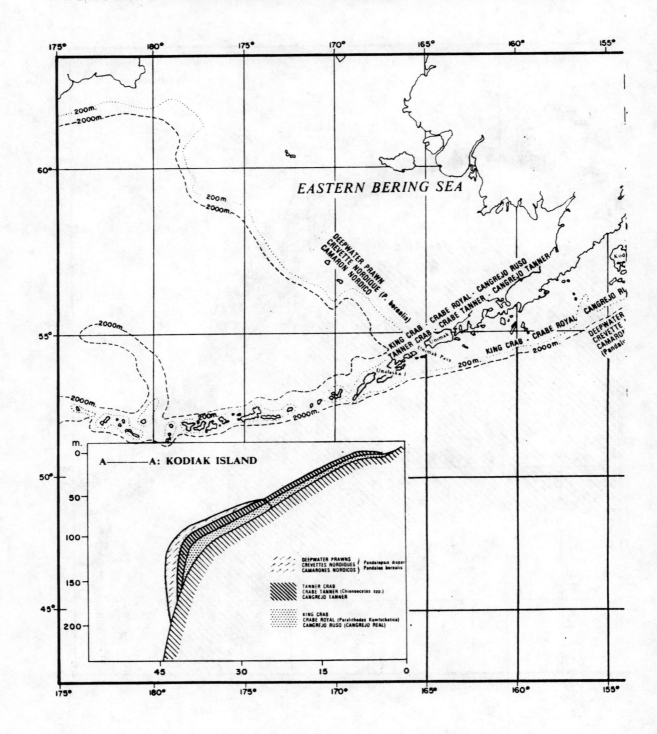

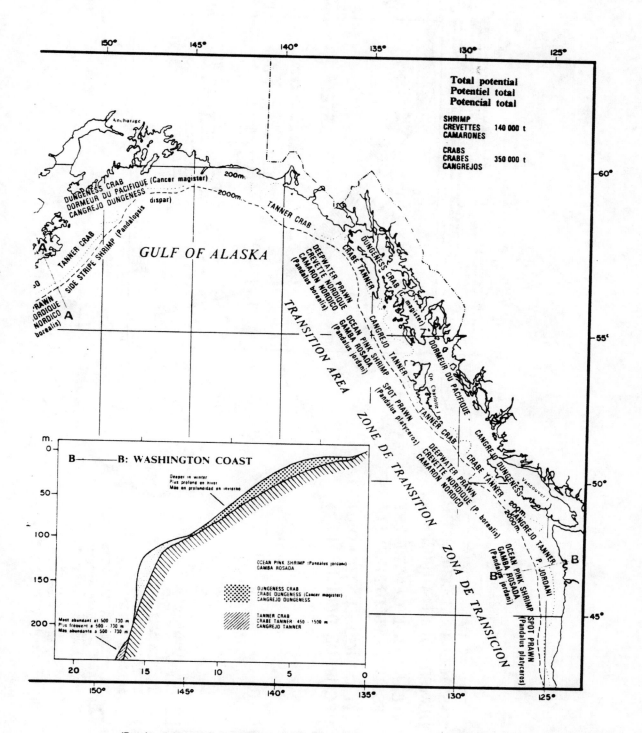

Figure 2.5–17

(Reprinted from FAO Department of Fisheries, *Atlas of the Living Resources of the Seas*, Rome, 1972, by permission of the Food and Agricultural Organization of the United Nations.)

Figure 2.5—18

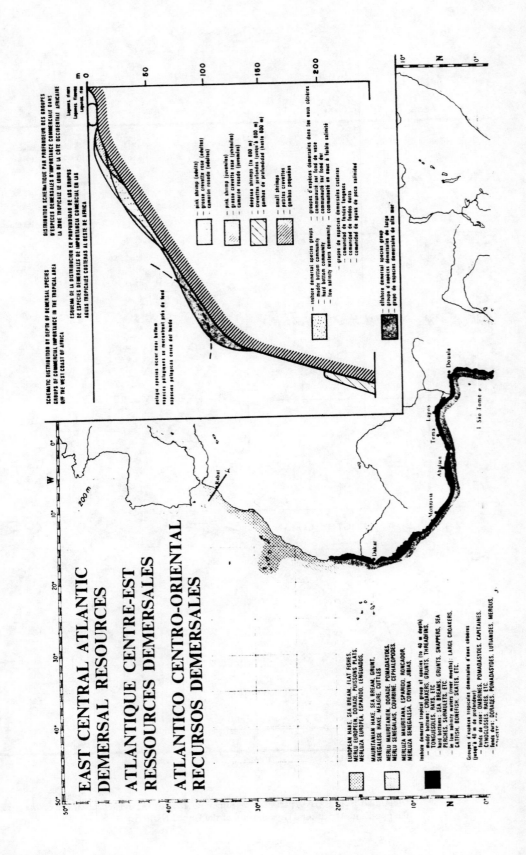

EAST CENTRAL ATLANTIC
DEMERSAL RESOURCES

ATLANTIQUE CENTRE-EST
RESSOURCES DEMERSALES

ATLANTICO CENTRO-ORIENTAL
RECURSOS DEMERSALES

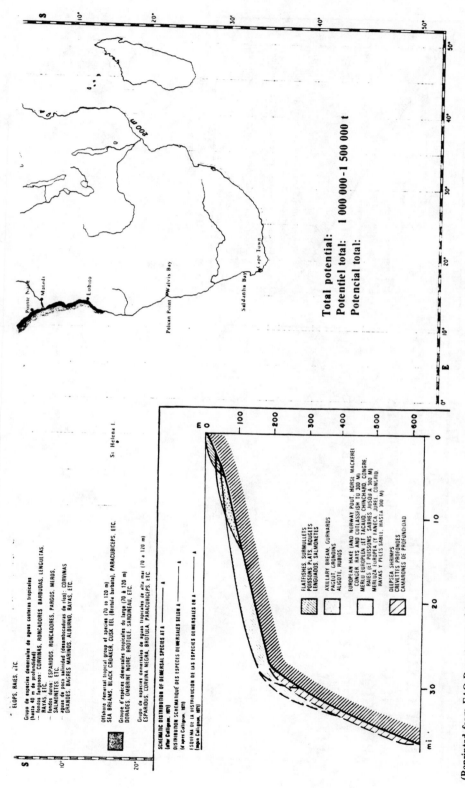

Figure 2.5—18

(Reprinted from FAO Department of Fisheries, *Atlas of the Living Resources of the Seas*, Rome, 1972, by permission of the Food and Agricultural Organization of the United Nations.)

137

Figure 2.5—19

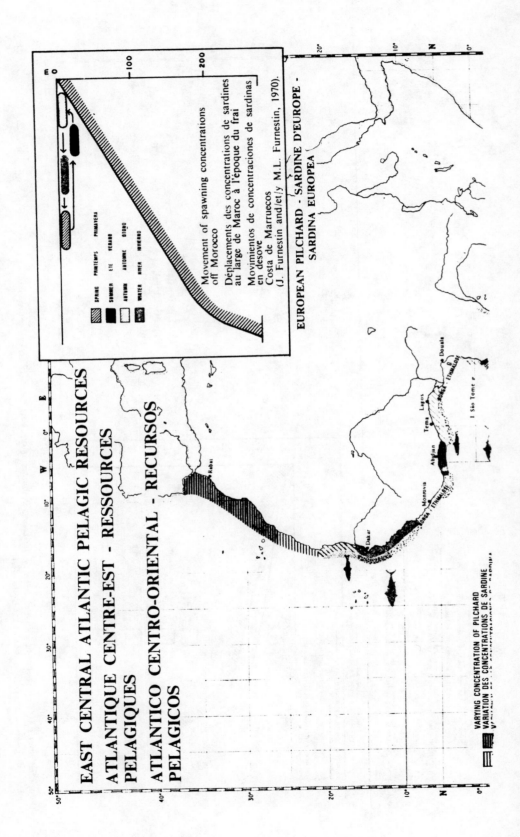

EAST CENTRAL ATLANTIC PELAGIC RESOURCES

ATLANTIQUE CENTRE-EST - RESSOURCES PELAGIQUES

ATLANTICO CENTRO-ORIENTAL - RECURSOS PELAGICOS

EUROPEAN PILCHARD - SARDINE D'EUROPE - SARDINA EUROPEA

Movement of spawning concentrations off Morocco

Déplacements des concentrations de sardines au large de Maroc à l'époque du frai

Movimientos de concentraciones de sardinas en desove

Costa de Marruecos
(J. Furnestin and/et/y M.L. Furnestin, 1970).

SPRING	PRINTEMPS	PRIMAVERA
SUMMER	ETE	VERANO
AUTUMN	AUTOMNE	OTOÑO
WATER	HIVER	INVIERNO

VARYING CONCENTRATION OF PILCHARD
VARIATION DES CONCENTRATIONS DE SARDINE

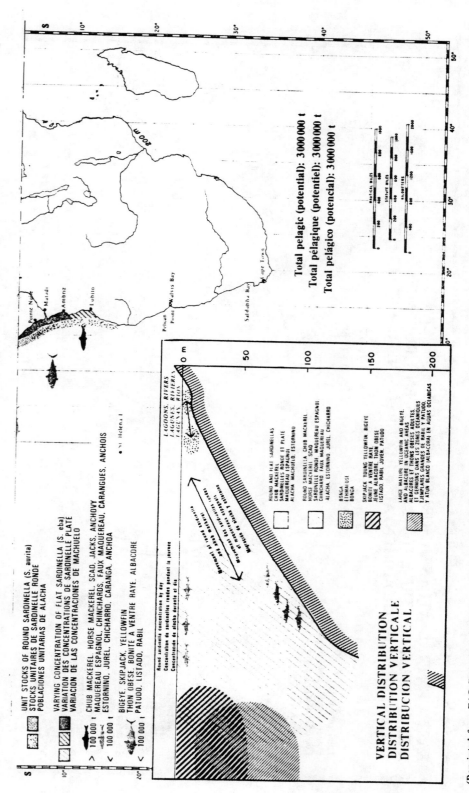

Figure 2.5—19

(Reprinted from FAO Department of Fisheries, *Atlas of the Living Resources of the Seas*, Rome, 1972, by permission of the Food and Agricultural Organization of the United Nations.)

Figure 2.5–20

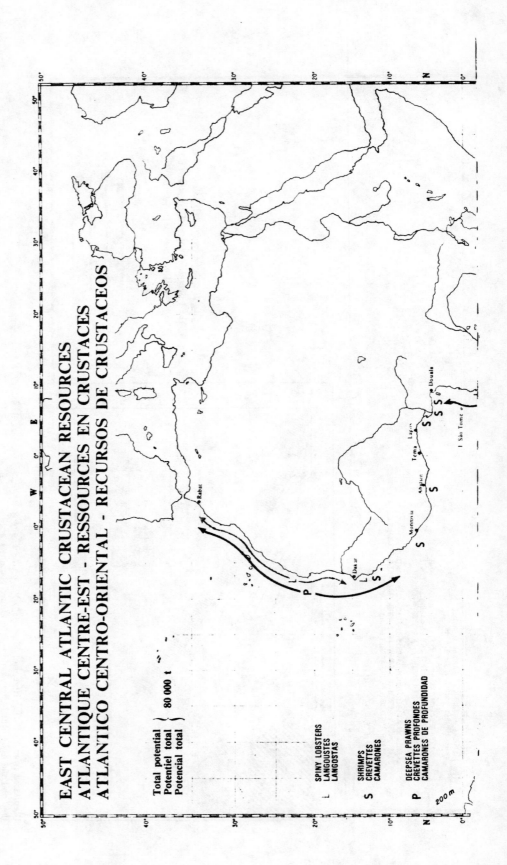

EAST CENTRAL ATLANTIC CRUSTACEAN RESOURCES
ATLANTIQUE CENTRE-EST - RESSOURCES EN CRUSTACES
ATLANTICO CENTRO-ORIENTAL - RECURSOS DE CRUSTACEOS

Total potential } 80 000 t
Potentiel total }
Potencial total }

L SPINY LOBSTERS
   LANGOUSTES
   LANGOSTAS

S SHRIMPS
   CREVETTES
   CAMARONES

P DEEPSEA PRAWNS
   CREVETTES PROFONDES
   CAMARONES DE PROFUNDIDAD

200 m

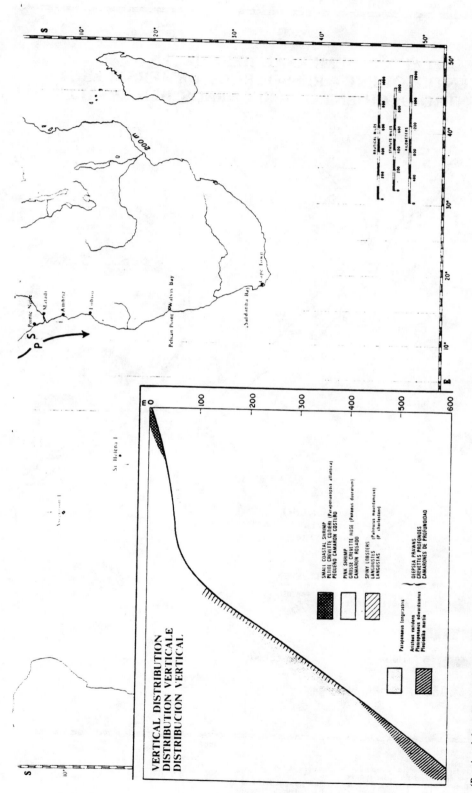

Figure 2.5—20

(Reprinted from FAO Department of Fisheries, *Atlas of the Living Resources of the Seas*, Rome, 1972, by permission of the Food and Agricultural Organization of the United Nations.)

141

Figure 2.5–21

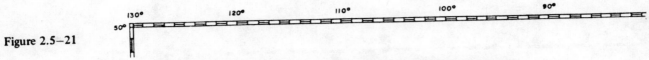

WEST CENTRAL ATLANTIC DEMERSAL RESOURCES
ATLANTIQUE CENTRE-OUEST - RESSOURCES DEMERSALES
ATLANTICO CENTRO-OCCIDENTAL - RECURSOS DEMERSALES

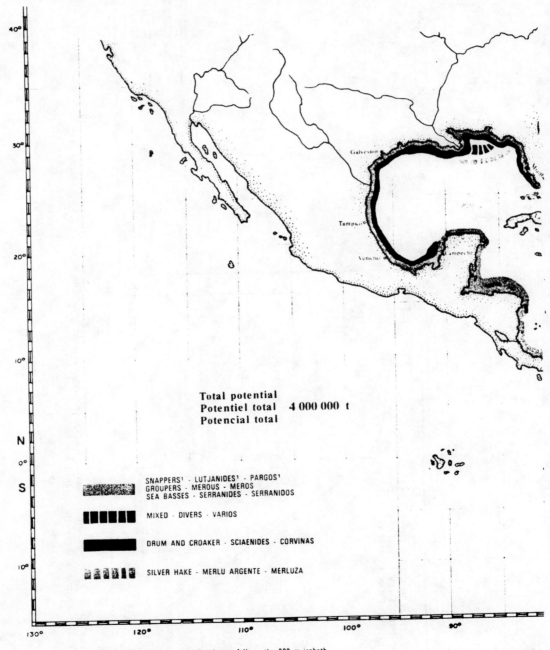

Total potential
Potentiel total   4 000 000 t
Potencial total

SNAPPERS[1] · LUTJANIDES[1] · PARGOS[1]
GROUPERS · MEROUS · MEROS
SEA BASSES · SERRANIDES · SERRANIDOS

MIXED · DIVERS · VARIOS

DRUM AND CROAKER · SCIAENIDES · CORVINAS

SILVER HAKE · MERLU ARGENTE · MERLUZA

[1] In the Caribbean Sea, snapper distribution follows the 200 m isobath
and does extend southward of 10°N.
[1] Dans la mer des Caraïbes, l'aire de répartition des lutjanidés suit
l'isobathe 200 m et ne s'étend pas au sud de la latitude 10°N.
[1] En el Mar de Caribe, el área de distribución de los pargos sigue la línea
isobática de 200 m y no se extiende hacia el sur más allá de los 10°N.

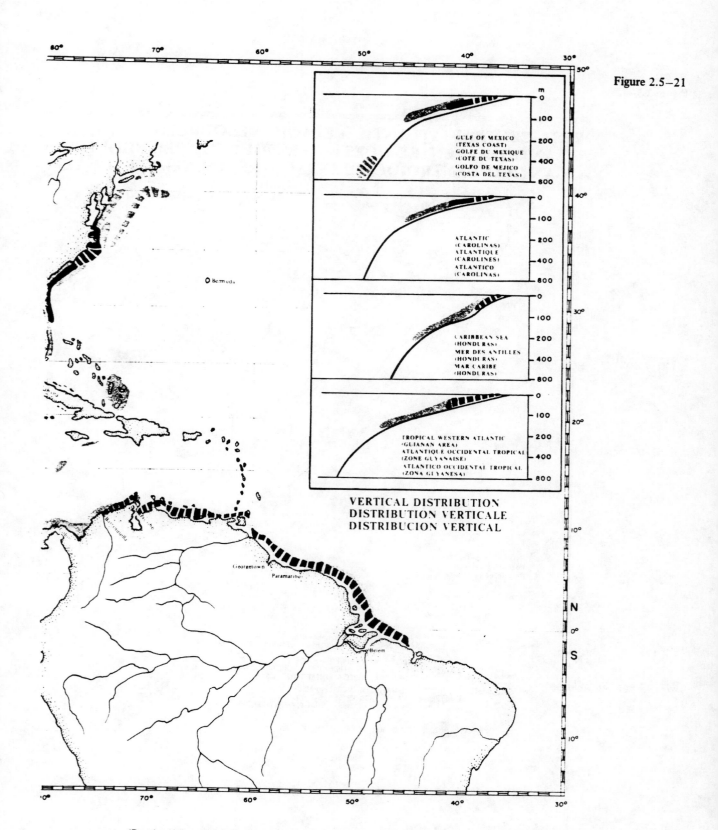

Figure 2.5—21

VERTICAL DISTRIBUTION
DISTRIBUTION VERTICALE
DISTRIBUCION VERTICAL

(Reprinted from FAO Department of Fisheries, *Atlas of the Living Resources of the Seas*, Rome, 1972, by permission of the Food and Agricultural Organization of the United Nations.)

Figure 2.5–22

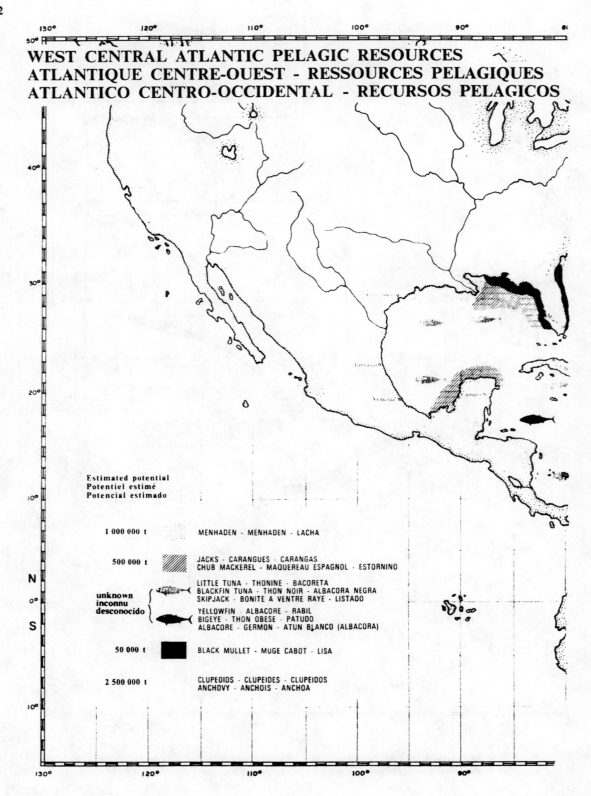

WEST CENTRAL ATLANTIC PELAGIC RESOURCES
ATLANTIQUE CENTRE-OUEST - RESSOURCES PELAGIQUES
ATLANTICO CENTRO-OCCIDENTAL - RECURSOS PELAGICOS

Estimated potential
Potentiel estimé
Potencial estimado

1 000 000 t MENHADEN - MENHADEN - LACHA

500 000 t JACKS - CARANGUES - CARANGAS
CHUB MACKEREL - MAQUEREAU ESPAGNOL - ESTORNINO

unknown
inconnu
desconocido

LITTLE TUNA - THONINE - BACORETA
BLACKFIN TUNA - THON NOIR - ALBACORA NEGRA
SKIPJACK - BONITE A VENTRE RAYE - LISTADO

YELLOWFIN ALBACORE - RABIL
BIGEYE - THON OBESE - PATUDO
ALBACORE - GERMON - ATUN BLANCO (ALBACORA)

50 000 t BLACK MULLET - MUGE CABOT - LISA

2 500 000 t CLUPEOIDS - CLUPEIDES - CLUPEIDOS
ANCHOVY - ANCHOIS - ANCHOA

Figure 2.5—22

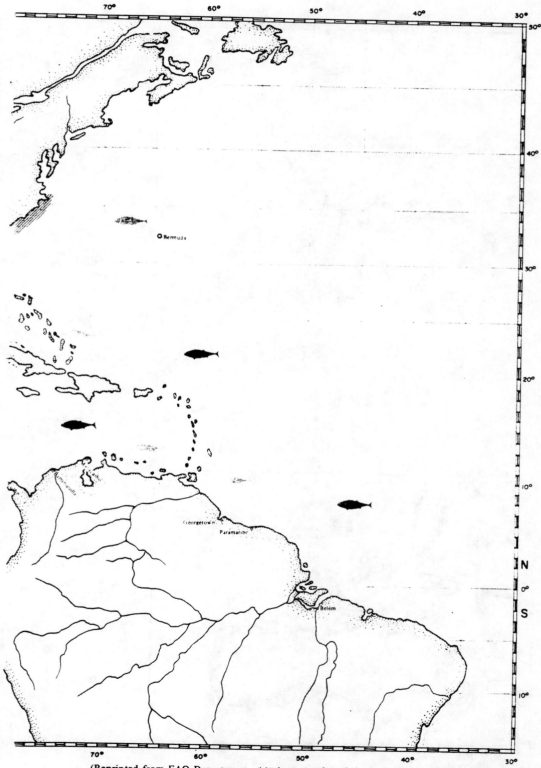

(Reprinted from FAO Department of Fisheries, *Atlas of the Living Resources of the Seas*, Rome, 1972, by permission of the Food and Agricultural Organization of the United Nations.)

Figure 2.5—23

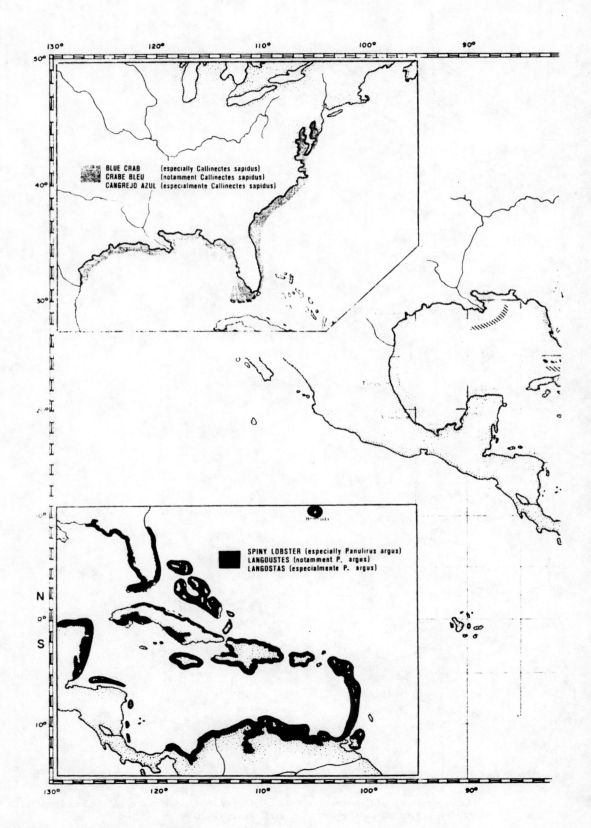

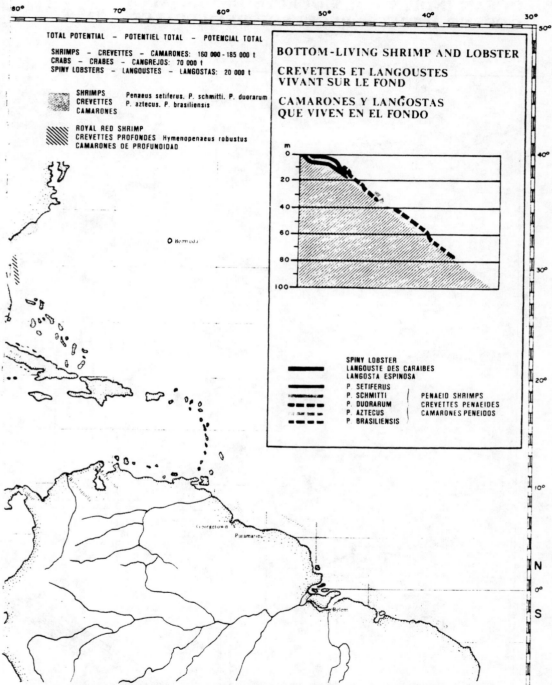

WEST CENTRAL ATLANTIC CRUSTACEAN RESOURCES
ATLANTIQUE CENTRE-OUEST - RESSOURCES EN CRUSTACES
ATLANTICO CENTRO-OCCIDENTAL - RECURSOS DE CRUSTACEOS

(Reprinted from FAO Department of Fisheries, *Atlas of the Living Resources of the Seas*, Rome, 1972, by permission of the Food and Agricultural Organization of the United Nations.)

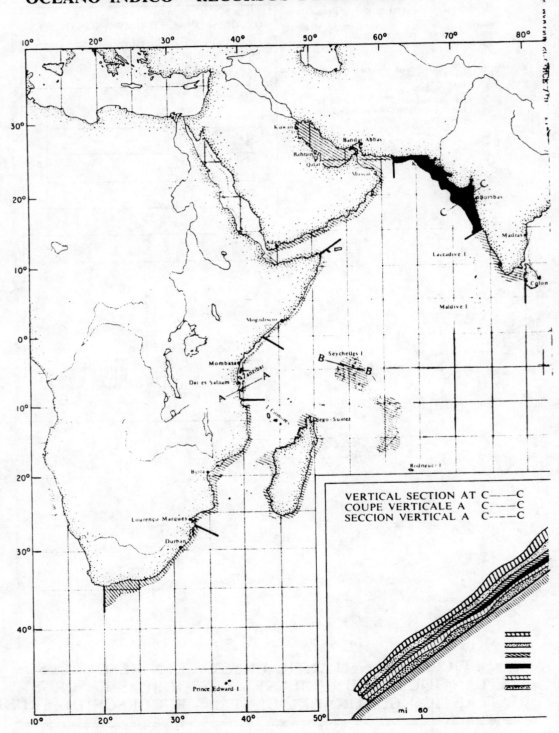

Figure 2.5–24

INDIAN OCEAN DEMERSAL RESOURCES
OCEAN INDIEN - RESSOURCES DEMERSALES
OCEANO INDICO - RECURSOS DEMERSALES

VERTICAL SECTION AT C——C
COUPE VERTICALE A   C——C
SECCION VERTICAL A C——C

Figure 2.5—24

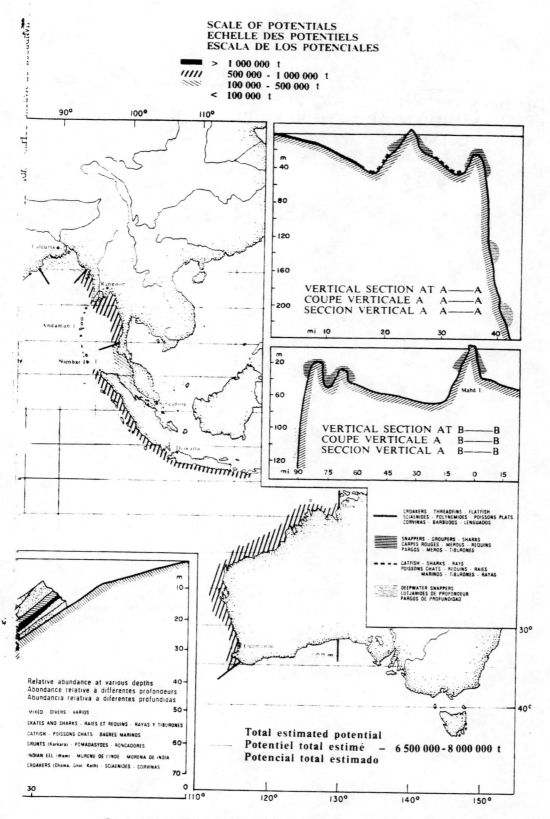

SCALE OF POTENTIALS
ECHELLE DES POTENTIELS
ESCALA DE LOS POTENCIALES

> 1 000 000 t
500 000 - 1 000 000 t
100 000 - 500 000 t
< 100 000 t

VERTICAL SECTION AT A——A
COUPE VERTICALE A   A——A
SECCION VERTICAL A   A——A

VERTICAL SECTION AT B——B
COUPE VERTICALE A   B——B
SECCION VERTICAL A   B——B

CROAKERS · THREADFINS · FLATFISH
SCIAENIDES · POLYNEMIDES · POISSONS PLATS
CORVINAS · BARBUDOS · LENGUADOS

SNAPPERS · GROUPERS · SHARKS
CARPES ROUGES · MEROUS · REQUINS
PARGOS · MEROS · TIBURONES

CATFISH · SHARKS · RAYS
POISSONS CHATS · REQUINS · RAIES
MARINOS · TIBURONES · RAYAS

DEEPWATER SNAPPERS
LUTJANIDES DE PROFONDEUR
PARGOS DE PROFUNDIDAD

Relative abundance at various depths
Abondance relative à différentes profondeurs
Abundancia relativa a diferentes profundidas

MIXED · DIVERS · VARIOS

SKATES AND SHARKS · RAIES ET REQUINS · RAYAS Y TIBURONES

CATFISH · POISSONS CHATS · BAGRES MARINOS

GRUNTS (Karkara) · POMADASYDES · RONCADORES

INDIAN EEL (Wam) · MURENE DE L'INDE · MORENA DE INDIA

CROAKERS (Dhoma, Ghol, Koth) · SCIAENIDES · CORVINAS

Total estimated potential
Potentiel total estimé   — 6 500 000 - 8 000 000 t
Potencial total estimado

(Reprinted from FAO Department of Fisheries, *Atlas of the Living Resources of the Seas*, Rome, 1972, by permission of the Food and Agricultural Organization of the United Nations.)

Figure 2.5–25

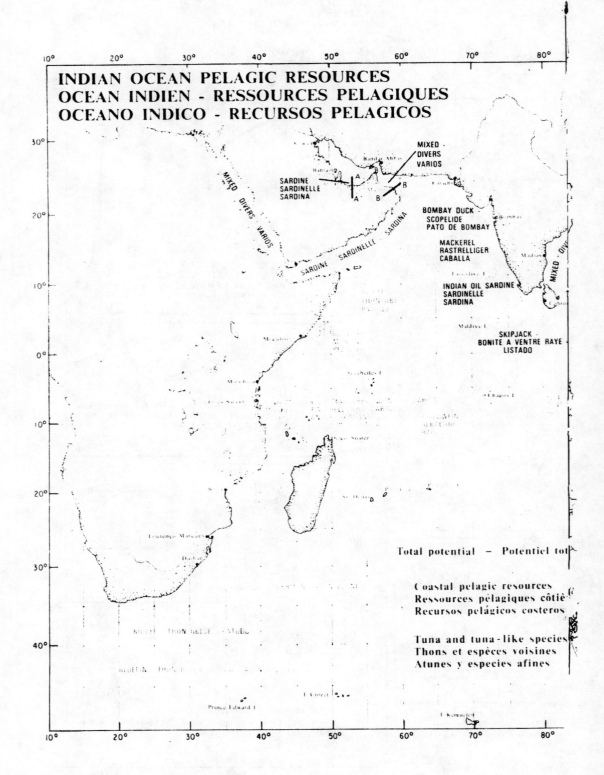

INDIAN OCEAN PELAGIC RESOURCES
OCEAN INDIEN - RESSOURCES PELAGIQUES
OCEANO INDICO - RECURSOS PELAGICOS

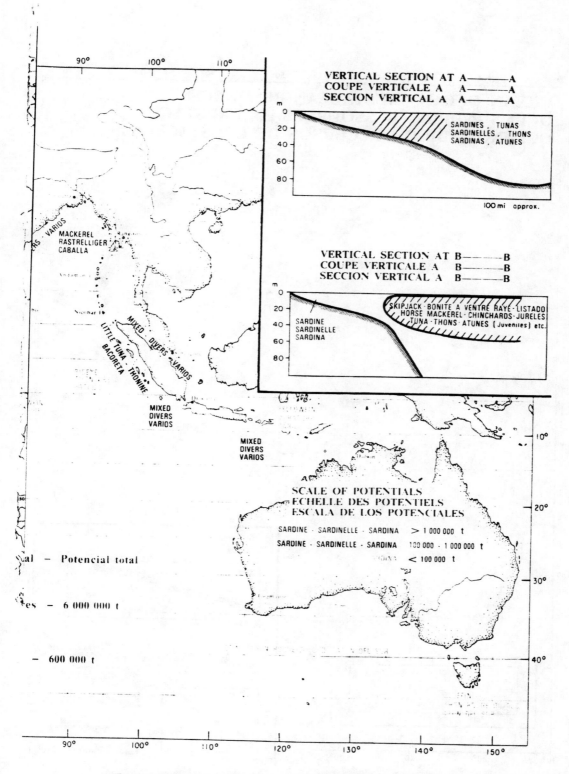

Figure 2.5-25

(Reprinted from FAO Department of Fisheries, *Atlas of the Living Resources of the Seas*, Rome, 1972, by permission of the Food and Agricultural Organization of the United Nations.)

Figure 2.5–26

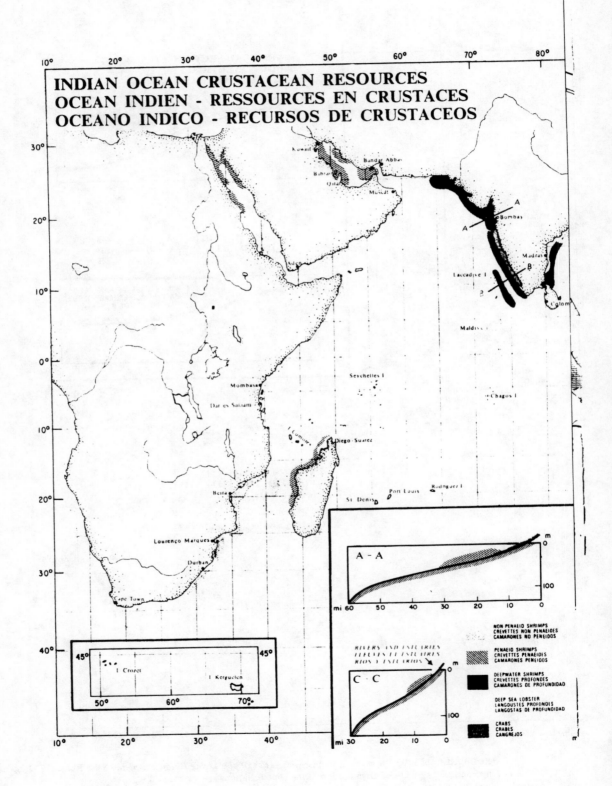

INDIAN OCEAN CRUSTACEAN RESOURCES
OCEAN INDIEN - RESSOURCES EN CRUSTACES
OCEANO INDICO - RECURSOS DE CRUSTACEOS

Figure 2.5—26

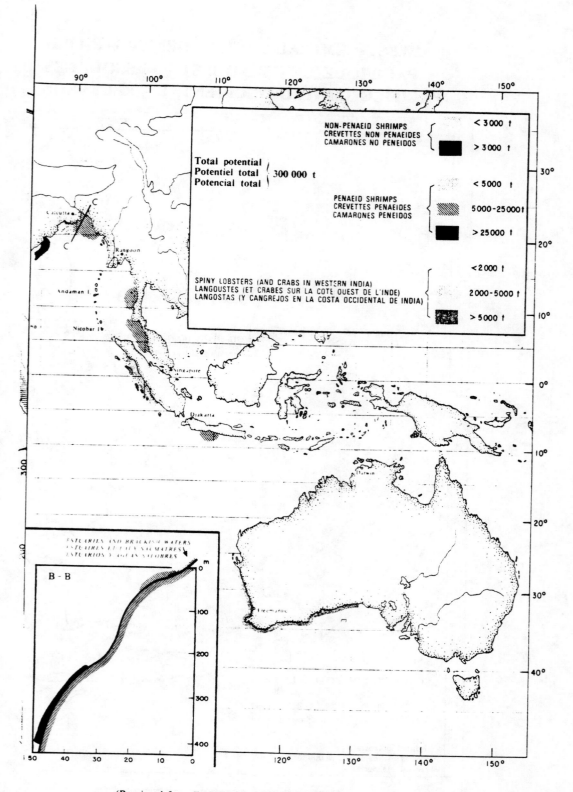

(Reprinted from FAO Department of Fisheries, *Atlas of the Living Resources of the Seas*, Rome, 1972, by permission of the Food and Agricultural Organization of the United Nations.)

Figure 2.5—27

WEST CENTRAL PACIFIC DEMERSAL FISH
PACIFIQUE CENTRE-OUEST - RESSOURCES
PACIFICO CENTRO-OCCIDENTAL - RECURSOS

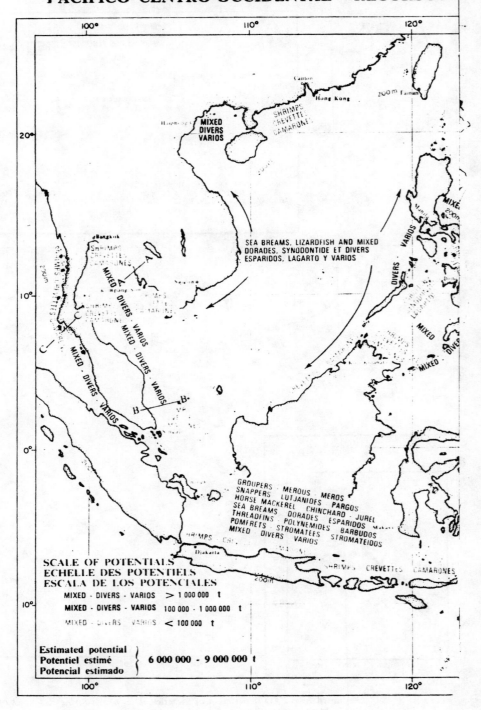

AND CRUSTACEAN RESOURCES
DEMERSALES: POISSONS ET CRUSTACES
DEMERSALES: PECES Y CRUSTACEOS

Figure 2.5–27

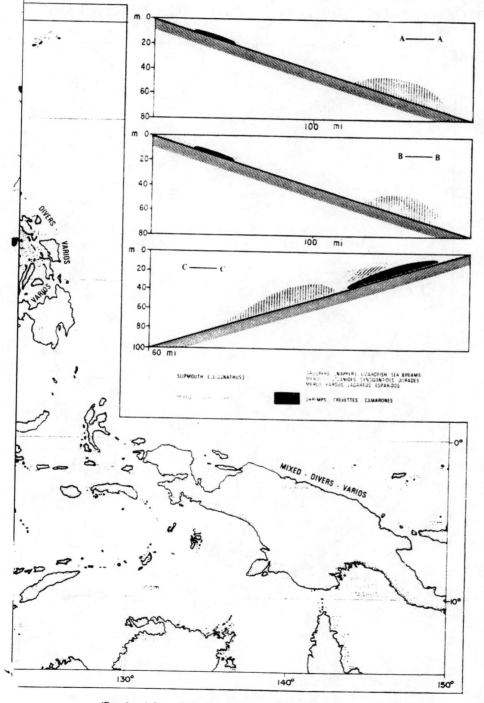

(Reprinted from FAO Department of Fisheries, *Atlas of the Living Resources of the Seas*, Rome, 1972, by permission of the Food and Agricultural Organization of the United Nations.)

Figure 2.5—28

WEST CENTRAL PACIFIC PELAGIC RESOURCES
PACIFIOUE CENTRE-OUEST - RESSOURCES PELAGIQUES
PACIFICO CENTRO-OCCIDENTAL - RECURSOS PELAGICOS

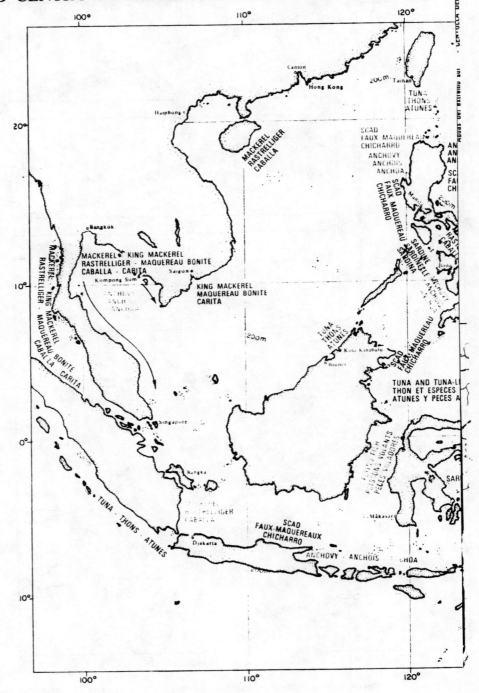

Figure 2.5—28

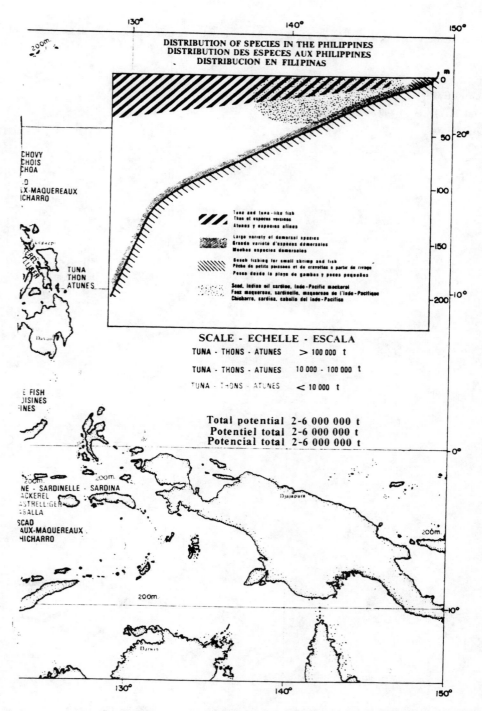

DISTRIBUTION OF SPECIES IN THE PHILIPPINES
DISTRIBUTION DES ESPECES AUX PHILIPPINES
DISTRIBUCION EN FILIPINAS

SCALE - ECHELLE - ESCALA

TUNA - THONS - ATUNES   > 100 000 t

TUNA - THONS - ATUNES   10 000 - 100 000 t

TUNA - THONS - ATUNES   < 10 000 t

Total potential 2-6 000 000 t
Potentiel total 2-6 000 000 t
Potencial total 2-6 000 000 t

(Reprinted from FAO Department of Fisheries, *Atlas of the Living Resources of the Seas*, Rome, 1972, by permission of the Food and Agricultural Organization of the United Nations.)

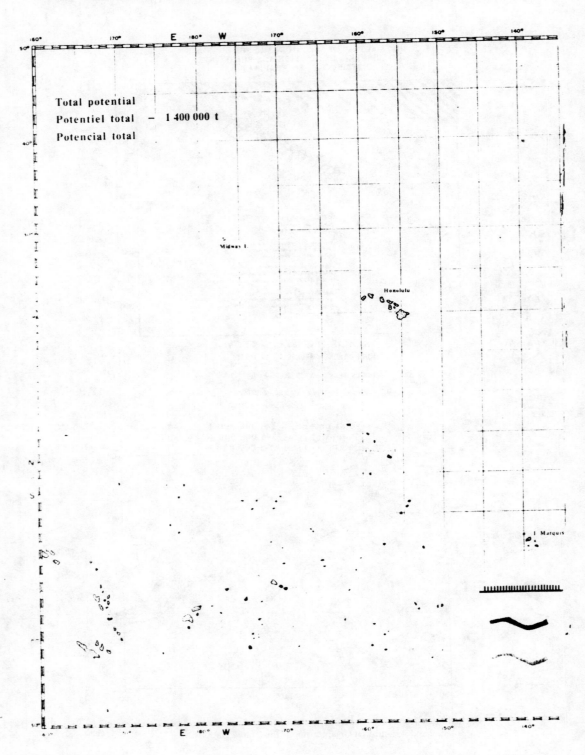

Figure 2.5—29

EAST CENTRAL PACIFIC DEMERSAL RESOURCES
PAFICIQUE CENTRE-EST - RESSOURCES DEMERSALES
PACIFICO CENTRO-ORIENTAL - RECURSOS DEMERSALES

Total potential
Potentiel total — 1 400 000 t
Potencial total

Figure 2.5—29

(Reprinted from FAO Department of Fisheries, *Atlas of the Living Resources of the Seas*, Rome, 1972, by permission of the Food and Agricultural Organization of the United Nations.)

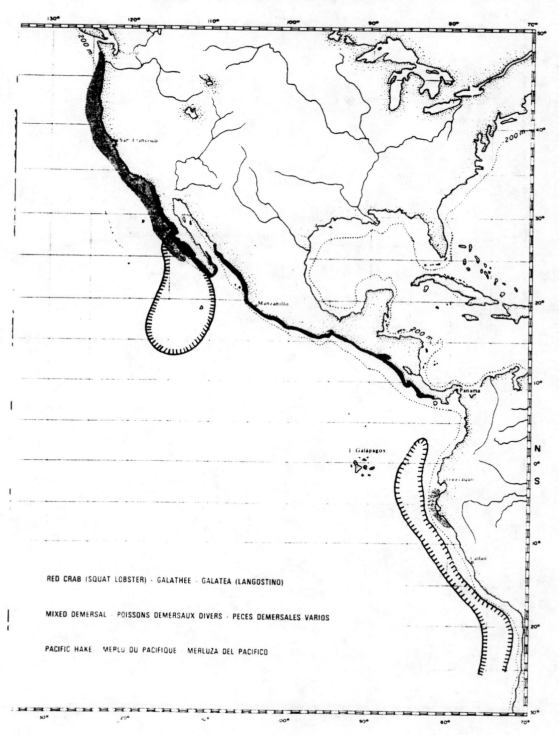

RED CRAB (SQUAT LOBSTER) · GALATHEE · GALATEA (LANGOSTINO)

MIXED DEMERSAL · POISSONS DEMERSAUX DIVERS · PECES DEMERSALES VARIOS

PACIFIC HAKE · MERLU DU PACIFIQUE · MERLUZA DEL PACIFICO

Figure 2.5–30

EAST CENTRAL PACIFIC PELAGIC RESOURCES
PACIFIQUE CENTRE-EST - RESSOURCES PELAGIQUES
PACIFICO CENTRO-ORIENTAL - RECURSOS PELAGICOS

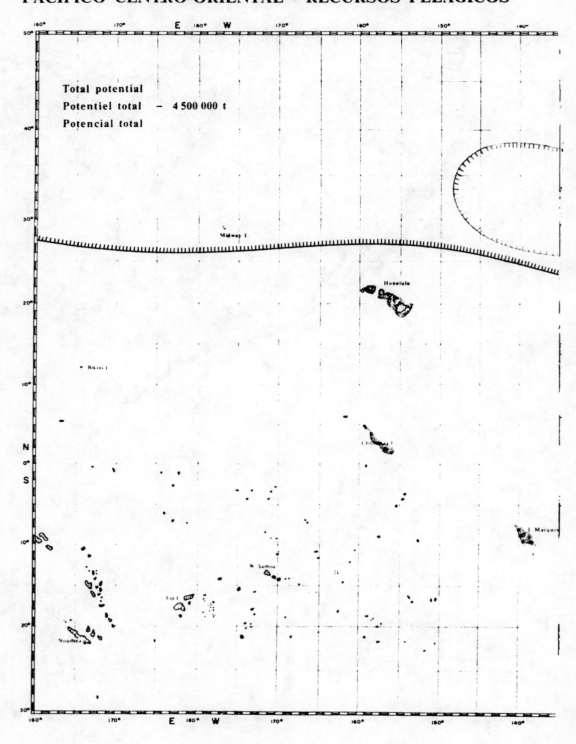

Total potential
Potentiel total   — 4 500 000 t
Potencial total

Figure 2.5–30

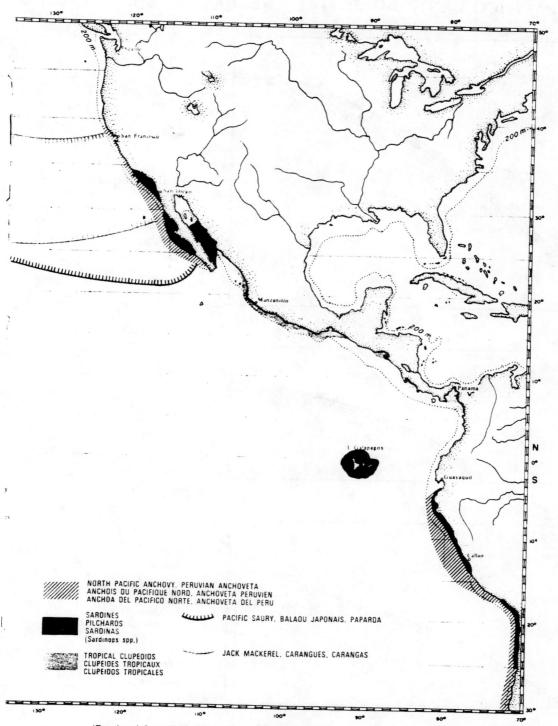

NORTH PACIFIC ANCHOVY, PERUVIAN ANCHOVETA
ANCHOIS DU PACIFIQUE NORD, ANCHOVETA PERUVIEN
ANCHOA DEL PACIFICO NORTE, ANCHOVETA DEL PERU

SARDINES
PILCHARDS
SARDINAS
(Sardinops spp.)

PACIFIC SAURY, BALAOU JAPONAIS, PAPARDA

TROPICAL CLUPEOIDS
CLUPEIDES TROPICAUX
CLUPEIDOS TROPICALES

JACK MACKEREL, CARANGUES, CARANGAS

(Reprinted from FAO Department of Fisheries, *Atlas of the Living Resources of the Seas*, Rome, 1972, by permission of the Food and Agricultural Organization of the United Nations.)

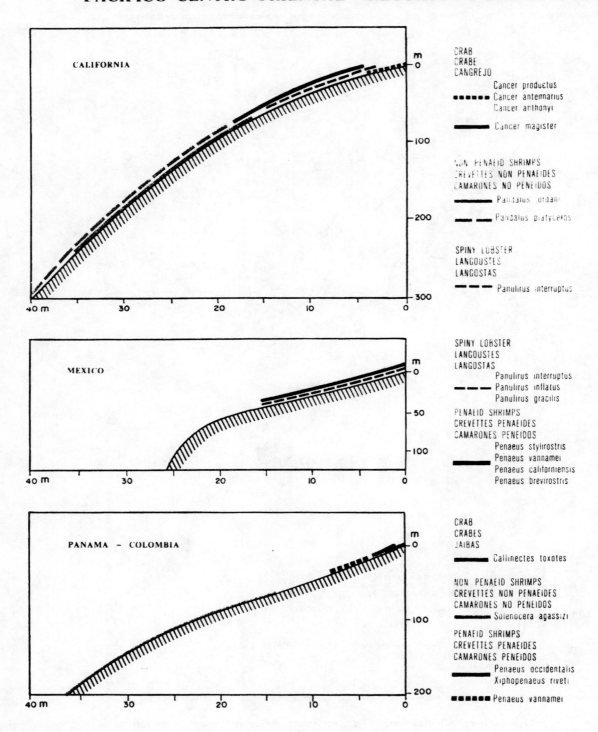

Figure 2.5–31

EAST CENTRAL PACIFIC CRUSTACEAN RESOURCES
PACIFIQUE CENTRE-EST - RESSOURCES EN CRUSTACES
PACIFICO CENTRO-ORIENTAL - RECURSOS DE CRUSTACEOS

CALIFORNIA

MEXICO

PANAMA – COLOMBIA

CRAB
CRABE
CANGREJO
    Cancer productus
•••••• Cancer antennarius
    Cancer anthonyi
    Cancer magister

NON PENAEID SHRIMPS
CREVETTES NON PENAEIDES
CAMARONES NO PENEIDOS
    Pandalus jordani
    Pandalus platyceros

SPINY LOBSTER
LANGOUSTES
LANGOSTAS
    Panulirus interruptus

SPINY LOBSTER
LANGOUSTES
LANGOSTAS
    Panulirus interruptus
    Panulirus inflatus
    Panulirus gracilis
PENAEID SHRIMPS
CREVETTES PENAEIDES
CAMARONES PENEIDOS
    Penaeus stylirostris
    Penaeus vannamei
    Penaeus californiensis
    Penaeus brevirostris

CRAB
CRABES
JAIBAS
    Callinectes toxotes

NON PENAEID SHRIMPS
CREVETTES NON PENAEIDES
CAMARONES NO PENEIDOS
    Solenocera agassizi
PENAEID SHRIMPS
CREVETTES PENAEIDES
CAMARONES PENEIDOS
    Penaeus occidentalis
    Xiphopenaeus riveti
•••••• Penaeus vannamei

Figure 2.5–31

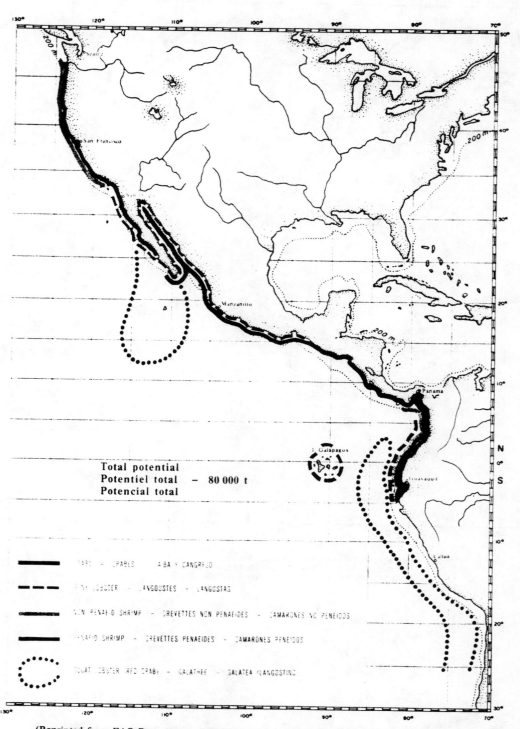

Total potential
Potentiel total   –   80 000 t
Potencial total

(Reprinted from FAO Department of Fisheries, *Atlas of the Living Resources of the Seas*, Rome, 1972, by permission of the Food and Agricultural Organization of the United Nations.)

Figure 2.5–32

SOUTH WEST PACIFIC DEMERSAL FISH AND CRUSTACEAN
PACIFIQUE SUD-OUEST - RESSOURCES DEMERSALES:
PACIFICO SUDOCCIDENTAL - RECURSOS DEMERSALES:

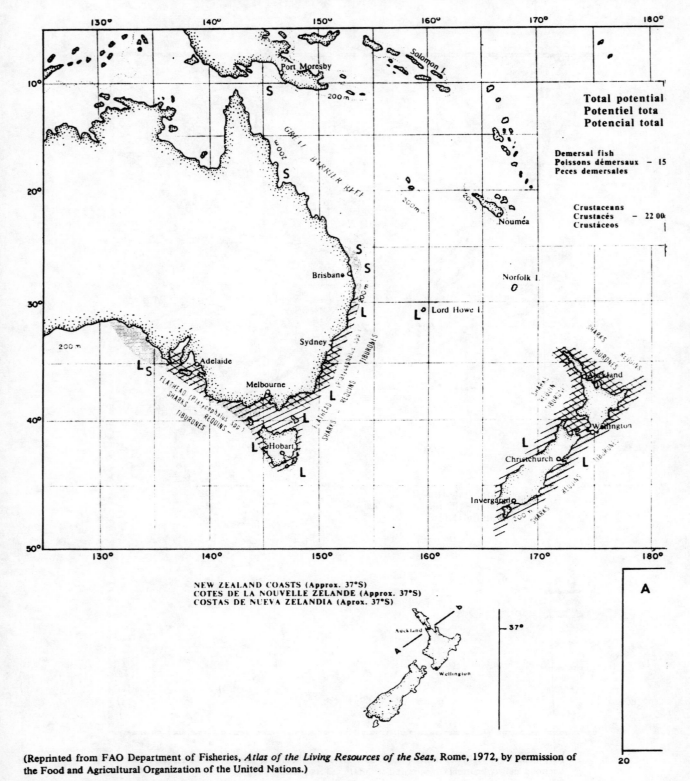

NEW ZEALAND COASTS (Approx. 37°S)
COTES DE LA NOUVELLE ZELANDE (Approx. 37°S)
COSTAS DE NUEVA ZELANDIA (Aprox. 37°S)

RESOURCES
POISSONS ET CRUSTACES
PECES Y CRUSTACEOS

Figure 2.5–32

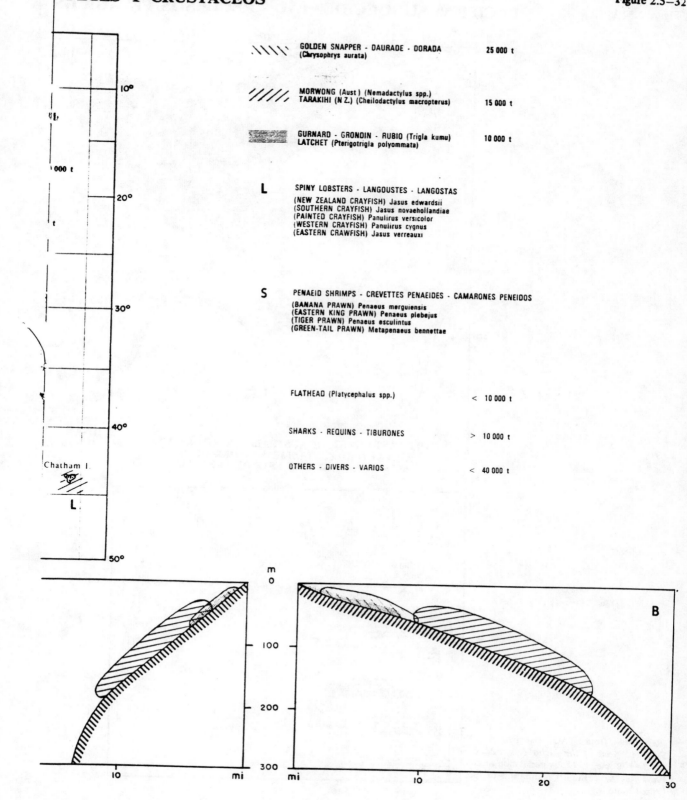

`\\\\\`	GOLDEN SNAPPER - DAURADE - DORADA (Chrysophrys aurata)	25 000 t
`/////`	MORWONG (Aust) (Nemadactylus spp.) TARAKIHI (N Z.) (Cheilodactylus macropterus)	15 000 t
▒▒▒	GURNARD - GRONDIN - RUBIO (Trigla kumu) LATCHET (Pterigotrigla polyommata)	10 000 t

L   SPINY LOBSTERS - LANGOUSTES - LANGOSTAS

(NEW ZEALAND CRAYFISH) Jasus edwardsii
(SOUTHERN CRAYFISH) Jasus novaehollandiae
(PAINTED CRAYFISH) Panulirus versicolor
(WESTERN CRAYFISH) Panulirus cygnus
(EASTERN CRAWFISH) Jasus verreauxi

S   PENAEID SHRIMPS - CREVETTES PENAEIDES - CAMARONES PENEIDOS

(BANANA PRAWN) Penaeus merguiensis
(EASTERN KING PRAWN) Penaeus plebejus
(TIGER PRAWN) Penaeus esculintus
(GREEN-TAIL PRAWN) Metapenaeus bennettae

FLATHEAD (Platycephalus spp.)	< 10 000 t
SHARKS - REQUINS - TIBURONES	> 10 000 t
OTHERS - DIVERS - VARIOS	< 40 000 t

165

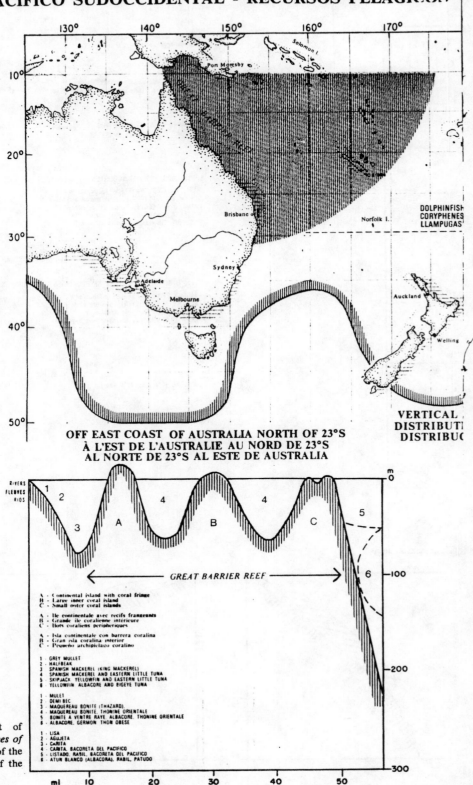

SOUTH WEST PACIFIC PELAGIC RESOURCES
PACIFIQUE SUD-OUEST - RESSOURCES PELAGIQUES
PACIFICO SUDOCCIDENTAL - RECURSOS PELAGICOS

Figure 2.5–33

(Reprinted from FAO Department of Fisheries, *Atlas of the Living Resources of the Seas*, Rome, 1972, by permission of the Food and Agricultural Organization of the United Nations.)

Figure 2.5—33

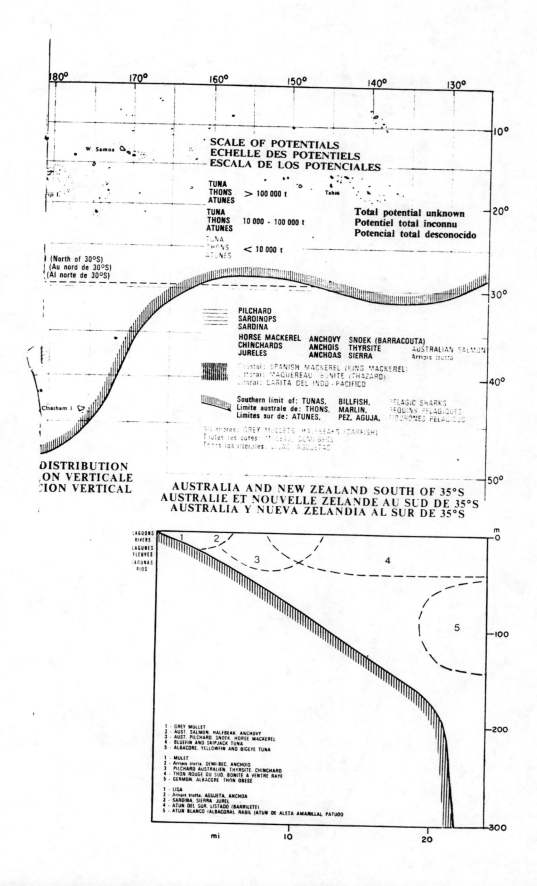

SCALE OF POTENTIALS
ECHELLE DES POTENTIELS
ESCALA DE LOS POTENCIALES

TUNA THONS ATUNES > 100 000 t

TUNA THONS ATUNES 10 000 - 100 000 t

TUNA THONS ATUNES < 10 000 t

Total potential unknown
Potentiel total inconnu
Potencial total desconocido

(North of 30ºS)
(Au nord de 30ºS)
(Al norte de 30ºS)

PILCHARD
SARDINOPS
SARDINA

HORSE MACKEREL ANCHOVY SNOEK (BARRACOUTA)
CHINCHARDS ANCHOIS THYRSITE AUSTRALIAN SALMON
JURELES ANCHOAS SIERRA Arripis trutta

Coastal: SPANISH MACKEREL (KING MACKEREL)
Litoral: MAQUEREAU BONITE (THAZARD)
Litoral: CARITA DEL INDO - PACIFICO

Southern limit of: TUNAS. BILLFISH. PELAGIC SHARKS
Limite australe de: THONS. MARLIN. REQUINS PELAGIQUES
Limites sur de: ATUNES. PEZ. AGUJA. TIBURONES PELAGICOS

All shores: GREY MULLETS HALFBEAKS (GARFISH)
Toutes les côtes: MULETS, DEMI-BECS
Todas las litorales: LISAS, AGUJETAS

DISTRIBUTION
ON VERTICALE
CION VERTICAL

AUSTRALIA AND NEW ZEALAND SOUTH OF 35ºS
AUSTRALIE ET NOUVELLE ZELANDE AU SUD DE 35ºS
AUSTRALIA Y NUEVA ZELANDIA AL SUR DE 35ºS

1 - GREY MULLET
2 - AUST. SALMON, HALFBEAK, ANCHOVY
3 - AUST. PILCHARD, SNOEK, HORSE MACKEREL
4 - BLUEFIN AND SKIPJACK TUNA
5 - ALBACORE, YELLOWFIN AND BIGEYE TUNA

1 - MULET
2 - Arripis trutta, DEMI-BEC, ANCHOIS
3 - PILCHARD AUSTRALIEN, THYRSITE, CHINCHARD
4 - THON ROUGE DU SUD, BONITE A VENTRE RAYE
5 - GERMON, ALBACORE THON OBESE

1 - LISA
2 - Arripis trutta, AGUJETA, ANCHOA
3 - SARDINA, SIERRA, JUREL
4 - ATUN DEL SUR, LISTADO (BARRILETE)
5 - ATUN BLANCO (ALBACORA), RABIL (ATUN DE ALETA AMARILLA), PATUDO

Figure 2.5—34

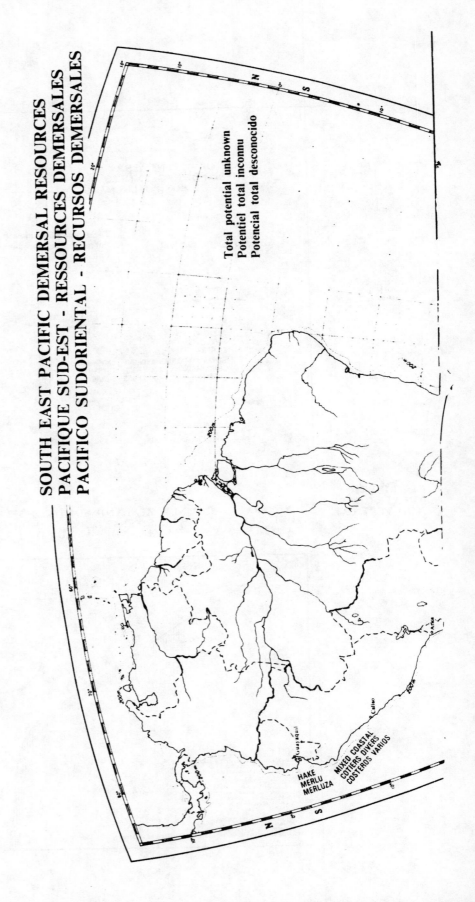

SOUTH EAST PACIFIC DEMERSAL RESOURCES
PACIFIQUE SUD-EST - RESSOURCES DEMERSALES
PACIFICO SUDORIENTAL - RECURSOS DEMERSALES

Total potential unknown
Potentiel total inconnu
Potencial total desconocido

HAKE
MERLU
MERLUZA

MIXED COASTAL
COTIERS DIVERS
COSTEROS VARIOS

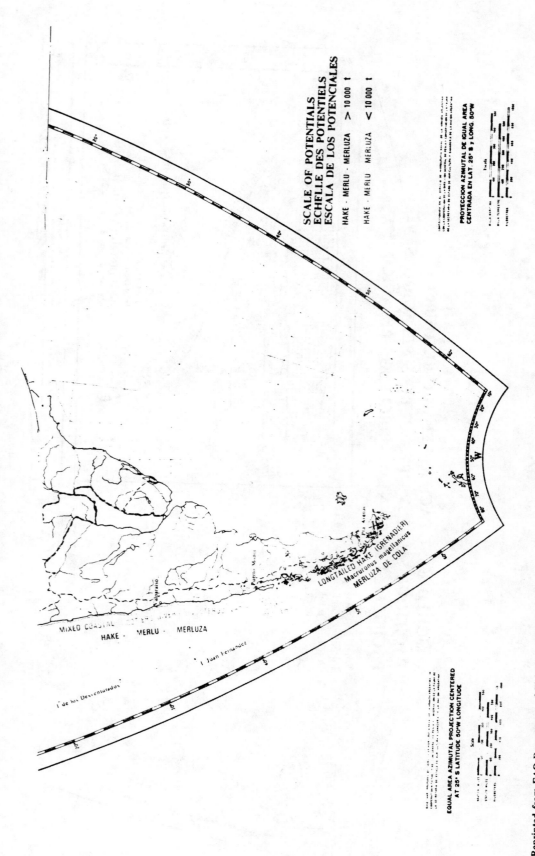

Figure 2.5—34

SCALE OF POTENTIALS
ECHELLE DES POTENTIELS
ESCALA DE LOS POTENCIALES

HAKE - MERLU - MERLUZA   > 10 000 t

HAKE - MERLU MERLUZA   < 10 000 t

LONGTAILED HAKE (GRENADIER)
Macruronus magellanicus
MERLUZA DE COLA

MIXED COASTAL
HAKE - MERLU - MERLUZA

PROYECCION AZIMUTAL DE IGUAL AREA
CENTRADA EN LAT. 25° S y LONG. 50°W

EQUAL AREA AZIMUTAL PROJECTION CENTERED
AT 25° S LATITUDE 50°W LONGITUDE

(Reprinted from FAO Department of Fisheries, *Atlas of the Living Resources of the Seas*, Rome, 1972, by permission of the Food and Agricultural Organization of the United Nations.)

Figure 2.5–35

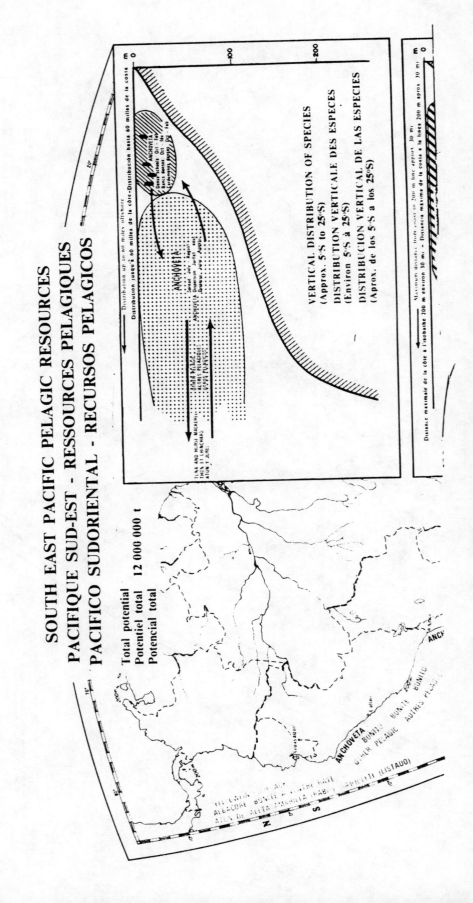

SOUTH EAST PACIFIC PELAGIC RESOURCES
PACIFIQUE SUD-EST - RESSOURCES PELAGIQUES
PACIFICO SUDORIENTAL - RECURSOS PELAGICOS

Total potential
Potentiel total   12 000 000 t
Potencial total

VERTICAL DISTRIBUTION OF SPECIES
(Approx. 5°S to 25°S)
DISTRIBUTION VERTICALE DES ESPECES
(Environ 5°S à 25°S)
DISTRIBUCION VERTICAL DE LAS ESPECIES
(Aprox. de los 5°S a los 25°S)

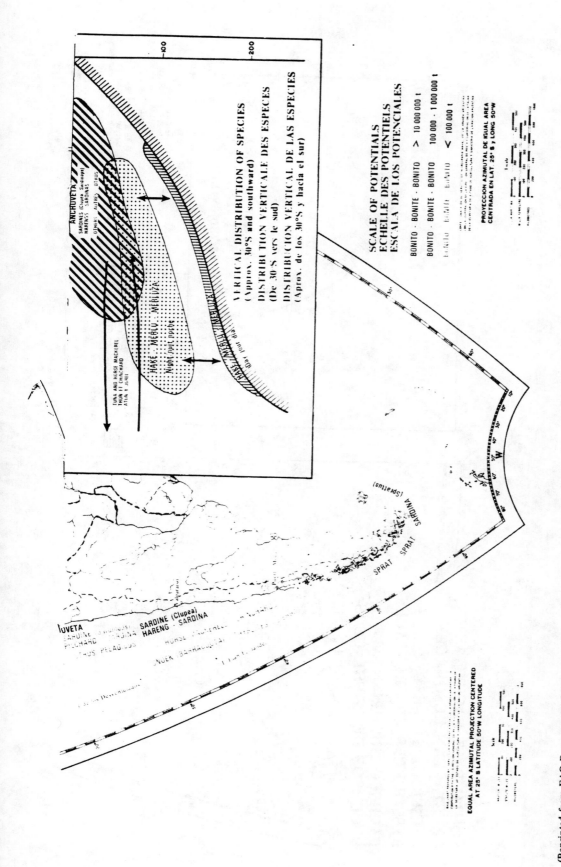

Figure 2.5–35

(Reprinted from FAO Department of Fisheries, *Atlas of the Living Resources of the Seas*, Rome, 1972, by permission of the Food and Agricultural Organization of the United Nations.)

Figure 2.5—36

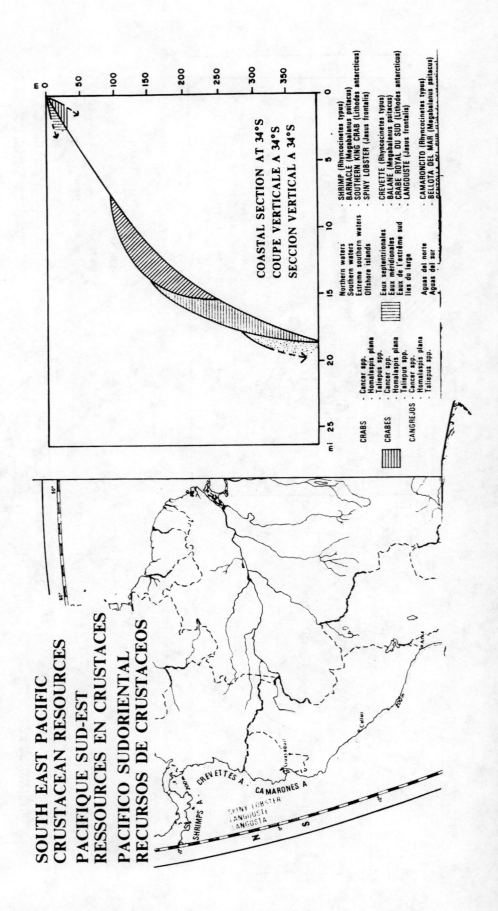

SOUTH EAST PACIFIC
CRUSTACEAN RESOURCES
PACIFIQUE SUD-EST
RESSOURCES EN CRUSTACES
PACIFICO SUDORIENTAL
RECURSOS DE CRUSTACEOS

COASTAL SECTION AT 34°S
COUPE VERTICALE A 34°S
SECCION VERTICAL A 34°S

Northern waters		SHRIMP (Rhyncocinetes typus)	
Southern waters		BARNACLE (Megabalanus psittacus)	
Extreme southern waters		SOUTHERN KING CRAB (Lithodes antarcticus)	
Offshore islands		SPINY LOBSTER (Jasus frontalis)	
Eaux septentrionales		CREVETTE (Rhyncocinetes typus)	
Eaux méridionales		BALANE (Megabalanus psittacus)	
Eaux de l'extrême sud		CRABE ROYAL DU SUD (Lithodes antarcticus)	
Iles du large		LANGOUSTE (Jasus frontalis)	
Aguas del norte		CAMARONCITO (Rhyncocinetes typus)	
Aguas del sur		BELLOTA DEL MAR (Megabalanus psittacus)	

CRABS — Cancer spp.
— Homalaspis plana
— Taliepus spp.

CRABES — Cancer spp.
— Homalaspis plana
— Taliepus spp.

CANGREJOS — Cancer spp.
— Homalaspis plana
— Taliepus spp.

CREVETTES A
SHRIMP'S A
CAMARONES A
SPINY LOBSTER
LANGOUSTE
LANGOSTA

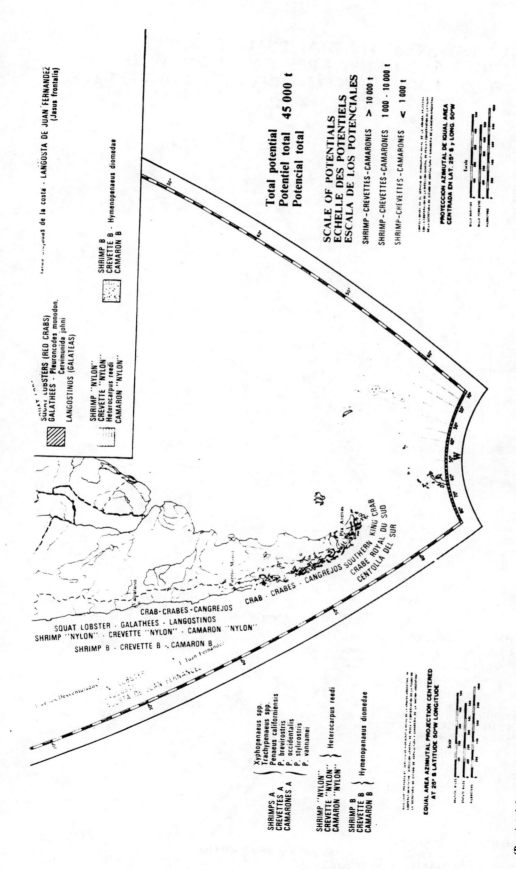

Figure 2.5–36

Total potential
Potentiel total 45 000 t
Potencial total

SCALE OF POTENTIALS
ECHELLE DES POTENTIELS
ESCALA DE LOS POTENCIALES

SHRIMP - CREVETTES - CAMARONES > 10 000 t

SHRIMP - CREVETTES - CAMARONES 1 000 - 10 000 t

SHRIMP - CREVETTES - CAMARONES < 1 000 t

PROYECCION AZIMUTAL DE IGUAL AREA
CENTRADA EN LAT. 25° S y LONG. 80°W

EQUAL AREA AZIMUTAL PROJECTION CENTERED
AT 25° S LATITUDE 80°W LONGITUDE

SHRIMPS A
CREVETTES A
CAMARONES A
— Xyphopenaeus spp.
Trachypenaeus spp.
Penaeus californiensis
P. brevirostris
P. occidentalis
P. stylirostris
P. vannamei

SHRIMP "NYLON"
CREVETTE "NYLON"
CAMARON "NYLON"
— Heterocarpus reedi

SHRIMP B
CREVETTE B
CAMARON B
— Hymenopenaeus diomedae

(Reprinted from FAO Department of Fisheries, *Atlas of the Living Resources of the Seas*, Rome, 1972, by permission of the Food and Agricultural Organization of the United Nations.)

SOUTH WEST ATLANTIC DEMERSAL RESOURCES
ATLANTIQUE SUD-OUEST - RESSOURCES DEMERSALES
ATLANTICO SUDOCCIDENTAL - RECURSOS DEMERSALES

Figure 2.5-37

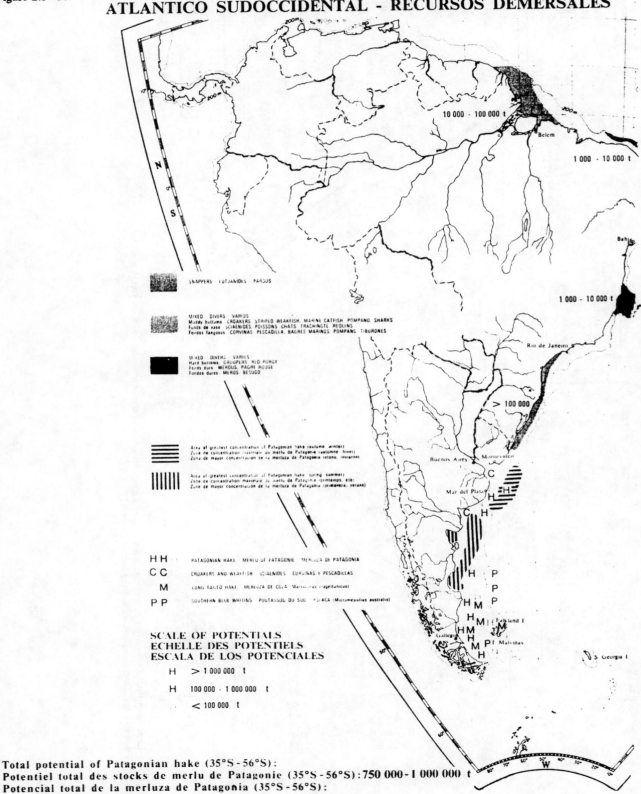

Total potential of Patagonian hake (35°S - 56°S):
Potentiel total des stocks de merlu de Patagonie (35°S - 56°S) : 750 000 - 1 000 000 t
Potencial total de la merluza de Patagonia (35°S - 56°S):

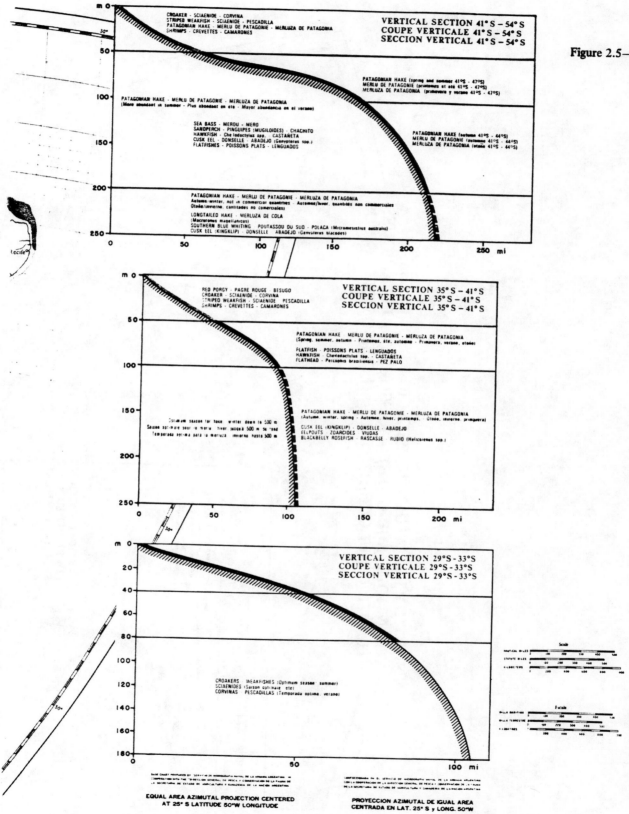

Figure 2.5–37

(Reprinted from FAO Department of Fisheries, *Atlas of the Living Resources of the Seas*, Rome, 1972, by permission of the Food and Agricultural Organization of the United Nations.)

Figure 2.5—38

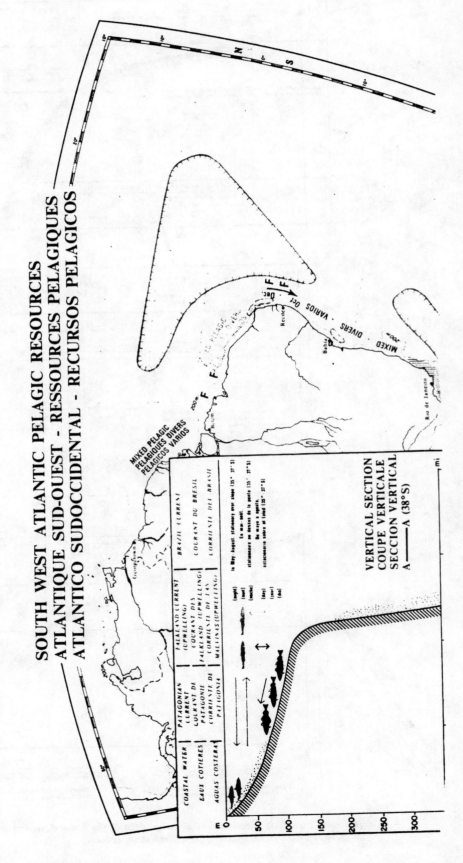

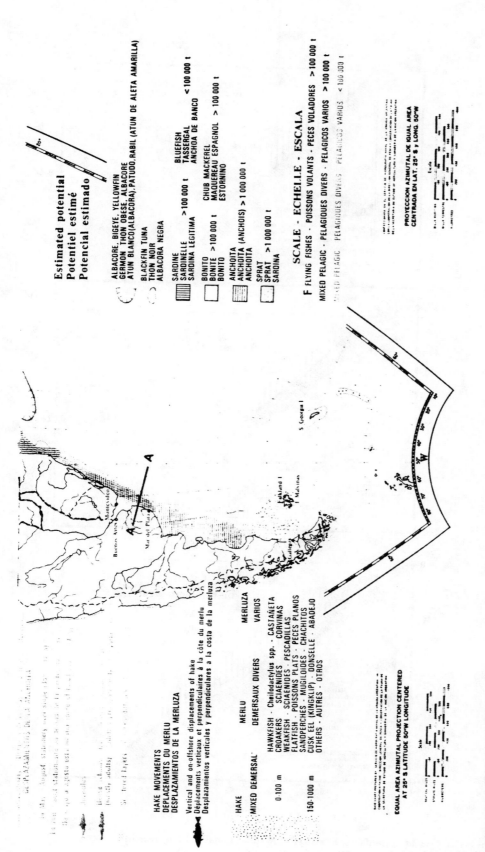

Figure 2.5–38

(Reprinted from FAO Department of Fisheries, *Atlas of the Living Resources of the Seas*, Rome, 1972, by permission of the Food and Agricultural Organization of the United Nations.)

Figure 2.5—39

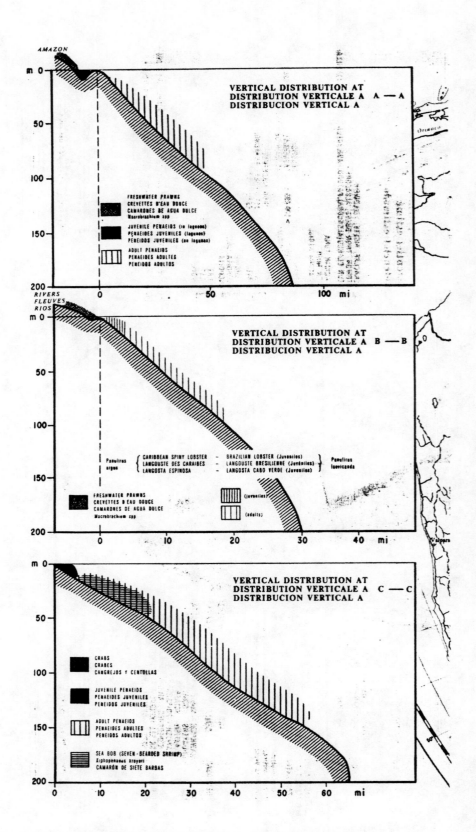

(Reprinted from FAO Department of Fisheries, *Atlas of the Living Resources of the Seas*, Rome, 1972, by permission of the Food and Agricultural Organization of the United Nations.)

SOUTH WEST ATLANTIC CRUSTACEAN RESOURCES
ATLANTIQUE SUD-OUEST - RESSOURCES EN CRUSTACES
ATLANTICO SUDOCCIDENTAL - RECURSOS DE CRUSTACEOS

Figure 2.5–39

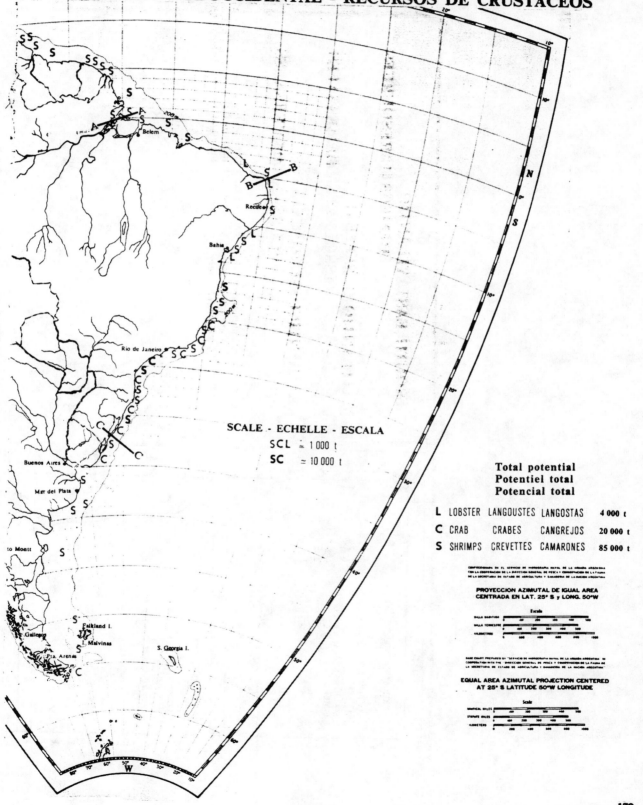

SCALE - ECHELLE - ESCALA

SCL ≃ 1 000 t

SC ≃ 10 000 t

Total potential
Potentiel total
Potencial total

L	LOBSTER	LANGOUSTES	LANGOSTAS	4 000 t
C	CRAB	CRABES	CANGREJOS	20 000 t
S	SHRIMPS	CREVETTES	CAMARONES	85 000 t

PROYECCION AZIMUTAL DE IGUAL AREA
CENTRADA EN LAT. 25° S y LONG. 50°W

EQUAL AREA AZIMUTAL PROJECTION CENTERED
AT 25° S LATITUDE 50°W LONGITUDE

Figure 2.5—40

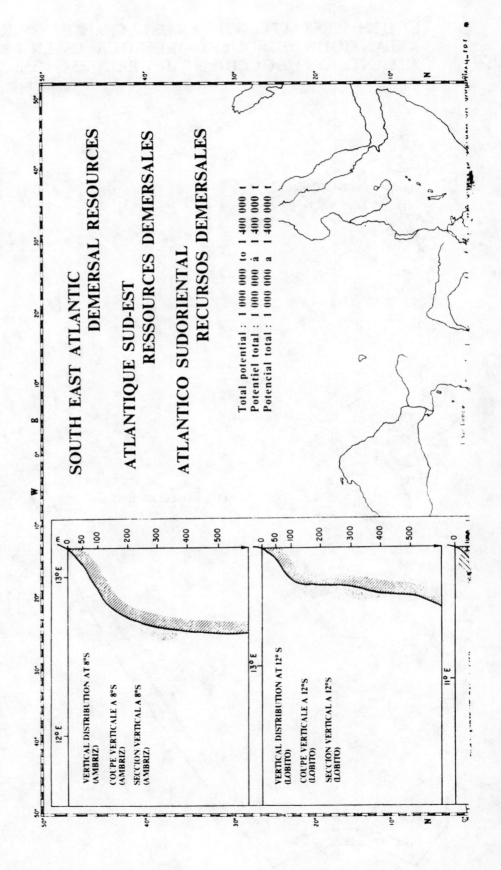

SOUTH EAST ATLANTIC DEMERSAL RESOURCES

ATLANTIQUE SUD-EST RESSOURCES DEMERSALES

ATLANTICO SUDORIENTAL RECURSOS DEMERSALES

Total potential : 1 000 000 to 1 400 000 t
Potentiel total : 1 000 000 à 1 400 000 t
Potencial total : 1 000 000 a 1 400 000 t

VERTICAL DISTRIBUTION AT 8°S (AMBRIZ)
COUPE VERTICALE A 8°S (AMBRIZ)
SECCION VERTICAL A 8°S (AMBRIZ)

VERTICAL DISTRIBUTION AT 12°S (LOBITO)
COUPE VERTICALE A 12°S (LOBITO)
SECCION VERTICAL A 12°S (LOBITO)

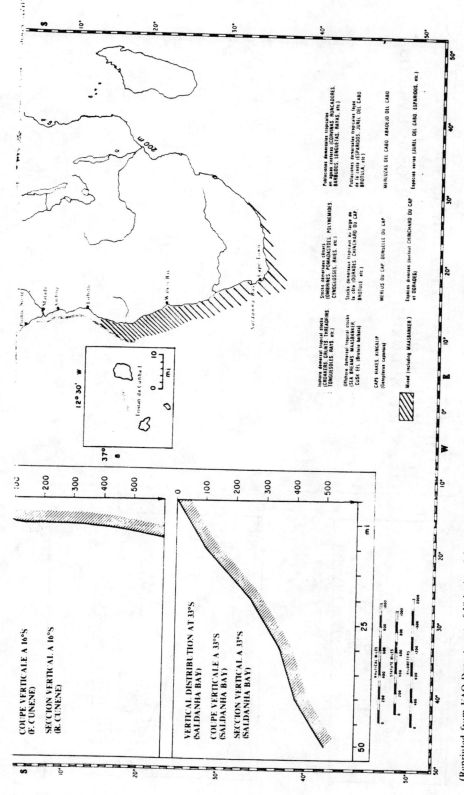

Figure 2.5–40

(Reprinted from FAO Department of Fisheries, *Atlas of the Living Resources of the Seas*, Rome, 1972, by permission of the Food and Agricultural Organization of the United Nations.)

Figure 2.5–41

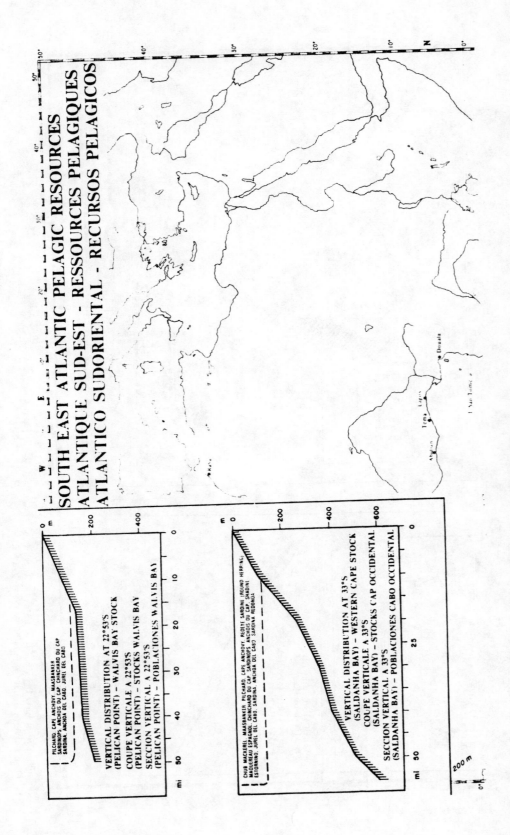

SOUTH EAST ATLANTIC PELAGIC RESOURCES
ATLANTIQUE SUD-EST - RESSOURCES PELAGIQUES
ATLANTICO SUDORIENTAL - RECURSOS PELAGICOS

PILCHARD CAPE ANCHOVY MAASBANKER
SARDINOPS ANCHOIS DU CAP CHINCHARD DU CAP
SARDINA. ANCHOA DEL CABO JUREL DEL CABO

VERTICAL DISTRIBUTION AT 22°53'S
(PELICAN POINT) – WALVIS BAY STOCK
COUPE VERTICALE A 22°53'S
(PELICAN POINT) – STOCKS WALVIS BAY
SECCION VERTICAL A 22°53'S
(PELICAN POINT) – POBLACIONES WALVIS BAY

CHUB MACKEREL MAASBANKER PILCHARD CAPE ANCHOVY REDEYE SARDINE (ROUND HERRING)
MAQUEREAU ESPAGNOL CHINCHARD DU CAP SARDINOPS ANCHOIS DU CAP SHADINE
ESTORNINO JUREL DEL CABO SARDINA. ANCHOA DEL CABO SARDINA REDONDA

VERTICAL DISTRIBUTION AT 33°S
(SALDANHA BAY) – WESTERN CAPE STOCK
COUPE VERTICALE A 33°S
(SALDANHA BAY) – STOCKS CAP OCCIDENTAL
SECCION VERTICAL A 33°S
(SALDANHA BAY) – POBLACIONES CABO OCCIDENTAL

Figure 2.5—41

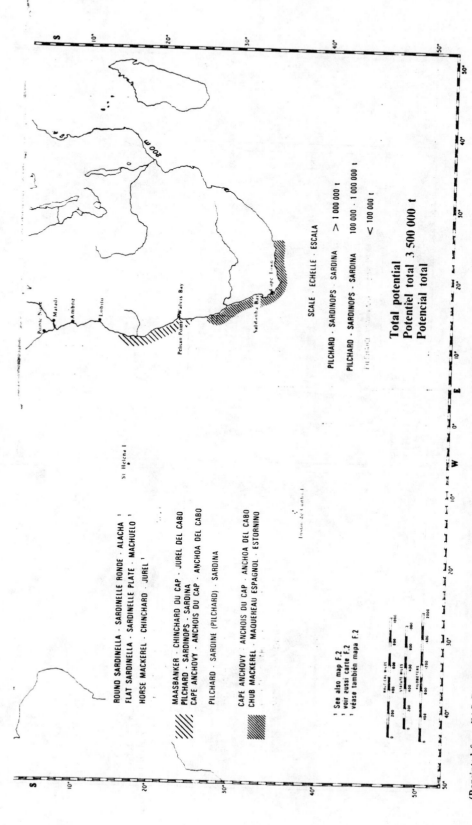

ROUND SARDINELLA - SARDINELLE RONDE - ALACHA [1]
FLAT SARDINELLA - SARDINELLE PLATE - MACHUELO [1]
HORSE MACKEREL - CHINCHARD - JUREL [1]

MAASBANKER - CHINCHARD DU CAP - JUREL DEL CABO
PILCHARD - SARDINOPS - SARDINA
CAPE ANCHOVY - ANCHOIS DU CAP - ANCHOA DEL CABO

PILCHARD - SARDINE (PILCHARD) - SARDINA

CAPE ANCHOVY - ANCHOIS DU CAP - ANCHOA DEL CABO
CHUB MACKEREL - MAQUEREAU ESPAGNOL - ESTORNINO

[1] See also map F.2
[1] voir aussi carte F.2
[1] véase también mapa F.2

SCALE - ECHELLE - ESCALA

PILCHARD - SARDINOPS - SARDINA   > 1 000 000 t

PILCHARD - SARDINOPS - SARDINA   100 000 - 1 000 000 t

< 100 000 t

Total potential
Potentiel total 3 500 000 t
Potencial total

(Reprinted from FAO Department of Fisheries, *Atlas of the Living Resources of the Seas*, Rome, 1972, by permission of the Food and Agricultural Organization of the United Nations.)

183

Figure 2.5—42

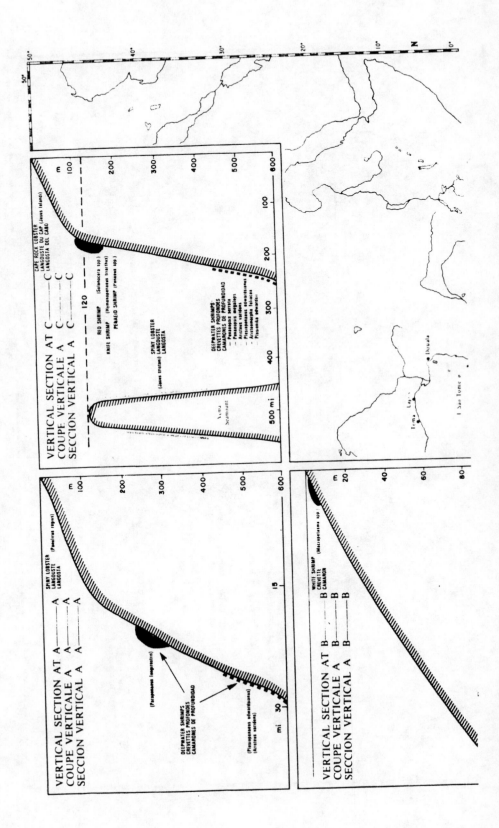

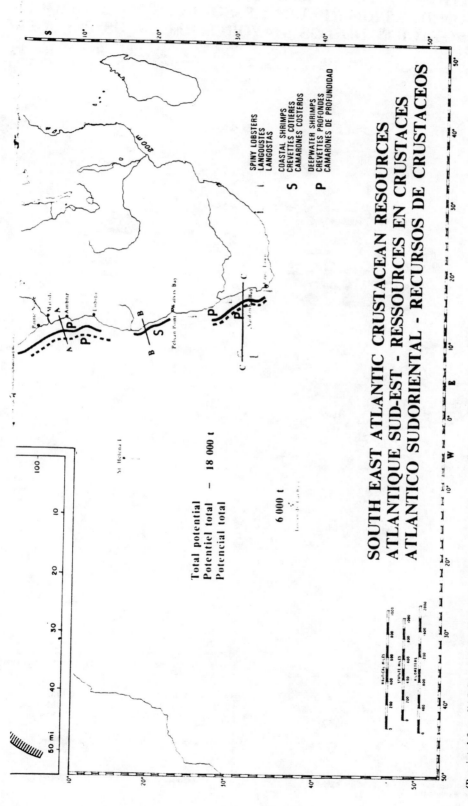

SOUTH EAST ATLANTIC CRUSTACEAN RESOURCES
ATLANTIQUE SUD-EST - RESSOURCES EN CRUSTACES
ATLANTICO SUDORIENTAL - RECURSOS DE CRUSTACEOS

SPINY LOBSTERS
LANGOUSTES
LANGOSTAS

S COASTAL SHRIMPS
CREVETTES COTIERES
CAMARONES COSTEROS

P DEEPWATER SHRIMPS
CREVETTES PROFONDES
CAMARONES DE PROFUNDIDAD

Total potential   — 18 000 t
Potentiel total
Potencial total

6 000 t

(Reprinted from FAO Department of Fisheries, *Atlas of the Living Resources of the Seas*, Rome, 1972, by permission of the Food and Agricultural Organization of the United Nations.)

Figure 2.5—42

STATE OF EXPLOITATION OF THE MAJOR STOCKS
ETAT D'EXPLOITATION DES PRINCIPALES RESSOURCES HALIEUTIQUES
ESTADO DE LA EXPLOTACION DE LOS MAYORES RECURSOS PESQUERO

Figure 2.5–43

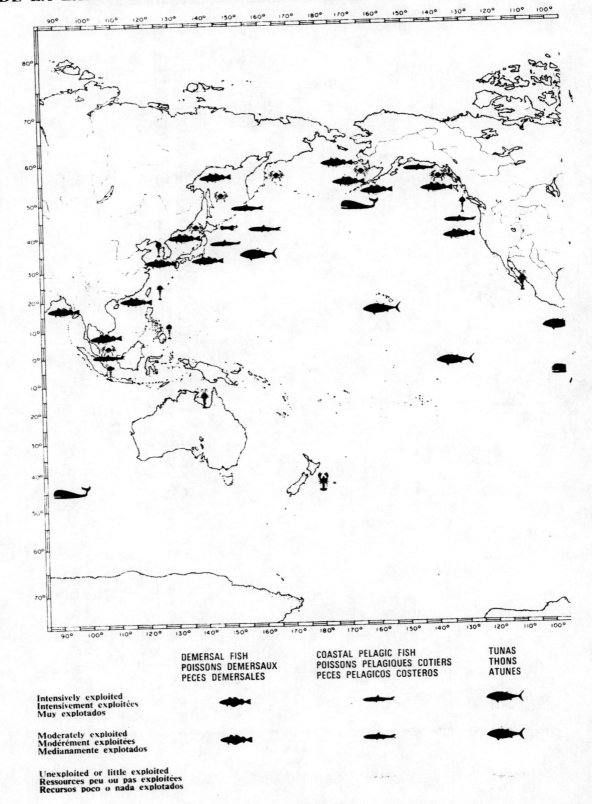

	DEMERSAL FISH POISSONS DEMERSAUX PECES DEMERSALES	COASTAL PELAGIC FISH POISSONS PELAGIQUES COTIERS PECES PELAGICOS COSTEROS	TUNAS THONS ATUNES
Intensively exploited Intensivement exploitées Muy explotados			
Moderately exploited Modérément exploitées Medianamente explotados			
Unexploited or little exploited Ressources peu ou pas exploitées Recursos poco o nada explotados			

S

Figure 2.5–43

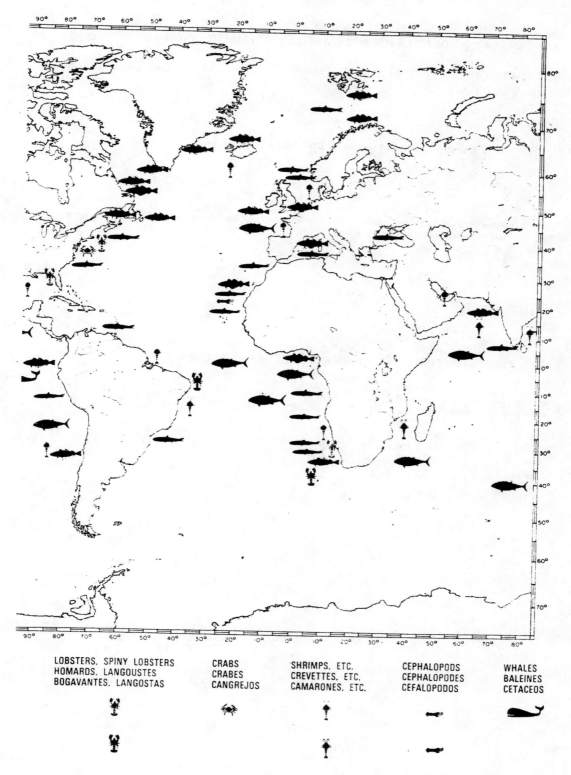

| LOBSTERS, SPINY LOBSTERS HOMARDS, LANGOUSTES BOGAVANTES, LANGOSTAS | CRABS CRABES CANGREJOS | SHRIMPS, ETC. CREVETTES, ETC. CAMARONES, ETC. | CEPHALOPODS CEPHALOPODES CEFALOPODOS | WHALES BALEINES CETACEOS |

187

2.6 Potential Yields Related to Trophic Levels and Values

Figure 2.6–1   VALUE OF WORLD CATCH (1968) AND ESTIMATED POTENTIALS BY REGION

Letters Refer To Suffix Designations Of Regional Maps In Sections 2.2, 2.3, 2.5 And Corresponding Areas In The Composite Map Of Figure 2.1–1

See special foldout section at back of the book, p. 382.

(Reprinted from FAO Department of Fisheries, *Atlas of the Living Resources of the Seas*, Rome, 1972, by permission of the Food and Agricultural Organization of the United Nations.)

Table 2.6–1
ESTIMATED FISH PRODUCTION IN THREE OCEAN PROVINCES

Province	Primary production [tons (organic carbon)]	Trophic levels	Efficiency (%)	Fish production [tons (fresh wt)]
Oceanic	16.3×10^9	5	10	16×10^5
Coastal	3.6×10^9	3	15	12×10^7
Upwelling	0.1×10^9	1½	20	12×10^7
Total				24×10^7

Table 2.6–2

**MEAN ANNUAL STANDING CROPS AND MEAN DAILY
PRODUCTION AND CONSUMPTION OF ORGANISMS IN
LONG ISLAND SOUND**

	Standing crop gC/m^2	Production $mgC/m^2/day$	Consumption $mgC/m^2/day$
Phytoplankton	8	530 (net)	–
Zooplankton	1	27	140
Other pelagic organisms	–	–	230
Benthic invertebrates	5	36 (14)	45
Benthic microbenthos	–	–	130
Benthic fishes	0.2	0.5	

Table 2.6–3

ESTIMATES OF POTENTIAL YIELDS* PER YEAR AT VARIOUS TROPHIC LEVELS

In Metric Tons

	Ecological efficiency factor					
	10%		15%		20%	
Trophic level	Carbon	Total wt	Carbon	Total wt	Carbon	Total wt
(0) Phytoplankton (net particulate production)	1.9×10^{10}		1.9×10^{10}		1.9×10^{10}	
(1) Herbivores	1.9×10^{9}	1.9×10^{10}	2.8×10^{9}	2.8×10^{10}	3.8×10^{9}	3.8×10^{10}
(2) 1st stage carnivores	1.9×10^{8}	1.9×10^{9}	4.2×10^{8}	4.2×10^{9}	7.6×10^{8}	7.6×10^{9}
(3) 2nd stage carnivores	1.9×10^{7}	$1.9 \times 10^{8^{2}}$	6.4×10^{7}	6.4×10^{8}	15.2×10^{7}	$15.2 \times 10^{8^{2}}$
(4) 3rd stage carnivores	1.9×10^{6}	1.9×10^{7}	9.6×10^{6}	9.6×10^{7}	30.4×10^{6}	30.4×10^{7}

* Output to predation at each trophic level.

Section 3

Zooplankton Population

3.1 Biomass as Related to Area, Depth, and Season

Table 3.1-1
RATIOS OF THE AVERAGE MONTHLY VALUES
OF THE MAXIMAL AND MINIMAL AMOUNTS (WEIGHT) OF PLANKTON
DURING A YEAR IN DIFFERENT REGIONS OF THE OCEAN

Region	0–100 m	0–1,000 m	Kind of data
North Atlantic-Norwegian Sea	1:406	1:2.36	Average of several samples per month
Northwestern Pacific and Bering Sea	1:12	1:3*	Average of trip stations
Northeastern Pacific	1:10	1:2	Average of single sample per month
Subantarctic	1:5.1	1:1.8	Average of many trip stations
Antarctic	1:3.4	1:1.9	Average of many trip stations

* Collections made at a depth of 0–500 m.

Figure 3.1–1

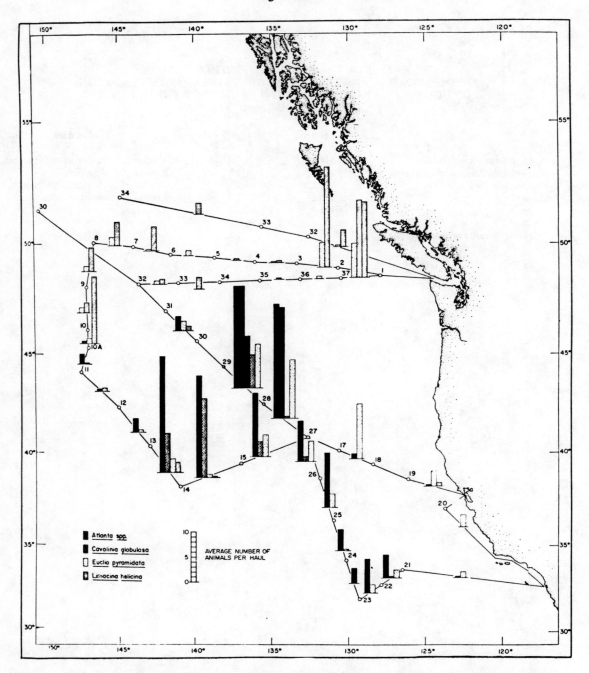

Figure 3.1–1   Cruise tracks showing position of hydrographic stations and the average catch of a heteropod and 3 pteropods taken at depths of 30, 60 and 225 m in the eastern north Pacific.

(From Aron, W., The distribution of animals in the eastern north Pacific and its relationship to physical and chemical conditions, *J. Fish. Res. Bd. Canada,* 19(2), 305, 1962. With permission.)

Figure 3.1–2

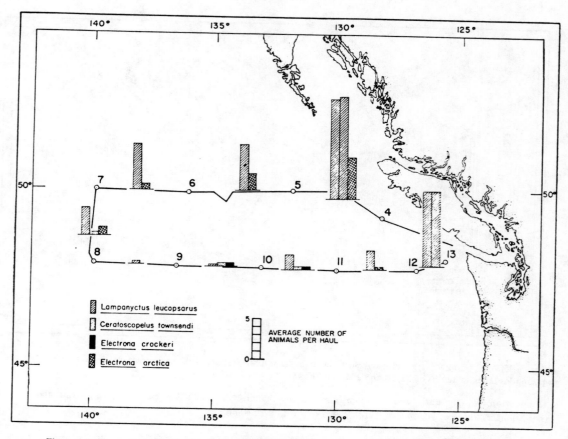

Figure 3.1–2   Cruise tracks showing position of hydrographic stations and the average catch of a heteropod and 3 pteropods taken at depths of 30, 60, and 225 m in the eastern north Pacific.

(From Aron, W., The distribution of animals in the eastern north Pacific and its relationship to physical and chemical conditions, *J. Fish. Res. Bd. Canada*, 19(2), 306, 1962.)

Figure 3.1—3

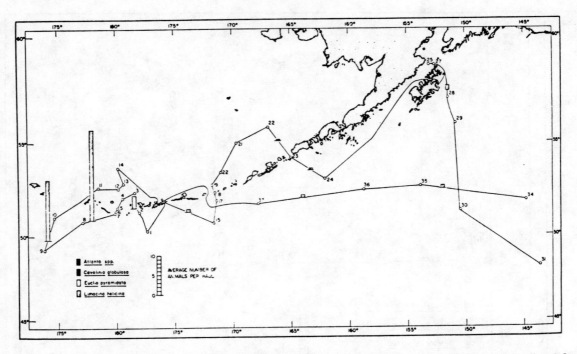

Figure 3.1—3   Cruise tracks showing position of hydrographic stations and the average catch of a heteropod and 3 pteropods taken at depths of 30, 60 and 225 m in the eastern north Pacific.

(From Aron, W., The distribution of animals in the eastern north Pacific and its relationship to physical and chemical conditions, *J. Fish Res. Bd. Canada,* 19(2), 304, 1962. With permission.)

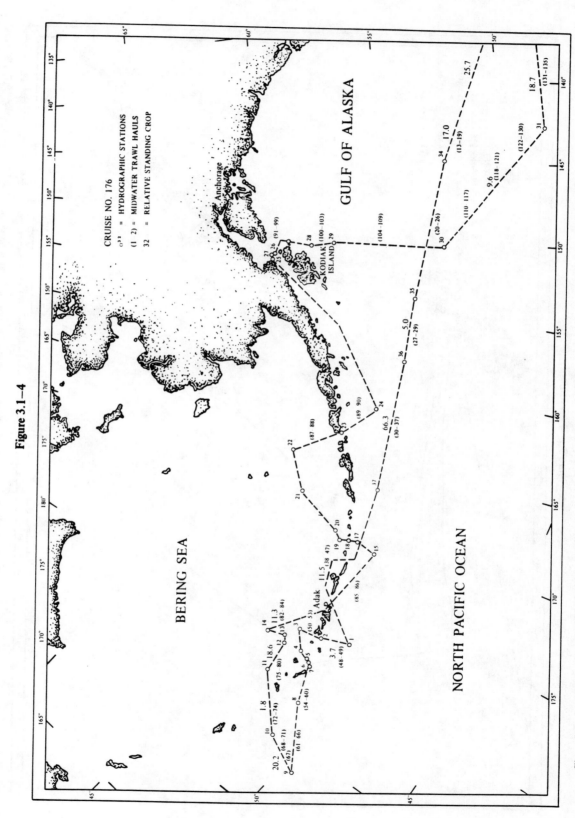

Figure 3.1-4

Figure 3.1-4   Cruise tracks showing position of hydrographic stations, position of hauls, and relative standing crop of plankton in the eastern north Pacific.

197

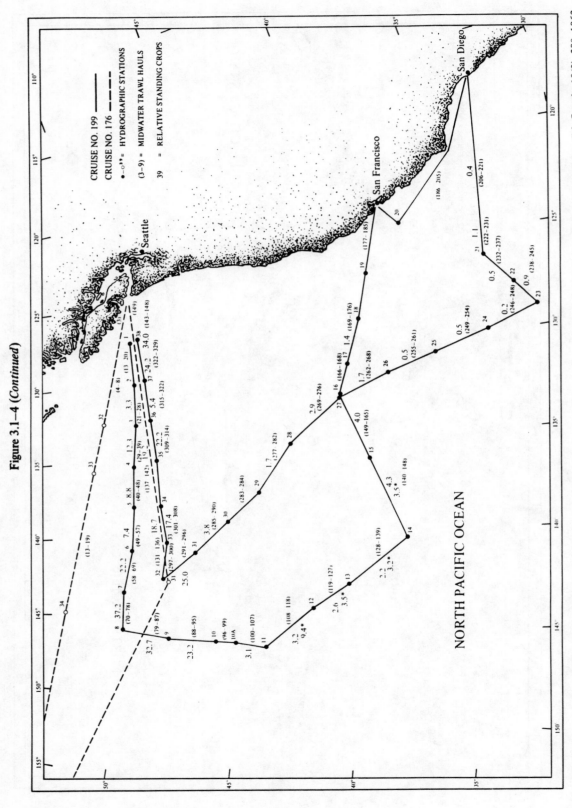

Figure 3.1–4 *(Continued)*

NORTH PACIFIC OCEAN

CRUISE NO. 199 ————
CRUISE NO. 176 – – – – = HYDROGRAPHIC STATIONS

(3–9) • ○ ** = MIDWATER TRAWL HAULS

39 = RELATIVE STANDING CROPS

Seattle

San Francisco

San Diego

(From, Aron, W., The distribution of animals in the eastern north Pacific and its relationship to physical and chemical conditions, *J. Fish. Res. Bd. Canada*, 19(2), 290, 1962. With permission.)

Figure 3.1–5

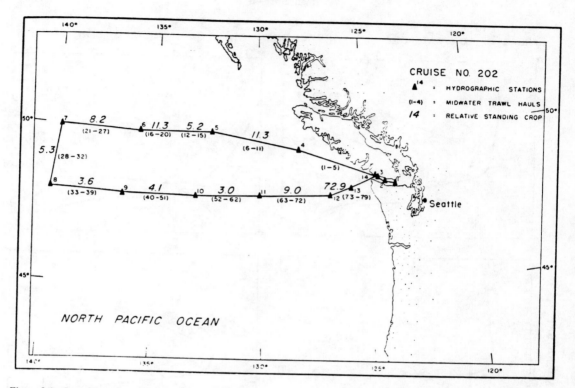

Figure 3.1–5   Cruise tracks showing position of hydrographic stations, position of hauls, and relative standing crop of plankton, in the eastern north Pacific.

(From Aron, W.. The distribution of animals in the eastern north Pacific and its relationship to physical and chemical conditions, *J. Fish. Res. Bd. Canada*, 19(2), 291, 1962. With permission.)

Figure 3.1−6

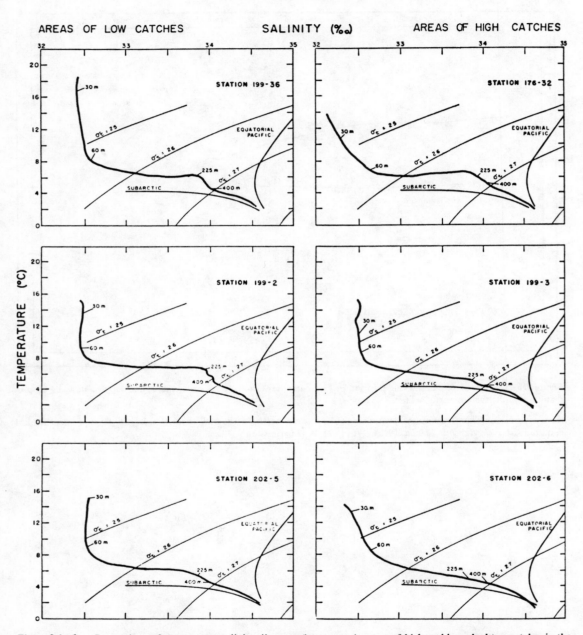

Figure 3.1−6   Comparison of temperature-salinity diagrams from oceanic areas of high and low plankton catches in the north eastern Pacific.

(From Aron, W., The distribution of animals in the eastern north Pacific and its relationship to physical and chemical conditions, *J. Fish. Res. Bd. Canada*, 19(2), 294, 1962. With permission.)

Figure 3.1–7

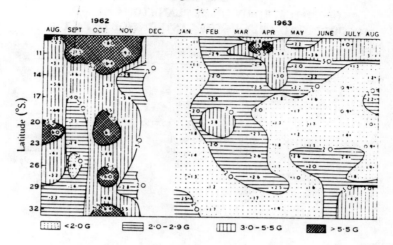

Figure 3.1–7   Contours of dry weight (g) per haul as a function of latitude and season for plankton catches in the central Indian Ocean, 1962 – 1963.

(From Legand, M., Seasonal variations in the Indian Ocean along 110° E. VI. Macroplankton and micronekton biomass, *Aust. J. Mar. Freshwater Res.*, 20, 98, 1969. With permission.)

Figure 3.1–8

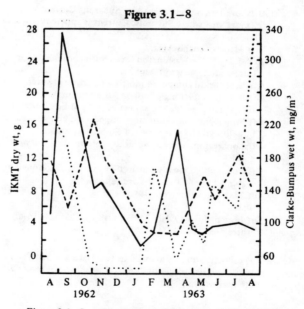

Figure 3.1–8   Comparison of the seasonal development of IKMT* plankton (● —— ●), IKMT macroplankton (X ----- X), and Clarke-Bumpus night plankton (△······△) at stations of maximal abundance in the north central Indian Ocean.

* IKMT — Isaacs-Kidd midwater trawl.

(From Legand, M., Seasonal variations in the Indian Ocean along 110° E. VI. Macroplankton and micronekton biomass, *Aust. J. Mar. Freshwater Res.*, 20, 102, 1969. With permission.)

Table 3.1−2
DEPTH DISTRIBUTION OF PLANKTON CATCHES
ACCORDING TO AREA IN THE EASTERN NORTH PACIFIC*

Area	Average volume of catch (*ml/min*)				
	20−30	31−60	61−120	121−250	251−400
A	43.7	12.7	4.3	5.8	No hauls
B	12.9	28.2	17.8	10.8	No hauls
C	18.3	7.7	No hauls	6.1	No hauls
D	0.6	6.1	No hauls	3.2	No hauls
E	28.6	22.1	No hauls	7.8	No hauls
F	1.8	3.4	6.2	5.1	No hauls
G	0.9	1.0	4.0	1.7	1.2
H	8.6	9.5	6.8	4.4	3.4
I	22.1	33.5	10.1	4.7	1.4
J	3.2	13.6	5.3	9.5	4.6
K	8.7	9.5	14.2	9.7	8.4

Note: Average volumes are 3 times average volumes caught by 3 ft net.

*Area code and description of area

Area	Description
A	Offshore area not influenced by Subarctic Current.
B	Offshore area where Subarctic Current is apparent south of 49°N latitude.
C	The Aleutian Island Area.
D	Area just off the Washington Coast with considerable Intermediate water present.
E	Offshore region composed mostly of Subarctic water.
F	Offshore region strongly influenced by the presence of Intermediate water and some Central and Equatorial water.
G	Offshore region similar in most respects to the previous region, but with greater quantities of Equatorial water.
H	Offshore region, oceanographically intermediate between E and F.
I	Offshore region, mostly Subarctic water, but Intermediate water is present along the cruise track.
J	Offshore region between 49°N, 135°W, and some Subarctic water
K	Offshore region (49°N, 135°W) with considerable Intermediate water present.

(From Aron, W., The distribution of animals in the eastern north Pacific and its relationship to physical and chemical conditions, *J. Fish. Res. Board Can.*, 19(2), 297, 299, 1962. With permission.)

Table 3.1–3
ESTIMATES OF ZOOPLANKTON PRODUCTION

Reference	Organism or group	Area and period	Production mgC/m²/day	Daily production ratios x/y standing crop	Production net primary production ratios x/y
Kamshilov, 1958	*Calanus finmarchicus*	E. Barents Sea; year	7.8	0.002	0.03
Mednikov, 1960	*Calanus cristatus*	N.W. Pacific; summer	5.6	0.012	—
	Calanus plumchrus	N.W. Pacific; summer	4.6	0.010	—
	Eucalanus bungii	N.W. Pacific; summer	3.5	0.014	—
Heinrich, 1962	*Calanus glacialis*	N. Bering Sea; year	0.7	—	0.005
	Calanus plumchrus	W. Bering Sea; year	3.1	—	0.012
	Calanus cristatus	W. Bering Sea; year	3.8	—	0.015
	Eucalanus bungii	W. Bering Sea; year	7.3	—	0.03
Yablonskaya, 1962	*Diaptomus salinus*	Aral Sea; year	0.66	0.007	—
Greze and Baldina, 1964	*Acartia clausi*	Black Sea; year	0.38	0.035	0.001
Lasker, 1966	*Centropages kröyeri*	Black Sea; summer	0.19	0.077	0.0002
Heinle, 1966	*Euphasia pacifica*	N.E. Pacific; year	0.9	0.008	0.0048
	Acartia tonsa	Chesapeake Bay estuary; summer	77	0.50	0.05
Petipa, 1967	*Acartia clausi*	Black Sea bay; June	15	0.17	—
	Acartia clausi	Black Sea, open sea; June	6.6	0.23	0.08
	Calanus helgolandicus	Black Sea, open sea; June	28	0.15	0.07
Riley, 1947	zooplankton	Georges Bank; year	200	0.03	0.25
Harvey, 1950	zooplankton	English Channel; year	75	0.10	0.30
Conover, 1956	zooplankton	Long Island Sound; year	166	0.17	0.30
Steele, 1958	zooplankton	N. North Sea; Apr.–Sept.	180	0.048	0.58
Cushing, 1959	herbivorous copepods	North Sea; Jan.–June	4.9	0.08	0.14
Cushing and Vucetic, 1963	copepods (mainly *Calanus*)	North Sea; Mar.–June	46 (author's calculation)	0.10 (author's calculation)	0.20 (author's calculation)
Smayda, 1966	zooplankton	Gulf of Panama; Jan.–April	70 or 234	0.29 or 0.98	0.09 or 0.31

(From Mullin, M., Production of zooplankton in the ocean: the present status and problems, *Oceanogr. Mar. Biol. Annu. Rev.*, 7, 308, 1970. With permission by George Allen and Unwin Ltd., London.)

Figure 3.1—9   DISTRIBUTION OF THE ABUNDANCE OF ZOOPLANKTON

See special foldout section at back of the book, p. 383.

(Reprinted from FAO Department of Fisheries, *Atlas of the Living Resources of the Seas,* Rome, 1972, by permission of the Food and Agricultural Organization of the United Nations.)

Figure 3.1—10

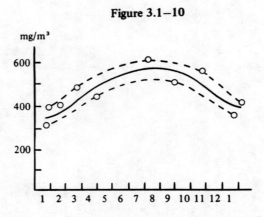

Figure 3.1—10.   Seasonal changes in plankton biomass on the southwest part of the Florida shelf.

Figure 3.1–11

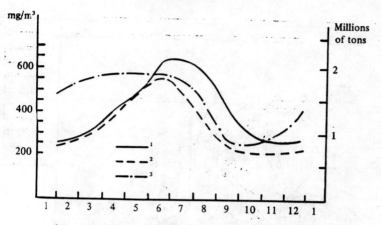

¹ Total quantity of plankton (millions of tons) on Bank.
² Average plankton biomass on eastern part of Bank.
³ Average plankton biomass on western part of Bank.

Figure 3.1–11. Seasonal changes in biomass and quantity of plankton on the Campeche Bank.

Figure 3.1–12

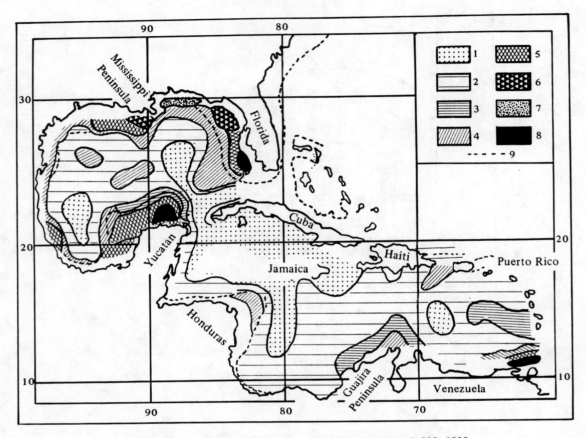

1) 30–100; 2) 50–150; 3) 100–200; 4) 100–300; 5) 200–600; 6) 200–1000;
7) 100–3000; 8) 300–1000; 9) edge of shelf.

Figure 3.1–12.   Mean plankton distributions in the upper 100 meter layer of the Caribbean and Gulf of Mexico in 1962 to 1966. Values are in mg/m³.

Table 3.1—4
RELATIVE RICHNESS OF VARIOUS POPULATIONS OFF THE AFRICAN COAST, IN LONG ISLAND SOUND, AND IN THE SARGASSO SEA

Populations	Richest	Less rich	Least rich
Primary production	Africa	> Long Island	> Sargasso Sea
Zooplankton (surface layer)	Africa	> Long Island	> Sargasso Sea
Zooplankton (total)	Africa	> Sargasso Sea	> Long Island
Ultraplankton (surface layer)	Long Island	> Africa	> Sargasso Sea
Ultraplankton (total)	Africa	> Long Island	> Sargasso Sea

Table 3.1—5
BALANCE SHEET OF ESTIMATED PRODUCTION AND CONSUMPTION IN THE SARGASSO SEA OFF BERMUDA

Process	Depth (m)	Component	Process rate $(mgC/m^2/day)$
Production	0–100	Uncorrected C^{14} uptake	200
		Estimated total production	350–400
Consumption	0–300	Zooplankton	135
		Ultraplankton	200
	300–900	Zooplankton + ultraplankton	40
	900–bottom	Zooplankton + ultraplankton	6

Figure 3.1–13

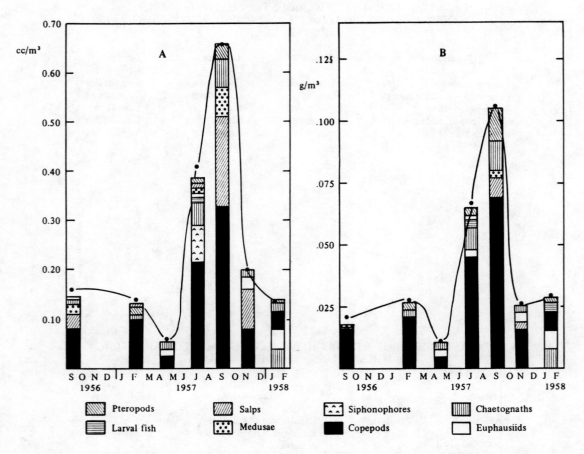

Figure 3.1–13. Average volumes (A) and dry weights (B) of zooplankton by group during 1956 to 1958 on the continental shelf south of New York.

Figure 3.1–14

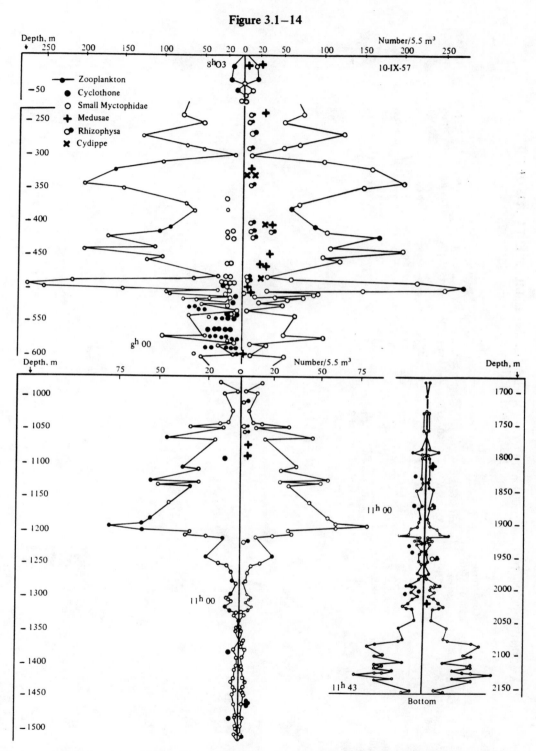

Figure 3.1–14. Zooplankton and nekton observed during a bathyscaphe dive in the Mediterranean, off Toulon. The scale on the abscissa changes at 1000 m.

(From Banse, K., On the vertical distribution of zooplankton in the sea, in *Progress Oceanography*, Vol. 2, Sears, M., Ed., Pergamon Press, New York, 1964, 57. With permission.)

Table 3.1–6

AMOUNT OF PLANKTON (cm³/1,000 m³) AT VARIOUS DEPTHS AND IN DIFFERENT PARTS OF THE NORTH ATLANTIC

Depth (m)	Gulf Stream 40–43°N	Continental slope of North America 38–41°N	Gulf Stream 35–37°N	Sargasso Sea 20–37°N	North Equatorial Current 16–19°N	African Littoral 15–22°N	Canaries Current 27–36°N	32°29'N 20°09'W	32°34'N 16°19'W
0–50	74.5	199.0	54.9	64.7	89.6	214.0	78.6	26.0	28.0
50–100	42.3	94.0	35.9	59.9	80.7	94.7	55.0		
100–200	26.2	35.1	23.3	32.1	30.6	59.9	27.0		
200–500	25.4	30.8	11.1	10.2	15.9	27.2	15.6		
500–1,000	15.7	19.0	6.0	4.1	5.4	–	6.5	5.0	4.0
1,000–2,000	5.0	7.8	2.8	1.2	1.6	–	2.1	1.0	0.7
2,000–3,000	–	–	2.1	0.6	1.0	–	0.8	0.6	–
3,000–4,000	–	–	–	0.3	0.3	–	0.2	–	–
4,000–5,000	–	–	–	0.1	–	–	–	–	–

Table 3.1—7

COMPARISONS OF DISPLACEMENT VOLUMES AND WET WEIGHTS OF ZOOPLANKTON IN VARIOUS REGIONS OF THE NORTH ATLANTIC

(1 = Summer; 2 = Fall; 3 = Winter; 4 = Spring; 5 = Yearly Mean)

Region	Displ. vol. ml/1,000 m³	Wet wt. mg/m³	Net mesh aperture (mm)	Depth range (m)	Reference
I. *Boreal North Atlantic and Neritic Waters*					
Iceland Coast					Jespersen, 1940
(1)	250	–	stramin	0–50	Jespersen, 1940
(4)	450	–	stramin	0–25	Jespersen, 1940
Southern Norwegian Sea					
(1)	340	–	0.366	0–100	Wiborg, 1954
(3)	10	–	0.366	0–100	Wiborg, 1954
(4)	230	–	0.366	0–100	Wiborg, 1954
Labrador Current					
(2) 43° N	–	1072	0.17	0–200	Yashnov, 1961
Norwegian Sea (5)	–	>500	0.17	0–100	Kanaeva, 1965
Cold-temperate Subarctic waters (5)	>100	>100	0.202	0–300	
Western North Atlantic					
Coastal	8100	–	0.158	0–25	Riley, 1939
Slope water (4)	4300	–	0.158	0–50	Riley, 1939
Slope water (1)	–	430–1600	0.158	0–400	Riley and Gorgy, 1948
Coastal (5)	540	–	10 strands/cm	0–85	Clarke, 1940
Offshore (5)	400	–	–	–	Clarke, 1940
Gulf of Maine (3)	120	–	Front: 29–38 meshes/inch	variable	Bigelow and Sears, 1939
(1)	260	–	Rear: 48–54 meshes/inch	variable	Bigelow and Sears, 1939
Cape Cod Chesapeake Bay coastal shelf					
(1)	700–800	–	–	variable	Bigelow and Sears, 1939
(3)	400	–	–	variable	Bigelow and Sears, 1939
Georges Bank					
(1)	1500	–	0.366	0–25	Clarke and Bishop, 1948
(3)	200	–	0.366	0–25	Clarke and Bishop, 1948

COMPARISONS OF DISPLACEMENT VOLUMES AND WET WEIGHTS OF ZOOPLANKTON IN VARIOUS REGIONS OF THE NORTH ATLANTIC

Region	Displ. vol. ml/1,000 m³	Wet wt. mg/m³	Net mesh aperture (mm)	Depth range (m)	Reference
Cape Hatteras-Cape Fear (1)	280	–	0.360	Variable over shelf	St. John, 1958
Continental Slope 38°−41°N (2)	328	–	0.170	0−200	Yashnov, 1961
New York-Bermuda Coastal water (5)	1070	–	0.230	0−200 or less	Grice and Hart, 1962
Slope water (5)	270	–	0.230	0−200	Grice and Hart, 1962
II. *Central Waters (Sargasso Sea)*					
Sargasso Sea (5)	20	–	0.230	0−200	Grice and Hart, 1962
Sargasso Sea Bermuda (5)	28	–	0.203−0.366	0−500	Menzel and Ryther, 1961
S. W. Sargasso Sea 20°−37°N (2)	–	156.7	0.17	0−200	Yashnov, 1961
Sargasso Sea (1)	–	45	0.158	0−400	Riley and Gorgy, 1948
Sargasso Sea (5)	–	50−100	0.170	0−100	Kanaeva, 1965
Sargasso Sea (5)	–	<50−100	–	0−300	Hela and Laevastu, 1961
Sargasso Sea (5)	10−25	<10−25	0.202	0−300	
III. *Boundary Currents*					
Florida Strait (4)	20	–	0.158	0−150	Riley, 1939
Florida Strait (4)	20	–	0.158	150−300	Riley, 1939
Gulf Stream off Florida (4)	50	–	0.158	0−100	Riley, 1939
Gulf Stream off Georgia (4)	70	–	0.158	0−150	Riley, 1939
Gulf Stream		137	0.158	0−400	Riley and Gorgy, 1948
Gulf Stream	–	250−500	0.170	0−100	Kanaeva, 1965
Gulf Stream betw. N.Y.-Bermuda	30	–	0.230	0−200	Grice and Hart, 1962

Table 3.1–7 (*Continued*)

COMPARISONS OF DISPLACEMENT VOLUMES AND WET WEIGHTS OF ZOOPLANKTON IN
VARIOUS REGIONS OF THE NORTH ATLANTIC

Region	Displ. vol. ml/1,000 m³	Wet wt. mg/m³	Net mesh aperture (mm)	Depth range (m)	Reference
Gulf Stream (north) 40°–43°N	–	143	0.17	0–200	Yashnov, 1961
Gulf Stream (south) 35°–37°N	–	114	0.17	0–200	Yashnov, 1961
N. Equat. Current 16°–19°N	–	201	0.17	0–200	Yashnov, 1961
Canary Current 27°–36°N	–	161	0.17	0–200	Yashnov, 1961
Equat. Current (1)	100–>300	–	0.336	0–~50	Mahnken, 1969
Equat. Current region (5)	–	100–500	0.17	0–100	Kanaeva, 1965
Equat. Current region (5)	–	100–300	–	0–300	Hela and Laevastu, 1961
All boundary Currents around Sargasso Sea (5)	25–100	25–100	0.202	0–300	

Table 3.1–8

PLANKTONIC BIOMASS AT DIFFERENT DEPTHS IN THE NORTHERN PART OF THE INDIAN OCEAN (mg/m³)

Depth (m)	10°18'N 110°23'E Sept. 1, 1956	10°03'S 108°00'E Nov. 5, 1959	6°31'S 108°08'E Nov. 21, 1959	6°20'S 90°02'E Dec. 18, 1959	6°03'S 90°10'E Dec. 22, 1959	1°59'S 86°41'E Jan. 11, 1960	9°56'S 86°27'E Jan. 14–15, 1960	3°11'S 57°02'E Feb. 9–10, 1960	2°46'S 65°41'E Feb. 12–13, 1960	30'S 71°19'E Dec. 24, 1960	2°03'S 91°28'E Sept. 5–7, 1962	3°59'S 77°00'E Oct. 6–7, 1962	16°48'S 96°54'E Apr. 26, 1957
0–50	162	34.0	20.5	18.7	53.9	17.1	45.5	88.0	67.5	3.5	26.4	13.8	16.1
50–100	50.0 (204)	}	20.0	}	41.3	29.2	43.6	71.0	52.0	18.0	12.6	–	25.0
100–200	28.9	32.0	14.0	18.7	17.3	14.7	22.4	14.8	27.1	14.7	6.2 (9.3)	–	10.5
200–500	19.9	11.2	7.3	4.6	5.0	10.7	7.3	8.5 (45.0)	14.2 (19.9)	5.8	2.3	6.5 (9.6)	4.4
500–1,000	15.1 (18.9)	7.5 (9.6)	7.1 (9.7)	3.9	2.7 (3.8)	6.1	4.6 (5.3)	6.5 (8.9)	5.2 (11.0)	4.5	3.3 (5.1)	4.2 (4.5)	4.2
1,000–1,500	–	–	–	–	–	2.6	–	3.3 (6.4)	–	–	–	–	–
1,500–2,000	–	–	–	–	–	0.99	0.97	1.5 (2.3)	–	–	–	–	–
1,000–2,000	3.73	1.7	0.78 (52.0) / 0.50 (25.5)	0.93	0.82 (4.1)	–	–	–	1.8 (2.2)	1.85	0.98 (4.9)	0.98 (2.1)	1.3
2,000–3,000	–	–	0.22	–	0.23	–	0.31	0.42	0.38	0.23[d]	0.37	–	0.57
2,000–4,000	–	–	0.14	0.12[a]	0.13–0.16	–	0.27[c]	–	–	–	0.03	0.41	–
2,000–4,000	0.36 (1.00)	0.36 (1.00)	–	–	–	0.26[b] (0.34)	–	–	–	–	–	–	–

The table shows the values of mesoplankton. Biomass of coelenterates and salps is not included.

( ) Biomass of the whole net plankton, including animals measuring more than 3 cm.

a Haul from 3,000–4,500 m.
b Haul from 2,000–4,500 m.
c Haul from 2,960–4,400 m.
d Haul from 2,000–3,500 m.

Figure 3.1—15

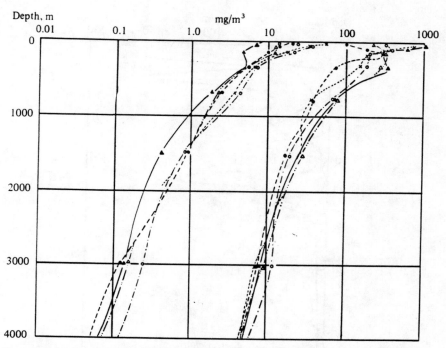

Figure 3.1—15 Vertical distribution of net plankton biomass (wet weight) in the Pacific, from divided hauls. The right family of curves is for stations off the Kurile Islands. The left family is for tropical stations (− − ● − − = average of two stations).

(From Banse, K., On the vertical distribution of zooplankton in the sea, in *Progress in Oceanography*, Vol. 2, Sears, M., Ed., Pergamon Press, New York, 1964. 82. With permission.)

Figure 3.1–16

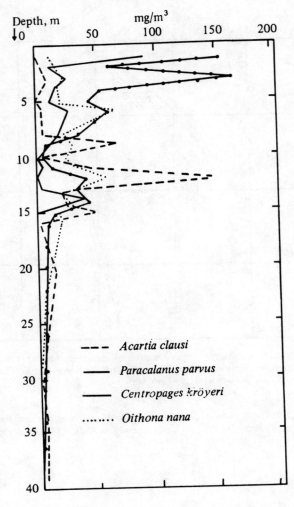

Figure 3.1–16   Vertical distribution of zooplankton biomass at two stations in the Black Sea.

(From Banse, K., On the vertical distribution of zooplankton in the sea, in *Progress in Oceanography,* Vol. 2, Sears, M., Ed., Pergamon Press, New York, 1964, 69. With permission.)

Table 3.1–9

PLANKTON BIOMASS AT DIFFERENT DEPTHS IN TROPICAL AREAS OF THE PACIFIC OCEAN (mg/m³)

Depth (m)	30°53'N 153°09'E Oct. 20–21, 1954	38°11'N 143°56'E Oct. 25–26, 1954	27°49'N 130°41'E Oct. 27–28, 1955	19°49'N 154°06'E July 9, 1957	6°17'S 153°45'E July 26–27, 1957	13°53'N 147°01'E Aug. 13–15, 1957	11°17'N 142°10'E Aug. 18, 1957	4°57'S 172°36'W Dec. 3–4, 1957	11°49'S 172°31'W Dec. 6–7, 1957	23°32'S 175°18'W Dec. 29, 1957
0–50	22.6		80.2	47.2	127	28.8	20.0	54.5	7.0	13.3
50–100	42.5	–	108	30.5	107	8.4	21.3	32.3		36.6
100–200	18.0	–	24.3	11.6	32.8	6.2	15.4	17.0	4.6	21.4
200–500	10.0	–	11.9	13.3	13.3	4.9	6.1	6.4	5.2	7.0
500–1,000	13.3	–	7.6	5.5	9.4	1.6	3.3	4.5	1.9	4.4
								3.8	1.7	
( )			(28.1)					(7.9)	(36.0)	
1,000–2,000	5.2	21.3	–	3.1	2.4	1.6	0.48	0.72	0.42	0.83
( )								(36.3)		
2,000–3,000	–	–	–	–	–	–	–	–	–	–
3,000–4,000	–	–	–	–	–	0.12	–	–	–	–
2,000–4,000	2.4	–	1.1[b]	0.67	0.40	–	–	–	–	0.25
4,000–6,000	–	–	0.48[c]	–	0.085	–	–	–	0.14	–
4,000–8,000	–	–	–	–	–	0.012	–	–	–	–
6,000–8,000	–	–	–	–	0.015	–	–	–	–	–

The table shows the mesoplanktonic biomass. The weight of coelenterates and salps is not included.

[b] Catch at a depth of 1,000–4,000 m.
[c] Catch at a depth of 4,000–5,500 m.
( ) Total biomass of net plankton, including animals larger than 3 cm.

Table 3.1–9 (Continued)

PLANKTON BIOMASS AT DIFFERENT DEPTHS IN TROPICAL AREAS OF THE PACIFIC OCEAN (mg/m³)

Depth (m)	30°32'S 176°39'W Jan. 4, 1958	19°57'N 172°37'E Feb. 14–15, 1958	25°02'N 39°46'W Jan. 4, 1959	10°26'N 140°00'W Oct. 13, 1961	5°58'N 139°57'W Sept. 16, 1961	14°11'S 140°05'W Sept. 28, 1961	6°59'S 154°00'W Oct. 17, 1961	7°56'N 153°45'W Oct. 26, 1961	13°00'N 176°00'W Nov. 1, 1961	5°58'N 176°04'W Nov. 5–6, 1961	7°58'S 175°57'W Nov. 13, 1961
0–50	–	23.1	–	71.6	119	31.4	20.7	–	–	–	–
50–100	23.4	12.1	–	28.0	231	21.2	34.8	–	5.4	15.9	–
100–200	25.0	12.8	–	12.1	56.1	4.8	8.7	–	–	9.5	–
200–500	10.9 (20.3)	4.9	–	15.5	9.6 (10.8)	2.05 (7.1)	4.1	2.5	4.0 (5.5)	2.3	2.7
500–1,000	9.6	2.4	3.9	7.51	3.7 (4.7)	2.3 (8.8)	4.1	2.9 (6.5)	1.1 (1.9)	1.8	–
1,000–2,000	2.05	1.04	0.56	0.40	1.54	–	0.4 (1.1)	1.0 (1.4)	0.51 (0.84)	0.55 (0.85)	0.45 (1.35)
2,000–3,000	–	–	0.42	0.19ᵃ	–	0.085	0.09	–	–	0.23	–
3,000–4,000	0.61 (1.05)	0.17	0.22	–	0.29	–	–	0.11	–	0.064	0.10
2,000–4,000	–	–	–	–	–	–	–	–	–	–	–
4,000–6,000	0.37	–	–	–	–	–	–	–	–	–	–
4,000–8,000	–	–	–	–	–	–	–	–	–	–	–
6,000–8,000	0.085	–	–	–	–	–	–	–	–	–	–

ᵃ Catch at a depth of 1,500–3,000 m.

3.2 Taxon Diversity as a Function of Area and Depth

Table 3.2–1

CHANGES IN THE RELATIVE NUMBERS OF CALANOIDA WITH DEPTH IN VARIOUS REGIONS OF THE OCEAN

Depth (m)	Open ocean			Isolated basin		Mediterranean Sea (Ionian Sea)		Sea of Japan (winter)
	Northwestern Pacific Ocean (spring)	Tropical zone Pacific Ocean	Tropical zone Indian Ocean	Norwegian Sea (annual average)	Central North Polar Basin	Winter	Summer	
0–50								
50–100	63.5	93.1	–	29.2	92.5	91.5	53.0	58.4
100–200	30.6	67.0	64.5	33.6	105	71.5	49.3	35.0
200–500	80.5	52.3	31.0	33.6	27.1	37.1	26.1	137.0
500–1000	16.2	40.0	61.0	160	27.1	8.8	27.0	56.4
1000–2000	59.0	28.5	18.7	5.1	5.8	3.5	4.7	3.1

(Numbers in each layer are expressed in % of numbers in the overlying layer.)

Table 3.2–2
MEAN ANNUAL TOTAL IN NUMBERS/m^3 (A) AND MEAN PERCENTAGES (B) OF MAJOR ZOOPLANKTON GROUPS IN THE SARGASSO SEA

	0–500 m		500–1000 m		1000–1500 m		1500–2000 m		0–2000 m
	A	B	A*	B	A*	B	A*	B	A
Calanoids	89.8	42.1	12.0	43.1	5.4	45.7	2.8	48.8	27.5
Other copepods	60.9	28.0	10.4	37.7	4.2	38.3	2.0	36.4	19.4
Total copepods	150.7	70.1	22.4	80.8	9.6	84.0	4.8	85.2	46.9
Ostracods	14.8	7.0	2.2	7.7	530	4.8	190	2.8	4.4
Other Crustacea (including larval forms)	7.1	3.3	1.1	3.5	370	3.4	270	4.4	2.2
Total Crustacea	172.6	80.4	25.7	92.0	10.5	92.2	5.2	92.4	53.5
Tunicates	15.1	6.5	185	0.6	20	0	20	0.5	3.8
Chaetognaths	6.5	3.0	312	1.1	25	2.2	35	0.6	1.8
Coelenterates	6.8	3.1	309	1.0	100	0.8	13	0.4	1.8
Larval forms (noncrustacean)	4.4	2.1	540	1.9	115	1.0	32	0.6	1.3
Protozoa	5.1	2.6	690	2.5	320	3.4	160	5.0	1.6
Miscellaneous	4.9	2.3	103	0.9	20	0.4	0		1.3
Total No./m³	215.3		27.9		11.3		5.5		65.0
Total No./m²	107,650		13,952.5		5672.5		2750		130,025

*Italicized numbers are ×10^{-3}, i.e., No./1000 m^3.

(From Deevey, G. B. and Brooks, A. L., The annual cycle in quantity and composition of the zooplankton of the Sargasso Sea off Bermuda II Surface to 2000 m, *Limnol. Oceanogr.*, 16 (6), 933, 1971. With permission.)

Table 3.2—3
ANALYSES OF CONTINENTAL SHELF ZOOPLANKTON SOUTH OF NEW YORK

| Organism | Water (wet weight) | Organic weight (ash free dry weight) | Constituent % of dry weight | | | |
			Ash weight	C (Total)	N (Kjeldahl)	P (Total)
Cnidaria						
Pelagia noctiluca	95.9	31.0	69.0	8.2−9.9	1.4	0.14−0.16
Aequorea vitrina	99.1	50.9	49.1	17.8−22.5	0.5	0.021
Cyanea capillata	95.4	37.0	63.0	11.6	0.4−1.4	0.0−0.4
Ctenophora						
Beroe cucumis	95.3	29.7	70.3	−	1.1	0.16
Mnemiopsis sp.	95.0	25.0	75.0	6.4	0.2	0.12
Mollusca (Pteropoda)						
Limacina retroversa and sp.	81.3	35.8	64.2	28.3	4.1	0.58
Clione limacina	91.0	66.7	33.3	26.3	2.2−5.0	0.26−0.35
Corolla sp.	96.5	75.7	24.3	−	−	−
Arthropoda						
Calanus finmarchicus	89.8	82.4	17.6	35.7−41.7	4.7−5.9	0.39−0.68
Centropages typicus and *hamatus*	84.2	77.2	22.8	32.5−38.5	5.2−7.1	0.72−0.84
Euphausia krohnia	80.0	81.4	18.6	35.8	6.8	0.94
Meganyctiphanes norvegica	81.0	77.6	22.4	33.4−37.0	5.2−7.1	1.16
Lophogaster sp.	−	−	−	46.8	−	−
Idotea metallica	−	−	−	33.2	6.0	4.07
Chaetognatha						
Sagitta elegans	89.4	78.4	21.6	−	7.8	0.20−0.57
Tunicata						
Salpa fusiformis	96.0	22.9	77.1	7.2−10.6	0.5−1.5	0.19−0.28
Pyrosoma sp.	95.9	31.0	71.2	9.4	0.3−0.4	0.14

Figure 3.2–1

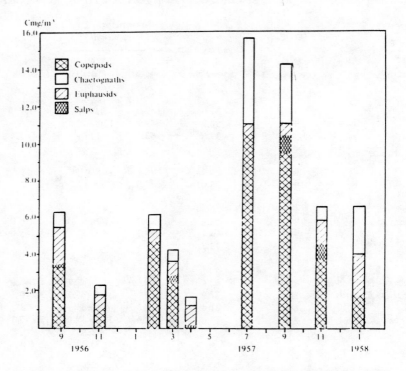

Figure 3.2–1. Contributions of copepods, salps, euphausiids and chaetognaths to standing crops of zooplankton 1956 to 1958 on the continental shelf south of New York.

Table 3.2—4
NUMBER OF EPIPLANKTONIC SPECIES IN
THE CALIFORNIA CURRENT REGION

Functional Group	Approx no. of species found in regions between the surface and 140 m depth	Source of estimate
Amphipoda	≥41	(Bowman, 1953)
Chaetognatha	24	(Alvarino, 1965)
Cladocera	5	(Baker, 1938; Strickland et al., 1968)
Copepoda	≥235	(Esterly, 1905; Olson, 1949; Fleminger, 1967)
Crustacean larvae	No estimate available	
Ctenophora	≥4	(Torrey, 1904)
Decapoda	No estimate available	
Euphausiacea	28	(Brinton, 1967a)
Heteropoda	13	(McGowan, 1967)
Larvacea	26	(Tokioka, 1960)
Medusae	24	(Alvarino, in press)
Mysidacea	≥14	(Banner, 1948; Tattersall, 1951; Clarke, W. D., pers. comm., 1969)
Ostracoda	≥16	(Juday, 1906; Miller, C., pers. comm., 1969)
Pteropoda	29	(McGowan, 1967)
Radiolaria, Tripylea	≥20	(Kling, 1966; Helms, P. B. and Milow, E. D., pers. comm., 1969)
Thaliacea	≥23	(Berner, 1967)
Siphonophora	44	(Alvarino, in press)
All Taxa Combined	≥546	

Table 3.2—5
NUMBER OF SPECIES OF CALANOIDA LIVING AT VARIOUS DEPTHS IN THE NORTHEAST

Atlantic Ocean (From 50° to 30°N)

Depth	0–200 m	200–500 m	500–1000 m	1000–2000 m	2000–3000 m	3000–4000 m	4000–5000 m
Total number of species	91	123	120	70	57	27	14
Number of appearing species	–	61	29	19	26	5	1
In % of total number of species at this depth	–	49	24	27	46	18	7
Number of disappearing species	29	32	69	39	36	14	–
In % of total number of species at this depth	32	26	58	56	61	52	–

Table 3.2—6
NUMBER OF SPECIES OF CALANOIDA LIVING AT VARIOUS DEPTHS IN THE NORTH ATLANTIC

(North of 50°N)

Depth	0–50 m	50–100 m	100–200 m	200–500 m	500–1000 m	1000–2000 m	Deeper than 2000 m
Total number of species	22	24	34	48	105	95	55
Number of disappearing species	–	7	10	17	59	21	3
Number of appearing species	5	0	3	2	31	43	–

Table 3.2—7
NUMBER OF SPECIES OF CALANOIDA AT VARIOUS DEPTHS OF THE ANTARCTIC AND SUBANTARCTIC REGIONS

Depth	0–50 m	50–100 m	100–200 m	250–500 m	500–750 m	750–1000 m
Total number of species	20	30	37	56	76	80
Number of appearing species	–	11	8	18	22	16
Number of disappearing species	0	1	0	2	12	–

Table 3.2–8

VERTICAL CHANGES IN THE DIVERSITY OF CALANID FAUNA IN THE EASTERN DEPRESSION OF THE NORTH POLAR BASIN

Depth (m)	No. of species	% of total calanid fauna	No. of appearing species	% of fauna of layer	No. of disappearing species	% of fauna of layer	Diversity index (α)	No. of specimens/m³
0–50	17	37			0	0	2.8	96.9
50–100	25	54	8	32	1	4	4.1	89.6
100–200 (300)	34	74	10	29	4	12	4.8	94.2
200 (300)–800 (1000)	40	87	10	25	9	22	6.1	26.0
800 (1000)–bottom	31	68	1	3	–	–	4.9	1.5

Table 3.2—9
VARIATION WITH DEPTH OF THE NUMBERS
(SPEC/m^3) OF COPEPODS (CALANOIDA)
IN TROPICAL AREAS OF THE PACIFIC OCEAN[a]
AND THE INDIAN OCEAN[b]

Hauls Made with BR 113/140 Nets

Depth (m)	Pacific Ocean (open waters)	Indian Ocean
0–50	12.4	18.6
50–100	11.3	–
100–200	6.9	12.0
200–500	3.5	3.7
500–1000	1.4	2.25
1000–2000	0.40	0.42
2000–3000	–	0.19
2000–4000	0.18	–
3000–4000	–	0.11
4000–6000	0.08	–
6000–8000	0.05	–

[a] Average of eight stations.
[b] Average of five stations.

Figure 3.2—2

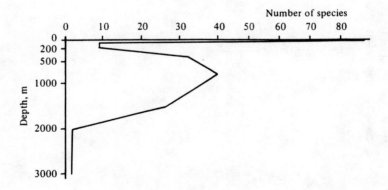

Figure 3.2—2.   Number of species of copepods (Calanoida) appearing in each of the layers from which collections were taken in the tropical regions of the Indian Ocean.

Table 3.2–10

NUMBER OF SPECIES OF CALANOIDA OCCUPYING VARIOUS DEPTHS IN THE BAY OF BISCAY, AND VERTICAL CHANGES IN THE DIVERSITY INDEX (α)

Depth	0–180 m	180–275 m	275–550 m	550–730 m	730–920 m	920–1370 m	1370–1830 m	1830–2300 m	2300–2750 m	2750–3700 m
Total number of species	44	57	68	67	65	70	67	24	14	11
Number of "appearing" species		24	19	10	4	12	8	1	0	0
Number of "disappearing" species	10	8	11	6	6	11	36	10	3	
Diversity index (α)		7.6	8.8	11.9	11.8	12.2	14.0	5.3	3.3	

Table 3.2–11
VARIATION WITH DEPTH IN THE SIGNIFICANCE OF VARIOUS TROPHIC GROUPS OF COPEPODS IN THE NORTHWEST PACIFIC OCEAN

In % of the Total Mass of Copepoda,
Except for Aphages

Depth (m)	Filter feeders	Predaceous carnivores	Species with mixed type of feeding
0–50	99.5	0	0.5
50–100	98.0	0	2.0
100–200	99.1	0.4	0.5
200–500	68.1	11.6	20.3
500–1000	15.1	36.9	48.0
1000–2000	4.0	68.0	28.0
2000–4000	9.7	66.0	22.3
4000–6000	0.34	21.3	78.3
6000–8000	>0.1	25.0	75.0

Table 3.2–12
VARIATION WITH DEPTH OF THE NUMBERS OF COPEPODS (CALANOIDA) IN THE NORTH PACIFIC OCEAN*

Hauls Made with Br 80/113 and BR 113/140 Nets

Depth (m)	Spec/m³
0–50	270
50–100	142
100–200	37
200–500	42
500–1000	9.5
1000–2000	3.6
2000–4000	1.1
6000–8500	0.13

* Average of five stations.

Table 3.2–13
BIOMASS OF COPEPODA AND ITS PART IN THE TOTAL MASS OF PLANKTON IN SUBARCTIC[a] AND TROPICAL[b] REGIONS OF THE PACIFIC OCEAN

Subarctic Region

Depth	0–50 m	50–100 m	100–200 m	200–300 m	300–500 m	500–750 m	750–1000 m	1000–1500 m
Mg/m³	461	78.3	38.9	202.6	152.2	57.4	33.5	17.6
% of total amount of plankton	82.8	57.7	39.8	76.8	70.3	61.1	66.1	65.6

Depth	1500–2000 m	2000–2500 m	2500–3000 m	3000–4000 m	4000–5000 m	5000–6000 m	6000–7000 m	7000–8700 m
Mg/m³	8.6	4.63	1.84	0.78	0.41	0.18	0.15	0.065
% of total amount of plankton	48.9	32.3	33.6	58.0	42.9	28.4	25.1	27.6

Tropical Region

Depth	0–100 m	100–200 m	200–500 m	500–1000 m	1000–2000 m	2000–4000 m
Mg/m³	20.2	10.4	2.6	1.3	0.31	0.1
% of total amount of plankton	42.5	62.2	40.1	43.1	52.3	51.0

[a] Average of nine stations.
[b] Average of five stations.

Table 3.2–14
BIOMASS OF CHAETOGNATHA AND THEIR ROLE IN THE TOTAL PLANKTONIC MASS IN THE SUBARCTIC[a] AND TROPICAL[b] REGIONS OF THE PACIFIC OCEAN

Subarctic Region

Depth	0–50 m	50–100 m	100–200 m	200–300 m	300–500 m	500–750 m	750–1000 m	1000–1500 m	1500–2000 m	2000–2500 m	2500–3000 m	3000–4000 m	4000–5000 m	5000–6000 m	6000–7000 m	7000–8000 m
Mg/m^3	36.8	33.6	38.7	33.1	28.4	15.4	6.4	1.4	5.6	7.8	2.0	0.058	0.012	0.002	<0.002	<0.002
% of total amount of plankton	8.7	28.7	43.9	14.5	13.2	15.3	12.7	5.4	30.1	43.6	37.1	4.5	0.9	0.6	0.4	<0.1

Tropical Region

Depth	0–100 m	100–200 m	200–500 m	500–1000 m	1000–2000 m	2000–4000 m	4000–8000 m
Mg/m^3	12.6	4.8	0.29	0.44	0.20	0.015	0
% of total amount of plankton	2.6	14	5.9	13.4	19.6	7.4	0

a Average of nine stations.
b Average of five stations.

Table 3.2–15
NUMBERS OF SPECIES OF CHAETOGNATHA OBSERVED IN VARIOUS AREAS OF THE INDIAN OCEAN AND SOUTHWESTERN PACIFIC

Authority	Number of valid species observed
Beraneck, 1895, Bay of Amboine	5
Burfield and Harvey, 1926, Indian Ocean	15
Doncaster, 1903, Maldive-Laccadive Archipelago	11
Fowler, 1906, Indian Ocean	18
George, 1952, Indian coastal waters	12
Lele and Gae, 1936, Bombay harbor	3
Oye, 1918, Java Sea	12
Rao, 1958, Lawson's Bay, Waltair, Bay of Bengal	13
Rao and Ganapati, 1958, off east coast of India and Ceylon	12
Ritter-Zahony, 1909, southern Indian Ocean	6
1910, southwest Australia	10
1911, Deutsche Sudpolar Expedition	17
Schilp, 1941, Indian Ocean	19
Tokioka, 1940, New South Wales	8
1955, NE Indian Ocean	13
1956, Central Indian Ocean	13
1956, Arafura Sea	9

Table 3.2–16
DISTRIBUTION OF CHAETOGNATHA IN THE WESTERN SUBTROPICAL ATLANTIC

SPECIES	SHELF	SLOPE	GULF STREAM	SARGASSO SEA
SAGITTA SERRATODENTATA				
S. ENFLATA				
S. ELEGANS				
S. LYRA				
S. BIPUNCTATA				
S. DECIPIENS				
PTEROSAGITTA DRACO				
SAGITTA HEXAPTERA				
S. PLANCTONIS				
S. MINIMA				
S. MAXIMA				
EUKROHNIA HAMATA				
KROHNITTA SUBTILIS				
K. PACIFICA				
SAGITTA FRIDERICI				
S. FEROX ?				

NOT FOUND IN GULF STREAM IN PRESENT COLLECTIONS BUT REPORTED FROM STREAM OFF FLORIDA (OWRE, 1960) OR NORTH CAROLINA (PIERCE, 1953)

(From Grice, G. D. and Hart, A. D., The abundance, seasonal occurrence, and distribution of the epizooplankton between New York and Bermuda, *Ecol. Monogr.*, 32(4), 297, 1962.)

Table 3.2–17
NUMBER OF SPECIES OF EUPHAUSIACEA LIVING AT VARIOUS DEPTHS OF THE PACIFIC OCEAN

Depth	0–200 m	200–500 m	500–1000 m	1000–2000 m	2000–4000 m	4000–8000 m
Total number of species	45	46	43	17	7	4
Number of appearing species	–	5	3	0	0	0
Number of disappearing species	4	6	26	10	3	–

Figure 3.2–3

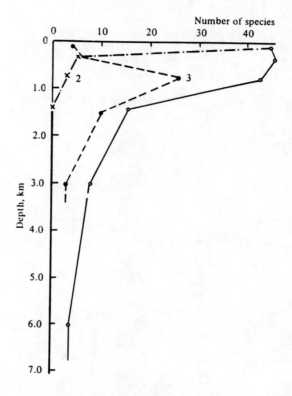

Figure 3.2–3. Variation with depth in numbers of species of euphausids in the central Pacific Ocean.

Table 3.2–18
ROLE OF ANIMALS OF VARIOUS TAXONOMIC GROUPS IN THE PLANKTON OF THE KURILE-KAMCHATKA AREA OF THE PACIFIC OCEAN[a]

Summer, 1956[b]

Depth (m)	Chaetognatha	Polychaeta	Ostracoda	Copepoda	Mysidacea	Amphipoda	Euphausiacea	Decapoda	Small fish
0–50	8.7	<0.1	0.1	82.8	+	2.2	3.6	+	+
50–100	28.7	0.2	0.3	57.7	+	3.5	7.0	0.2	+
100–200	43.9	0.6	0.5	39.8	+	5.2	5.0	1.1	+
200–300	14.5	0.2	0.4	76.8	+	3.1	1.3	+	0.3
300–500	13.2	0.7	1.8	70.3	0.4	2.4	0.9	0.2	0.3
500–750	15.3	1.2	1.2	61.1	3.8	1.0	0.2	1.2	3.6
750–1000	12.7	0.7	1.4	66.1	5.7	0.5	+	1.6	0.1
1000–1500	5.4	0.7	1.3	65.6	6.9	1.0	0.1	7.7	0.9
1500–2000	30.1	0.2	0.8	48.9	1.9	1.2	0.3	4.3	2.6
2000–2500	43.6	0.3	0.8	32.3	7.4	1.3	0.05	8.0	3.0
2500–3000	37.1	0.7	2.1	33.6	0.4	0.6	10.5	7.5	+
3000–4000	4.5	0.5	3.1	58.0	3.3	1.4	+	11.0	0
4000–5000	0.9	0.3	1.0	42.9	28.3	1.5	2.3	+	0
5000–6000	0.6		0.8	28.4	20.3	19.3	+	+	0
6000–7000	0.4	2.6	1.2	25.1	6.9	15.6	0	0	0
7000–8700	+	6.5	3.6	27.6	0	10.6	0	0	0

[a] In % of the total planktonic mass in each of the layers from which catches were made with BR nets.
[b] Average of nine stations.

Table 3.2–19
BIOMASS OF EUPHAUSIACEA AND THEIR PART IN THE TOTAL MASS OF PLANKTON IN THE SUBARCTIC[a] AND TROPICAL[b] REGIONS OF THE PACIFIC OCEAN

Subartic Region

Depth	0–50 m	50–100 m	100–200 m	200–300 m	300–500 m	500–750 m	750–1000 m	1000–1500 m
Mg/m³	14.8	4.6	4.6	3.5	2.1	0.2	~0	0.2
% of total amount of plankton in the layer	3.6	7.0	5.0	1.3	0.9	0.2	<0.1	0.1

Depth	1500–2000 m	2000–2500 m	2500–3000 m	3000–4000 m	4000–5000 m	5000–6000 m	6000–7000 m	Deeper than 7000 m
Mg/m³	0.06	0.02	0.70	~0	0.03	0	0	0
% of total amount of plankton in the layer	0.3	<0.1	10.5	<0.1	2.3	0	0	0

Tropical Region

Depth	0–200 m	200–500 m	500–1000 m	1000–2000 m	2000–4000 m
Mg/m³	1.9	2.4	0.08	0.025	0.03
% of total amount of plankton in the layer	7.2	25.0	2.4	2.4	13.4

[a] Average of nine stations.
[b] Average of five stations.

Table 3.2—20
TAXON DIVERSITY OF ZOOPLANKTON IN THE WESTERN PACIFIC

Sample number and species	Position		Date	Depth of sampling (m)
Pteropoda				
2 Euclio pyramidata (Linné)	30°02.2'N	142°00.0'E	Sept. 15, 1967	0—1000
3 Cavolinia longirostris longirostris (Lesueur)	34°33.6'N	140°56.1'E	Aug. 7, 1967	0—350
4 Limacina inflata (d'Orbigny)	0°03.9'N	148°38.8'E	Dec. 25, 1967	surface
5 Limacina helicina helicina (Phipps), juv.[a]	42°22.5'N	144°58.0'E	July 30, 1967	surface
Copepoda				
6a Calanus cristatus (Kroyer), V	42°21.3'N	145°00.4'E	July 30, 1967	0—550
6b Calanus cristatus (Kroyer), V	43°54.6'N	149°54.5'E	Dec. 6, 1967	180—210
6c Calanus cristatus (Kroyer), V	42°02.5'N	145°54.0'E	May 22, 1968	0—1000
7 Calanus plumchrus (Marukawa),♀	42°21.3'N	145°00.4'E	July 30, 1967	0—550
8a Calanus pacificus (Brodsky), V	34°41.2'N	139°58.0'E	Apr. 25, 1967	0—118
8b Calanus pacificus (Brodsky), V	34°53.8'N	138°38.5'E	Nov. 3, 1968	0—300
9 Calanus lighti (Bowman)	9°36.5'N	178°50.1'W	Jan. 12, 1968	surface
10 Eucalanus bungii bungii (Giesbrecht),♀	42°21.3'N	145°00.4'E	July 30, 1967	0—550
11 Rhincalanus nasutus (Giesbrecht)	34°51.2'N	140°09.9'E	Apr. 26, 1967	0—1420
12 Pareuchaeta birostrata (Brodsky),♀	34°41.2'N	139°58.2'E	Apr. 25, 1967	450—725
13 Pareuchaeta sarsi (Farran), egg	34°53.8'N	138°38.5'E	Nov. 3, 1968	0—1000
14 Metridia okhotensis (Brodsky),♀	42°21.3'N	145°00.4'E	July 30, 1967	0—550
15 Pleuromamma xiphias (Giesbrecht)	13°59.8'N	174°40.6'W	Jan. 18, 1968	0—770
16 Disseta palumboi (Giesbrecht)	9°36.3'N	178°50.4'W	Jan. 12, 1968	0—960
17 Candacia aetiopica (Dana)	18°12.2'N	174°34.9'W	Jan. 17, 1968	surface
18 Candacia columbiae (Campbell)	37°55.0'N	142°54.4'E	Aug. 6, 1967	0—370
19 Labidocera actifrons (Dana)	20°13.4'N	150°21.2'E	Dec. 18, 1967	surface
20 Pontellina plumata (Dana)	9°50.6'N	149°38.6'E	Dec. 22, 1967	surface
Amphipoda				
21 Parathemisto japonica (Bovallius), juv.[b]	42°22.7'N	144°47.2'E	July 31, 1967	30—50
22 Platyscelus serratulus (Stebbing)	2°27.2'N	156°01.5'E	Jan. 3, 1968	surface
23 Cyphocaris challengeri (Stebbing)	42°25.7'N	144°55.7'E	July 30, 1967	25—45
Mysidacea				
24 Siriella aequiremis (Hansen)	0°03.9'N	148°38.8'E	Dec. 25, 1967	surface
Euphausiacea				
25a Euphausia pacifica (Hansen)	42°25.7'N	144°55.7'E	July 30, 1967	25—45
25b Euphausia pacifica (Hansen), juv.[c]	42°22.5'N	144°54.0'E	July 30, 1967	surface
26 Tessarabrachion oculatus (Hansen)	42°22.9'N	144°54.3'E	July 31, 1967	0—520
Decapoda				
27 Lucifer renaudii (Bate), juv.[d]	30°02.8'N	142°02.4'E	Sept. 15, 1967	0—100
Insecta				
28 Halobates sericeus (Eschscholtz)	18°12.2'N	174°34.9'W	Jan. 17, 1968	surface
Chaetognatha				
29 Sagitta elegans (Verrill)	42°21.3'N	145°00.4'E	July 30, 1967	0—550
30 Sagitta nagae (Alvariño)	34°55.4'N	138°38.9'E	Nov. 4, 1968	0—100

[a] Shell diameter 0.6 to 1.3 mm.
[b] Body length 4.1 to 6.9 mm.
[c] Body length 8.9 to 12.2 mm.
[d] Body length 8.2 to 11.0 mm.

(From Omori, M., Weight and chemical composition of some important oceanic zooplankton in the North Pacific Ocean, *Mar. Biol.*, 3, 5, 1969. With permission.)

Table 3.2–21
VERTICAL DISTRIBUTION OF PACIFIC EUPHAUSIDS
BY FAUNAL ZONES

Epipelagic Zone

Subarctic epipelagic:

Thysanoessa raschii
T. inermis
T. spinifera   0–280 m
T. longipes
Euphausia pacifica

Transition-zone epipelagic:
Nematoscelis difficilis
Euphausia pacifica   0–280 m
Thysanoessa gregaria
Euphausia gibboides   0–700 m
Thysanopoda acutifrons

Central epipelagic:
Thysanopoda obtusifrons
T. aequalis, T. subaequalis
Euphausia brevis
E. mutica
E. recurva
E. hemigibba (North Pacific)
E. gibba (South Pacific)   0–700 m
Nematoscellis atlantica
N. microps
Stylocheiron carinatum
S. abbreviatum
S. suhmii
S. affine, Central Form
Nematobrachion fexipes

Equatorial epipelagic:
Euphausia tenera
E. distinguenda   0–280 m
Stylocheiron microphthalma
Thysanopoda tricuspidata
Euphausia diomedige
E. eximia
E. lamelligera
E. fallax
Nematoscelis gracilis   0–700 m
Stylocheiron affine
W. Equatorial Form
E. Equatorial Form
Indo-Australian Form

Table 3.2—21 (*Continued*)
**VERTICAL DISTRIBUTION OF PACIFIC EUPHAUSIDS
BY FAUNAL ZONES**

Mesopelagic Zone

Cosmopolitan mesopelagic
   Stylocheiron maximum   140—1000 m

Subarctic mesopelagic
   Tessarabrachion oculatus   0—1000 m

Central-Equatorial mesopelagic (40° N.—40° S.)
   Stylocheiron longicorne   140—700 m
   S. elongatum
   Thysanopoda pectinata
   T. orientalis
   T. monachantha   140—1000 m
   Nematoscelis tenella
   Nematobrachion boopis

Central mesopelagic:
   Thysanoessa parva
   Nematobrachion sexspinosus
   Stylocheiron robustum   280—1000 m
   Thysanopoda cristata

Bathypelagic Zone

   Thysanopoda cornuta
   T. egregia
   T. spinicaudata
   Bentheuphausia amblyops

(From Brinton, E., The distribution of Pacific, euphausiids, *Bull. Scripps Inst. Oceanogr., Univ. Calif.* 8, 196, 1962. With permission.)

Table 3.2–22

STATISTICS FOR THE FIVE SPECIES OF ZOOPLANKTON THAT COMPRISE THE SUBARCTIC GROUP

Tessarabracnion oculatus, Thysanoessa longipes, Euphausia pacifica, Sagitta elegans, and *Limacina helicina*

Species	Frequency[a]	Abundance[b]	Average rank[c]	Dominance[d]	Dispersion (aggregated)[e]	Fidelity[f]	Vitality[g]
T. oculatus	25/62	2 744 (8)	1.24	0/62	430	Aleutians S to 40° N in Pacific	All stages present
Th. longipes	59/62	4 9784 (413)	3.35	13/62	2275	Arctic Ocean S to 40° N in Pacific	All stages present
E. pacifica	55/62	4 22,278 (345)	3.36	23/62	6950	60° N in Bering Sea, S to 25° N in Calif. Current to 35° N in Central Pacific	All stages present
S. elegans	60/62	17 7900 (1405)	4.15	29/62	1830	N. Atlantic, Arctic Ocean, S to 40° N in Pacific	All stages present
I. helicina	58/62	20 75,700 (360)	2.90	7/62	52,400	N. Atlantic, Arctic Ocean, S to 30° N in Calif. Current to 35° N in Central Pacific	All stages present

a Proportion of samples in which the species was found (total number of samples, 62). *T. oculatus* occurred in deeper tows at nine additional stations.
b Range and median (in parentheses) of numbers of individuals/1000 cu m in samples in which species was found.
c Species were ranked within each sample on the basis of numbers of individuals. Ranks for each species were averaged over the 62 samples (1: least abundant; 5: most abundant).
d Proportion of samples in which the species was among those making up 50% of the individuals; summation in each sample was begun with the most abundant species.
e The ratio of variance to mean; the expected value for a random (Poisson) distribution is 1.0.
f Degree of restriction to the Subarctic Water Mass.
g 2 Proportion of life lived in the Subarctic Water Mass.

(From Fager, E. W. and McGowan, J. A., Zooplankton Species groups in the north Pacific, Science, 140(3536), 458, © 1963, American Association for the Advancement of Science. With permission.)

3.3 Chemical Compositions

Table 3.3–1
TOTAL CARBON, CARBONATE CARBON, AND ORGANIC CARBON CONCENTRATION IN PLANKTONIC MARINE ORGANISMS[a]

Organism			Total carbon ($^o/_o$ DW)	Carbonate carbon ($^o/_o$ DW)[b]	Organic Carbon	
					($^o/_o$ DW)	($^o/_o$ a-f DW)
Cnidaria						
7b	*Cyanea capillata*		13.8	–	13.8	36.0
11B	*Physalia physalis*		31.4	–	31.4	62.8
22b	*Pelagia noctiluca*		12.9	–	12.9	26.0
28a	*Pelagia noctiluca*		15.9	–	15.9	31.2
57	*Aequorea vitrina*		26.8	–	26.8	52.5
		Average	17.5	–	17.5	41.7
Ctenophora						
14	*Mnemiopsis* sp.		6.4	–	6.4	20.6
Arthropoda						
1a	*Euphausia krohnii*		35.8	–	35.8	43.9
2a	*Centropages hamatus* *C. typicus* 1:1		36.3	–	36.3	46.2
4	*Calanus finmarchicus*		41.7	–	41.7	50.5
28b	*Meganyctiphanes norvegica*		42.0	–	42.0	51.6
29c	*Lophogaster* sp.		46.8	–	46.8	57.4
29	*Centropages* sp.		38.5	–	38.5	49.7
30	*Centropages* sp. *Sagitta elegans* 1:1		38.7	–	38.7	50.0
36a	*Idotea metallica*		33.2	2.36	30.8	48.0
36c	*Calanus finmarchicus*		39.8	–	39.8	48.3
40	Mixed copepods		35.6	–	35.6	46.0
56a	*Calanus finmarchicus* (a small admixture of euphausiids and shell-less pteropods)		37.8	–	37.8	46.0
		Average	38.3	–	38.0	48.9
Mollusca						
7a	*Limacina* sp.		28.3	2.74	25.6	56.0
8	*Ommastrephes* sp.		45.1	–	45.1	48.8
24b	*Sthenoteuthis* sp.		37.2	–	37.2	40.4
32	*Clione limacina*		26.3	–	26.3	39.4
38	*Illex illecebrosus*		39.2	–	39.2	42.6
39	Squid eggs (*Loligo*)		21.7	–	21.7	45.0
		Average	33.1	–	32.7	45.4
Chordata						
5	*Salpa* sp.		10.6	–	10.6	46.1
6b	*Salpa* sp.		9.6	–	3.6	39.0
27	*Salpa fusiformis*		7.8	–	7.8	33.9
55	*Pyrosoma* sp.		9.4	–	9.4	41.0
		Average	9.4	–	9.4	40.0

[a] All figures are % of dry weight except last column, which is ash-free dry weight.
[b] Where no results are listed, the inorganic carbonate was not detectable.

Table 3.3–1 *(Continued)*
TOTAL CARBON, CARBONATE CARBON, AND ORGANIC CARBON
CONCENTRATION IN PLANKTONIC MARINE ORGANISMS[a]

Organism		Total carbon (°/₀ DW)	Carbonate carbon (°/₀ DW)[b]	Organic Carbon	
				(°/₀ DW)	(°/₀ a-f DW)
Mixed Samples					
6a	Mixed copepods and phyto-plankton	29.8	–	29.8	38.5
19	Copepods and phytoplankton	25.2	–	25.2	48.0
33b	Phytoplankton and fish	4.8	1.54	3.3	32.7
34a	Phytoplankton and copepods	6.6	–	6.6	56.0
34b	Copepods and phytoplankton	14.3	–	14.3	48.0
59	Mixed zooplankton	28.4	–	28.4	48.6
	Average	18.2	–	17.9	38.8

[a] All figures are % of dry weight except last column, which is ash-free dry weight.
[b] Where no results are listed, the inorganic carbonate was not detectable.

(From Curl, H. S., Jr., Analyses of carbon in marine plankton organisms, *J. Mar. Res.*, 20(3), 185, 1962. With permission.)

Table 3.3–2
ORGANIC CONTENT OF COPEPODS AND
SAGITTAE BASED ON DRY WEIGHTS

	Protein %	Fat %	Carbohydrate %	Ash %	P₂O₅ %	Nitrogen %
Copepods	70.9–77.0	4.6–19.2	0–4.4	4.2–6.4	0.9–2.6	11.1–12.0
Sagittae	69.6	1.9	13.9	16.3	3.6	10.9

Table 3.3—3
TOTAL CARBON, NITROGEN, HYDROGEN AND ASH CONTENTS (% DRY WEIGHT), AND CARBON:NITROGEN RATIO OF NORTH PACIFIC ZOOPLANKTON

Sample number	Method of preservation	Total dry wt. analyzed (mg)	Carbon (%)	Nitrogen (%)	Hydrogen (%)	Ash (%)	C/N (ratio)
Dinoflagellida							
1	Drying	13.419	43.9	5.8	6.4	2.8	7.6
Pteropoda							
2	Drying	26.813	20.3	2.9	2.1	42.8	7.0
3	Drying	24.847	22.0	3.5	2.4	39.8	6.3
4	Drying	7.904	17.0	1.5	1.1	46.6	11.1
5	Drying	7.540	29.0	6.0	3.8	28.6	4.9
Average	—	—	22.4	3.5	2.4	39.3	7.3
Copepoda							
6a	Freezing	20.385	60.9	6.3	9.8	2.4	9.7
6b	Freezing	12.385	39.9	7.6	7.0	3.4	5.1
6c	Freezing	48.805	59.0	5.9	10.1	2.9	10.0
7	Drying	9.456	61.8	7.0	9.7	1.9	8.8
8a	Freezing	8.810	46.4	11.2	7.0	4.4	4.1
8b	Freezing	13.637	58.4	7.8	9.2	2.9	7.5
9	Freezing	4.206	48.0	12.7	7.6	4.3	3.8
10	Drying	7.189	49.9	7.6	7.7	3.9	6.5
11	Drying	7.749	52.8	9.9	8.3	3.4	5.3
12	Drying	19.327	58.4	7.1	9.6	2.1	8.2
13	Freezing	7.748	66.6	5.1	10.3	2.1	13.2
14	Drying	6.521	63.5	5.8	10.0	2.7	11.0
15	Freezing	4.071	47.4	13.1	7.3	3.3	3.6
16	Freezing	10.214	51.0	10.7	8.0	2.8	4.8
17	Freezing	5.422	46.6	12.6	7.2	3.7	3.7
18	Drying	6.745	46.6	11.2	7.2	3.3	4.2
19	Freezing	7.526	45.8	12.9	7.2	5.7	3.5
20	Drying	8.180	44.3	12.2	6.7	6.4	3.6
Average	—	—	53.3	9.4	8.4	3.4	6.5
Amphipoda							
21	Drying	7.796	48.4	8.2	7.5	13.4	5.9
22	Drying	11.690	25.9	4.4	4.4	37.7	6.0
23	Drying	19.390	45.9	6.1	7.1	10.0	7.5
Average	—	—	40.0	6.2	6.3	20.4	6.5
Mysidacea							
24	Drying	9.539	42.4	11.0	6.7	10.2	3.9
Euphausiacea							
25a	Drying	42.000	38.7	10.7	7.3	8.0	3.6
25b	Drying	8.969	39.6	10.1	6.7	8.5	3.9
26	Drying	44.472	47.2	10.0	7.6	8.1	4.7
Average	—	—	41.8	10.3	7.2	8.2	4.1

Table 3.3–3 (*Continued*)
TOTAL CARBON, NITROGEN, HYDROGEN AND ASH CONTENTS (% DRY WEIGHT),
AND CARBON:NITROGEN RATIO OF NORTH PACIFIC ZOOPLANKTON

Sample number	Method of preservation	Total dry wt. analyzed (mg)	Carbon (%)	Nitrogen (%)	Hydrogen (%)	Ash (%)	C/N (ratio)
Decapoda							
27	Drying	5.844	41.1	9.3	6.7	11.9	4.4
Insecta							
28	Freezing	10.947	52.6	9.7	7.8	5.6	5.4
Chaetognatha							
29	Drying	13.542	47.7	10.7	7.6	4.8	4.4
30	Drying	14.706	43.5	11.1	7.2	4.2	3.9
Average	–	–	45.6	10.9	7.4	4.5	4.2
Pisces							
31	Drying	14.302	41.5	11.2	7.0	8.9	3.7
32	Freezing	34.108	37.9	9.8	5.8	12.9	3.9
33	Drying	10.296	46.5	12.6	7.2	6.8	3.7
Average	–	–	42.0	11.2	6.7	9.5	3.8

(From Omori, M., Weight and chemical composition of some important oceanic zooplankton in the north Pacific Ocean, *Mar. Biol.*, 3, 8, 1969. With permission.)

Table 3.3—4
WET AND DRY WEIGHT OF NORTH PACIFIC ZOOPLANKTON

Sample number	Average wet wt./individual (mg)	Average dry wt./individual (mg)	Dry wt./wet wt. (%)
Dinoflagellida			
1	0.10	0.0011	1.1
Pteropoda			
2	16.00	4.97	31.1
3	–	3.98	–
4	0.22	0.08	36.4
5	0.56	0.14	25.0
Copepoda			
6a	20.81	3.47	16.7
6b	16.59	2.57	15.5
6c	26.20	8.87	33.9
7	4.55	1.26	27.7
8a	1.28	0.25	19.7
8b	1.46	0.31	21.2
9	1.00	0.13	13.0
10	9.08	1.09	12.0
11	1.04	0.14	13.5
12	17.47	3.23	18.5
13	–	0.10	–
14	2.72	0.51	18.8
15	1.88	0.23	12.2
16	5.64	0.52	9.2
17	0.42	0.06	14.3
18	3.10	0.41	13.2
19	2.21	0.25	11.3
20	0.20	0.03	16.5
Amphipoda			
21	2.56	0.47	18.4
22	5.99	2.19	36.6
23	14.22	3.13	22.0
Mysidacea			
24	5.25	0.98	18.7
Euphausiacea			
25a	7.63	1.54	20.2
25b	61.81	14.00	20.7
26	67.69	14.45	21.3
Decapoda			
27	1.20	0.16	13.3
Insecta			
28	2.91	0.81	27.8
Chaetognatha			
29	10.14	1.43	14.1
30	11.49	1.33	11.6
Pisces			
31	31.80	5.93	18.6
32	96.97	18.60	21.1
33	10.40	2.19	21.1

(From Omori, M., Weight and chemical composition of some important oceanic zooplankton in the north Pacific Ocean, *Mar. Biol.*, 3, 7, 1969. With permission.)

Table 3.3—5
CARBON CONTENT OF FRESH AND PRESERVED *NEMATOSCELIS DIFFICILIS*

Preservation	Rinse	Size (cm)	Runs	Av. dry wt. (mg)	Carbon content[b] % dry wt.	C(mg)	Carbon loss
Fresh dried at 55°–60°C	No rinse	1.8[a]	10	7.66	42.8 (42.0–43.7)	3.28 (3.19–3.50)	–
		1.9	10	8.30	41.6 (40.3–44.0)	3.45 (3.30–3.33)	–
15 weeks in 10% buffered seawater formalin	1 hr rinse in seawater	1.8	10	6.73	40.4 (39.8–41.0)	2.72 (2.64–2.78)	3.28–2.72= 0.56 (17%)
15 weeks in 10% buffered freshwater formalin	1 hr rinse in distilled water	1.9	10	5.39	51.3 (50.8–51.7)	2.76 (2.69–2.82)	3.45–2.76= 0.69 (20%)
15 weeks in 70% ethanol	No rinse	1.9	10	6.90	44.2 (43.8–45.1)	2.66 (2.58–2.76)	3.45–2.66= 0.79 (23%)

[a] Anterior edge of eye to tip of telson.
[b] Range in parentheses.

(From Hopkins, T. L., Carbon and nitrogen content of fish and preserved *Nematoscelis difficilis*, a euphausiid crustacean, *J. Cons. Cons. Int. Explor. Mer.*, 31, 302, 1968. With permission.)

Table 3.3—6
NITROGEN CONTENT OF FRESH AND PRESERVED *NEMATOSCELIS DIFFICILIS*

Preservation	Rinse	Size (cm)	Runs	Av. dry wt. (mg)	Nitrogen content[b] % dry wt	N(mg)	Nitrogen loss
Fresh dried at 55°–60°C	No rinse	1.7	10	6.61	10.9 (10.6–11.0)	0.72 (0.67–0.76)	–
		2.1	10	10.39	10.6 (9.7–11.9)	1.10 (0.92–1.30)	–
		2.3	4	13.07	10.1 (9.2–11.0)	1.31 (1.02–1.74)	–
17 weeks in 10% buffered seawater formalin	1 hr rinse in seawater	2.0	9	8.17	10.3 (9.7–10.8)	0.84 (0.67–1.02)	1.04–0.84 = 0.20 (19.2%)
17 weeks in 10% buffered freshwater formalin	1 hr rinse in distilled water	1.7	10	4.13	12.4 (11.5–12.9)	0.51 (0.48–0.56)	0.72–0.51 = 0.21 (29.7%)
17 weeks in 70% ethanol	No rinse	2.2	9	7.53	13.1 (12.2–16.6)	0.98 (0.76–1.22)	1.24[a]–0.98 = 0.26 (21.0%)

[a] Fresh nitrogen values estimated by interpolation.
[b] Range in parentheses.

(From Hopkins, T. L., Carbon and nitrogen content of fish and preserved *Nematoscelis difficilis*, a euphausiid crustacean, *J. Cons. Cons. Int. Explor. Mer.*, 31, 303, 1968. With permission.)

Table 3.3—7
ANALYSIS OF MAJOR CONSTITUENTS OF
NEOMYSIS INTEGER

Mean Values and Totals of All Fractions Obtained
for Each Method of Preservation

Sample and preservation time	Dry weight	Ash	Chitin	Lipid	Protein	Carbohydrate	Total
Fresh	22 ± 1	12 ± 0	3 ± 0	11 3	73 ± 7	3 ± 0	103%
24 hr formalin	21 ± 0	12 ± 0	3 ± 0	8 1	6 ± 3	3 ± 0	32%
4 weeks formalin	20 ± 1	9 ± 2	3 ± 1	15 ± 2	5 ± 5	9 ± 6	41%
3 weeks TCA	36 ± 2	1 ± 0	2 ± 0	7 ± 2	44 ± 4	1 ± 0	55%
4 weeks ethanol	17 ± 2	10 ± 3	5 ± 1	9 ± 1	92 ± 11	2 ± 0	118%
5 weeks deep freeze	29 ± 2	11 ± 0	5 ± 1	15 ± 3	79 ± 7	3 ± 2	113%
4 weeks freeze dried	20 ± 1*	13 ± 0	4 ± 0	14 ± 2	74 ± 9	2 ± 0	107%

* The freeze dried material was, of course, anhydrous; the dry weight quoted here is the figure obtained for drying fresh material by freeze drying.

(From Fudge, H., Biochemical analysis of preserved zooplankton, *Nature*, 219, 381, 1968. With permission.)

Table 3.3—8
LIPID COMPOSITION OF DIATOMS AND *CALANUS HELGOLANDICUS*

	Calanus helgolandicus			Diatoms		
		Laboratory grown		*Lauderia borealis*	*Skeletonema costatum*	*Chaetoceros curvisetus*
Fraction	Wild	(b)	(c)	Weight (%)		
Hydrocarbon	tr	3	1	11	16	2
Wax ester	37	25	41	ND[b]	ND	ND
Triglyceride	5	3	12	16	14	12
Sterol	14	10	16	17	15	10
Free fatty acid	tr[a]	tr	tr	5	12	16
Phospholipid[c]	44	59	28	51	57	50
Total lipid (% dry weight)	12.4	18.6	24.2	13.2	8.6	9.1

[a] tr — trace.
[b] ND — Not detected.
[c] This fraction includes galactolipids in the case of the diatoms.
  (b) Fed 400 μg C/l of *Skeletonema costatum*.
  (c) Fed 600—800 μg C/l of *S. costatum* (90%) and *Chaetoceros curvisetus*.

(From Lee, R. F., Nerenzel, J. C., and Poffenhofer, G. A., Importance of wax esters and other lipids in the marine food chain: phytoplankton and copepods, *Mar. Biol.*, 9, 101, 1971. With permission.)

Figure 3.3–1

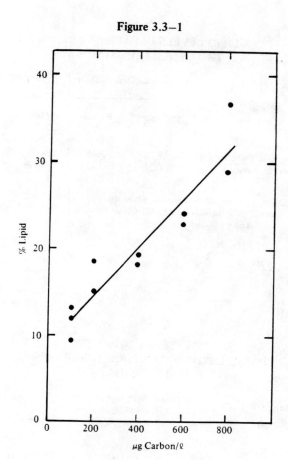

Figure 3.3–1. The amount of lipid in *Calanus helgolandicus* as a function of the concentration of *Skeletonema costatum* fed.

Note: The equation of the best straight line (drawn) through the points is $0.0276 \times (\mu gC/l) + 9.00 = \%$ lipid. The regression coefficient for the 11 points is → 0.9424.

(From Lee, R. F., Nerenzel, J. C., and Poffenhofer, G. A., Importance of wax esters and other lipids in the marine food chain: phytoplankton and copepods. *Mar. Biol.*, 9. 101, 1971. With permission.)

Table 3.3—9
LIPID LEVELS IN EACH SEX OF
TWO GENERA OF MYSIDS

Species sex	Mesopodopsis slabberi		Neomysis integer	
	Male	Female	Male	Female
No. of values	25	21	17	26
Means (mg lipid/g wet weight)	19.13	23.45	14.08	27.29
Standard error of means	0.236	0.451	0.499	0.625

(From Linford, E., Biochemical studies on marine zooplankton. II. Variations in the lipid content of some Mysidacea, *J. Cons. Cons. Int. Explor. Mer.*, 30, 19, 1965. With permission.)

Table 3.3—10
TOTAL LIPID CONTENTS OF THREE SPECIES
OF MYSIDS

Species	Mean % lipid/wet wt. (95% confidence level)	Mean % lipid/dry wt.
Mesopodopsis slabberi	1.98 ± 0.09	9.00
Neomysis integer	2.21 ± 0.21	10.05
Praunus neglectus	2.05 ± 0.30	9.32

(From Linford, E., Biochemical studies on marine zooplankton. II. Variations in the lipid content of some Mysidacea, *J. Cons. Cons. Int. Explor. Mer.*, 30, 19, 1965. With permission.)

Table 3.3—11
CARBOHYDRATE CONTENT OF CERTAIN ZOOPLANKTON

No. of animals	Carbohydrate content	
	μg/animal	% wet wt
Paraeuchaeta		
11 V	3.2	0.07
12, mostly ♀	9.2	0.10
30 V	3.7	0.07
25 ♀	8.0	0.06
15 ♀	13.3	0.11
Euthemisto		
7	>110	>0.25
4	128	0.33
4	166	0.35
10	28	0.62
Pleuromamma		
94 V and VI	1.5	0.10
Hyperia		
10	24	0.15
Conchoecia		
50	7.5	0.31

(From Raymont, J. E. G. and Conover, R. J., Further investigations on the carbohydrate content of marine zooplankton, *Limnol. Oceanogr.*, 6, 161, 1961. With permission.)

Table 3.3—12
CARBOHYDRATE CONTENT OF *NEOMYSIS INTEGER*

Fed Animals

Wet wt. (mg)	% carbohydrate/wet wt.	% carbohydrate/dry wt.
160.0	0.31	1.55
142.0	0.20	1.00
180.4	0.42	2.10
353.4	0.22	1.10
243.6	0.24	1.20
228.0	0.28	1.40
251.0	0.23	1.15
208.0	0.22	1.10
310.0	0.25	1.25
276.7	0.20	1.00
318.8	0.22	1.10
980.0	0.22	1.10
389.2	0.29	1.45
403.0	0.26	1.30
333.3	0.23	1.18
278.0	0.39	1.95

Unfed Animals

Wet wt. (mg)	% carbohydrate/wet wt.	% carbohydrate/dry wt.
190.0	0.15	0.75
161.6	0.28	1.40
444.2	0.16	0.80
389.2	0.20	1.00
258.7	0.20	1.00
251.0	0.25	1.25
334.2	0.21	1.05
305.5	0.24	1.20
392.0	0.17	0.85
398.4	0.27	1.35
455.5	0.20	1.00

Comparative Summary

Fed % carbohydrate/wet wt.	Unfed % carbohydrate/wet wt.
0.31	0.15
0.20	0.28
0.42	0.16
0.22	0.20
0.24	0.20
0.28	0.25
0.23	0.21
	0.24
0.22	0.17
0.25	0.27
0.20	0.20

(From Raymont, J. E. G. and Krishaswamy, S., Carbohydrates in some marine planktonic animals, *J. Mar. Biol. Assoc. U. K.*, 39, 242, 1960. With permission.)

Table 3.3—13
CARBOHYDRATE CONTENT, RESPIRATORY
RATE, AND GRAZING RATE FOR
NEOMYSIS AMERICANA

All Animals Obtained from Cape Cod Bay
25.VIII.1959

| No. of animals | Carbohydrate content | | Carbohydrate equivalent, μg glucose/animal day |
	μg/ animal	% wet wt.	
15	16.7	0.23	–
15	17.3	0.22	–
14	15.0	0.20	62.8
14	14.3	0.20	60.4
14	12.5	0.21	72.1
14	13.9	0.20	78.8
15	13.0	0.19	–
15	11.7	0.22	–

(From Raymont, J. E. G. and Conover, R. J., Further investigations on the carbohydrate content of marine zooplankton, *Limnol. Oceanogr.*, 6, 160, 1961. With permission.)

Table 3.3–14
CARBOHYDRATE CONTENT AND RESPIRATORY RATE FOR *CALANUS HYPERBOREUS*

No. and stage of copepods	Carbohydrate content		Respiratory rate		Carbohydrate equivalent, μg glucose/animal day	Grazing rate ml/animal day	Remarks
	μg/copepod	% wet wt.	μL/animal day	μL/mg wet wt. day			
15 V	11.7	0.21	—	—	—	—	Not fed
20 V	8.2	0.16	—	—	—	—	Starved 7 days, fed 1 day
17 ♀	29.4	0.29	—	—	—	—	Starved 7 days, fed 1 day
20 V	9.2	0.18	—	—	—	3.71 (*Skeletonema*)	Starved 7 days, fed 3 days
20 V	12.7	0.20	—	—	—	2.57 (*Skeletonema*)	Starved 7 days, fed 3 days
20 V	8.8	0.16	—	—	—	2.03 (*Skeletonema*)	Starved 7 days, fed 3 days
18 ♀	15.8	0.18	—	—	—	4.25 (*Skeletonema*)	Starved 7 days, fed 1 day
19 V	7.9	0.16	9.9	2.11	13.0	—	Starved 7 days, fed 1 day before expt., then starved 2 days
20 V	7.4	0.14	10.7	2.03	14.0	—	Starved 7 days, fed 1 day before expt., then starved 2 days
20 V	7.5	0.14	7.9	1.47	10.4	—	Starved 7 days, fed 1 day before expt., then starved 2 days
18 ♀	29.7	0.29	17.4	1.70	22.8	—	Starved 7 days, fed 1 day before expt., then starved 2 days

(From Raymont, J. E. G. and Conover, R. J., Further investigations on the carbohydrate content of marine zooplankton, *Limnol. Oceanogr.*, 6, 158, 1961. With permission.)

Table 3.3—15
CARBOHYDRATE CONTENT OF *CALANUS FINMARCHICUS*

No. of animals used	Carbohydrate/ copepod (μg)	% carbohydrate/wet wt.	Remarks
125	2.24	0.18	Fed
125	1.76	0.14	Unfed
200	0.41	0.03	Unfed
200	1.15	0.09	Unfed
150	1.10	0.09	Fed
250	0.72	0.06	Fed
130	0.63	0.05	Fed
200	<0.25	<0.02	Unfed
140	1.70	0.13	Fresh animals
200	1.40	0.11	Fresh animals

(From Raymont, J. E. G. and Krishaswamy, S., Carbohydrates in some marine planktonic animals, *J. Mar. Biol. Assoc. U. K.*, 39, 241, 1960. With permission.)

Table 3.3—16
CARBOHYDRATE CONTENT OF FED AND STARVED *CALANUS FINMARCHICUS*

	All experiments on fed and unfed *Calanus*				Experiments on *Calanus* kept in laboratory not more than 7 days			
	Carbohydrate % wet wt.		Carbohydrate/ copepod (μg)		Carbohydrate % wet wt.		Carbohydrate/ copepod (μg)	
	Range	Mean	Range	Mean	Range	Mean	Range	Mean
Fed	0.11—0.40	0.23	0.90—2.85	1.71	0.18—0.40	0.27	1.23—2.85	1.93
Starved	0.10—0.24	0.16	0.68—1.50	1.16	0.10—0.24	0.16	0.68—1.50	1.16

(From Raymont, J. E. G. and Conover, R. J., Further investigations on the carbohydrate content of marine zooplankton, *Limnol. Oceanogr.*, 6, 157, 1961. With permission.)

Table 3.3–17
CARBOHYDRATE CONTENT, GRAZING RATE, AND FECAL PELLET PRODUCTION OF
CALANUS FINMARCHICUS

Mainly Stage V Copepodites

Number of animals	Carbohydrate content		Grazing rate ml/ copepod day	No. fecal pellets produced	Remarks
	μg/ copepod	% wet wt.			
100 V	1.50	0.24	–	–	Initial, starved 24 hr
99 V	1.26	0.18	–	2200	Starved 24 hr, fed 20 hr
97 V	1.39	0.20	–	1932	Starved 24 hr, fed 21 hr
98 V	1.89	0.29	–	4730	Starved 24 hr, fed 22 hr
94 V	1.70	0.24	–	6110	Starved 24 hr, fed 23 hr
82 V	1.22	0.18	–	281	Starved 38 hr
100 V	(1.10)	0.16	–	318	Starved 39 hr
84 V	1.07	0.14	–	390	Starved 41 hr
100 V	(0.85)	0.14	–	252	Starved 42 hr
200 V	1.23	0.18	6.9	9160	Starved for 5 days, fed 16 hr
200 V	0.68	0.10		1000	Starved for 6 days
100 V	1.45	0.19	–	1000	Starved for 21 hr
100 V	2.45	0.28	29.8	6760	Fed for 21 hr
100 V	2.70	0.38	–	5760	Fed for 23 hr
100 V	2.85	0.40	–	9360	Fed for 47 hr
106 V	(2.10)	(0.26)	–	5320	Fed for 23 hr, then accidentally killed
100 V	0.90	0.11	av. 15.6	3041	Starved for 10 days, fed 18 hr
100 V	1.15	0.15	av. 17.0	2680	Starved for 10 days, fed 45 hr
100 V	1.25	0.17	av. 17.0	1998	Starved for 10 days, fed 24 hr, then starved 22 hr
117 IV and V	1.07	0.16	–	–	Starved for 1 day

(From Raymont, J. E. G. and Conover, R. J., Further investigations on the carbohydrate content of marine zooplankton, *Limnol. Oceanogr.*, 6, 156, 1961. With permission.)

Table 3.3–18

CARBOHYDRATE CONTENT, RESPIRATORY RATE AND GRAZING RATE FOR SOME EUPHAUSIIDS

Animal	Carbohydrate content		Respiratory rate		Carbohydrate equivalent, μg glucose/animal day	Grazing rate, ml/animal day		Remarks
	μg/animal	% wet wt.	μ/animal day	μ/mg wet wt. day				
Meganyctiphanes norvegica	153	0.05	—	—	—	—		Starved 5 days, fed 2 days
Meganyctiphanes norvegica	99	0.05	—	—	—	—		Starved 5 days, fed 3 days
Meganyctiphanes norvegica	233	0.06	—	—	—	—		Starved 5 days, fed 3 days
Meganyctiphanes norvegica	114	0.04*	—	—	—	—		Starved 5 days, fed 3 days
Meganyctiphanes norvegica	213	0.04*	—	—	—	—		Starved 5 days, fed 3 days
Meganyctiphanes norvegica	151	0.05	—	—	—	55.1	*Skeletonema*	Starved 5 days, fed 7 days
Meganyctiphanes norvegica	130	0.05	—	—	—	25.7	*Thalassiosira*	Starved 5 days, fed 7 days
Meganyctiphanes norvegica	128	0.08	—	—	—	—		Starved 3 days
Thysanoessa sp.	72	0.11	—	—	—	62.3 / 25.7	*Skeletonema* / *Thalassiosira*	Fed 7 days
Thysanoessa sp.	68	0.11	—	—	—	42.9 / 38.3	*Skeletonema* / *Thalassiosira*	Fed 7 days
Thysanoessa sp.	16	0.09	—	—	—	—		Starved 3 days
Nematoscelis megalops	135	0.13	—	—	—	—		Starved 5 days, fed 3 days
Nematoscelis megalops	53	0.12	—	—	—	—		—
Nematoscelis megalops	—	0.17	—	—	—	—		—
Nematoscelis megalops	50	0.15	—	—	—	—		—
Nematoscelis megalops	82	0.13	—	—	—	—		—
Nematoscelis megalops	50	0.13	—	—	—	—		—
Nematoscelis megalops	56	0.15	—	—	—	—		—
Nematoscelis megalops	51	0.11	—	—	—	—		—
Nematoscelis megalops	46	0.13	93.5	2.54	122.8	—		No food 10°C
Nematoscelis megalops	71	0.08	112.2	3.12	147.3	—		No food 10°C
Nematoscelis megalops	50	0.14	140.5	2.63	184.5	—		No food 10°C
Nematoscelis megalops	60	0.13	159.3	3.42	209.2	—		No food 4°C
Nematoscelis megalops	45	0.13	116.8	3.23	153.4	—		No food 4°C
Nematoscelis megalops	49	0.12	97.9	2.57	128.5	—		No food 4°C

* Animals dead.

(From Raymont, J. E. G. and Conover, R. J., Further investigations on the carbohydrate content of marine zooplankton, *Limnol. Oceanogr.,* 6, 159, 1961. With permission.)

Table 3.3—19
CARBOHYDRATE CONTENT OF
PLEUROBRACHIA PILEUS

% carbohydrate/wet wt.	% carbohydrate/dry wt.
0.011	0.275
0.057	1.425
0.002	0.050
0.007	0.175
0.005	0.125
<0.001	<0.025
0.006	0.150
0.001	0.025
0.009	0.225
0.007	0.175
0.001	0.025
0.001	0.025
0.001	0.025
<0.001	<0.025
0.046	1.150
0.003	0.075
0.004	0.100
0.002	0.050
0.003	0.075
0.001	0.025
0.002	0.050
0.007	0.175
*0.006	0.150
*0.009	0.225
*0.008	0.200
*0.009	0.225

* Determinations made on animals immediately following capture.

(From Raymont, J. E. G. and Krishaswamy, S., Carbohydrates in some marine planktonic animals, *J. Mar. Biol. Assoc. U. K.*, 39, 245, 1960. With permission.)

Table 3.3—20
CAROTENOID AND VITAMIN A CONTENT OF ZOOPLANKTON

	Vitamin A		Carotenoids	
	i.u.[a]/g animal	i.u./g oil	µg[b]/g animal	µg/g oil
Meganyctiphanes norvegica	15	680	42	1900
Thysanoessa raschii	32	495	33	500
Pandalus bonnieri	2.1	89	24	1000
Spirontocarus spinus	1.0	22	27	950
Crangon allmanni	0.4	30	5	390
Crangon vulgaris	0.2	21	5	550

[a] International unit.
[b] Microgram.

Table 3.3–21
COMPARISON OF NET PLANKTON AND SUSPENDED MATTER CONTAINING PROTEIN IN THE NORTH PACIFIC AND NORTH ATLANTIC OCEANS*

Meters	100	200	300	400	500	1000	1400	1500	2000
Vinogradov (net plankton)	$\vert$— 12.4 —$\vert\vert$— 10.8 —$\vert\vert$— 2.8 —$\vert\vert$— 1.0 —$\vert$								
McAllister et al.	48_2	38_2	34_2	–	–	–	–	–	–
Parsons and Strickland	–	–	60_1	–	73_2	–	–	54_3	–
Krey	10_{28}	9_{12}	10_7	7_2	7_4	–	6_{10}	–	–

* In mg/m^3 protein. The biomass of net plankton is converted into protein (albumen equivalents): dry weight 14% of wet weight; organic matter, 92% of dry weight; albumen equivalents, 37% of organic matter. Because McAllister et al. standardized their measurements against egg albumen, their values are used as reported, whereas the data of Parsons and Strickland have been reduced taking the albumen equivalents as 60% of 6.25 times the nitrogen content found by the Kjeldahl method. From pigment analyses, interference by plant matter is estimated to be of the order of 10 to 20% of protein in the upper layers. Subscripts give the number of samples.

(From Banse, K., On the vertical distribution of zooplankton in the sea, in *Progress in Oceanography*, Vol. 2, Sears, M., Ed., Pergamon Press, New York, 1964, 85. With permission.)

Table 3.4—1

THE AREAS OF THE NORTH PACIFIC IN WHICH THE LISTED SPECIES HAVE BEEN SHOWN TO OCCUR

Organism	Subarctic	Transitional	Central	Equatorial	Eastern Tropic Pacific	Warm water cosmopolites	Comments	References
PROTOZOA								
Foraminifera								
Globigerina quinqueloba	+							Bradshaw, 1959
Globigerinoides minuta	+							Bradshaw, 1959
Globigerina pachyderma	+							Bradshaw, 1959
Globorotalia truncatulinoides			+					Bradshaw, 1959
Pulleniatina obliquiloculata				+				Bradshaw, 1959
Sphaeroidinella dehiscens				+				Bradshaw, 1959
Globigerina conglomerata				+				Bradshaw, 1959
Globorotalia tumida				+				Bradshaw, 1959
Globorotalia hirsuta				+				Bradshaw, 1959
Globigerinella aequilateralis						+	Pure	Bradshaw, 1959
Globigerinoides conglobata						+	Pure	Bradshaw, 1959
Globigerinoides rubra						+	Pure	Bradshaw, 1959
Orbulina universa						+	Pure	Bradshaw, 1959
Globigerinoides sacculifera						+	Peak at equator	Bradshaw, 1959
Globorotalia menardii						+	Peak at equator	Bradshaw, 1959
Globigerina eggeri						+	Edge effect	Bradshaw, 1959
Hastigerina pelagica						+	Edge effect	Bradshaw, 1959
Radiolaria								
Castanidium apsteini	+							Kling, 1966 (E.N.P. only studied)
Castanidium variabile	+							Kling, 1966
Haeckeliana porcellana	+							Kling, 1966
Castanea amphora			+					Kling, 1966
Castanissa brevidentata			+					Kling, 1966
Castanella thomsoni			+				Doubtful, may be deep central too	Kling, 1966
Castanea henseni			+				Doubtful, may be deep central too	Kling, 1966
Castanea globosa							T. zone w/upwelled water?	Kling, 1966
Castanidium longispinum				+			T. zone w/upwelled water?	Kling, 1966
Castanella aculeata				+			T. zone w/upwelled water?	Kling, 1966

E.N.P. – Eastern North Pacific.

T zone – Transitional zone.

Table 3.4–1 (Continued)
THE AREAS OF THE NORTH PACIFIC IN WHICH THE LISTED SPECIES HAVE BEEN SHOWN TO OCCUR

Organism	Subarctic	Transitional	Central	Equatorial	Eastern Tropic Pacific	Warm water cosmopolites	Comments	References
CHAETOGNATHA								
Sagitta elegans	+							Bieri, 1959
Eukrohnia hamata	+							Alvarino, 1962
Sagitta scrippsae		+						Alvarino, 1962
Sagitta pseudoserratodentata			+				Crossing W.T.P.	Bieri, 1959
Sagitta californica			+					Bieri, 1959
Sagitta ferox				+				Bieri, 1959, Alvarino, 1962
Sagitta robusta				+			Patchy	Bieri, 1959, Alvarino, 1962
Sagitta regularis				+				Bieri, 1959
Sagitta hexaptera						+	Peak at equator	Bieri, 1959
Sagitta enflata						+	Peak at equator	Bieri, 1959
Pterosagitta draco						+	Peak at equator	Bieri, 1959
Sagitta pacifica						+	Edge effect	Bieri, 1959
Sagitta minima								Bieri, 1959
ANNELIDA								
Tomopteris septentrionalis						+		Tebble, 1962
Tomopteris pacifica	+							Tebble, 1962
Poeobius meseres	+							McGowan, 1960
ARTHROPODA								
Copepoda								
Calanus pacificus	+						May be T. zone	Brodsky, 1965
Calanus plumchrus	+							Brodsky, 1960
Calanus tonsus	+							Johnson and Brinton, 1963
Calanus cristatus	+							Johnson and Brinton, 1963
Eucalanus bungii bungii	+							Johnson and Brinton, 1963
Eucalanus elongatus hyalinus		+					South Pacific also	Lang, 1965
Eucalanus bungii californicus		+						Lang, 1965
Clausocalanus pergens		+						Frost and Fleminger
Clausocalanus lividus			+					Frost and Fleminger
Eucalanus subcrassus				+				Lang, 1964

W.T.P. – Western Tropical Pacific.
T zone – Transitional zone.

Table 3.4–1 (Continued)

THE AREAS OF THE NORTH PACIFIC IN WHICH THE LISTED SPECIES HAVE BEEN SHOWN TO OCCUR

Organism	Subarctic	Transitional	Central	Equatorial	Eastern Tropic Pacific	Warm water cosmopolites	Comments	References
Rhincalanus cornutus				+				Lang, 1964
Eucalanus inermis					+			Lang, 1964
Eucalanus crassus						+	Patchy, 'pure'	Lang, 1964
Rhincalanus nasutus						+	Very patchy, almost pure equatorial	Lang, 1964
Eucalanus attenuatus						+	Peak at equator; some edge effect	Lang, 1964
Eucalanus subtenuis						+	Patchy, peak at equator; some edge effect	Lang, 1964
Clausocalanus arcuicornis						+		Frost and Fleminger
Eucalanus longiceps								Lang, 1964
Rhincalanus gigas								Lang, 1964
Clausocalanus laticeps								Frost and Fleminger
Euphausiacea								Brinton, 1962
Thysanoessa longipes	+							Brinton, 1962
Euphausia pacifica	+							Brinton, 1962
Thysanopoda acutifrons		+						Brinton, 1962
Thysanoessa gregaria		+						Brinton, 1962
Euphausia gibboides		+						Brinton, 1962
Nematoscelis difficilismegalops			+					Brinton, 1962
Nematoscelis atlantica			+					Brinton, 1962
Euphausia brevis			+					Brinton, 1962
Euphausia hemigibba			+					Brinton, 1962
Euphausia gibba			+					Brinton, 1962
Euphausia mutica			+					Brinton, 1962
Stylocheiron suhmii							Crossing in W.T.P.	Brinton, 1962
Euphausia diomediae				+				Brinton, 1962
Euphausia distinguenda				+				Brinton, 1962
Nematoscelis gracilis				+				Brinton, 1962
Euphausia distinguenda					+			Brinton, 1962
Euphausia eximia					+			Brinton, 1962
Euphausia lamelligera					+	+	Peak at equator	Brinton, 1962
Euphausia tenera						+	Avoids E.T.P.	Brinton, 1962
Stylocheiron abbreviatum								Brinton, 1962
Euphausia superba								Marr, 1962

W.T.P. – Western Tropical Pacific.
E.T.P. – Eastern Tropical Pacific.

Table 3.4—1 (Continued)
THE AREAS OF THE NORTH PACIFIC IN WHICH THE LISTED SPECIES HAVE BEEN SHOWN TO OCCUR

Organism	Subarctic	Transitional	Central	Equatorial	Eastern Tropic Pacific	Warm water cosmopolites	Comments	References
Amphipoda								
Parathimisto pacifica	+							Bowman, 1960
MOLLUSCA								
Pteropoda								
Limacina helicina	+							McGowan, 1963
Clio polita	+							McGowan
Corolla pacifica		+						Beklemishev, 1961
Clio balantium		+						McGowan, 1960
Cavolinia inflexa			+					McGowan, 1960
Clio pyramidata			+					McGowan, 1960
Styliola subula			+					McGowan, 1960
Limacina lesueuri			+				Crossing in W.T.P.	McGowan, 1960
Clio n.sp.				+			Crossing in W.T.P.	McGowan
Cavolinia uncinata				+			Crossing in W.T.P.	McGowan, 1960
Limacina trochiformis					+		Very patchy	McGowan, 1960
Limacina inflata						+		McGowan, 1960
Cavolinia longirostris						+	Very patchy; almost pure equatorial	McGowan, 1960
Cavolinia gibbosa						+	Very patchy, avoids E.T.P.	McGowan, 1960
Hyalocylix striata						+		McGowan, 1960
Creseis virgula						+	Edge effect	McGowan, 1960
Creseis acicula						+		McGowan, 1960
Cavolinia tridentata						+	Peak at equator	McGowan, 1960
Diacria trispinosa						+	Peak at equator	McGowan, 1960
Limacina bulimoides						+	Avoids E.T.P.	McGowan, 1960
Clio antarctica						+		McGowan
Heteropoda								
Caranaria japonica		+						McGowan
Gymnosomata								
Clione limacina	+							McGowan

E.T.P. – Eastern Tropical Pacific.
W.T.P. – Western Tropical Pacific.
(From McGowan, J. A., "Oceanic Biogeography of the Pacific," in *The Micropalaeontology of Oceans*, Cambridge University Press (Eng.), 1971, p. 14 to 17. With permission.)

Table 3.4—2
TYPICAL COSMOPOLITAN OCEANIC SPECIES

Siphonophora

Physophora hydrostatica
Agalma elegans
Dimophyes arctica
Lensia conoidea
Chelopheys appendiculata
Sulculeolaria biloba

Medusae

Cosmetira pilosella
Laodicea undulata
Halicreas sp.
Periphylla periphylla

Mollusca

Euclio pyramidata
Euclio cuspidata
Diacria trispinosa
Pneumodermopsis ciliata
Taonidium pfefferi
Tracheloteuthis risei

Polychaeta

Travisiopsis lanceolata
Vanadis formosa
Rhynchonerella angelini
Tomopteris septentrionalis

Chaetognatha

Sagitta serratodentata
  f. tasmanica
Sagitta hexaptera

Thaliacea

Salpa fusiformis
Dolioletta gegenbauri

Copepoda

Rhincalanus nasutus
Eucalanus elongatus
Pleuromamma robusta
Euchirella rostrata
Euchirella curticaudata
Oithona spinirostris

Other Crustacea

Lepas sp.
Munnopsis murrayi
Brachyscelus crusulum
Meganyctiphane norvegica

Euphausia krohni
Anchialus agilis

Table 3.4—3
SOME PLANKTONIC SPECIES TYPICAL OF DEEP WATER

Gaetanus pileatus
Arietellus plumifer
Pontoptilus muticus
Centraugaptilus rattrayi
Augaptilus megalaurus
and many other copepods

Amalopenaeus elegans
Hymenodora elegans
Boreomysis microps
Eucopia unguiculata
Cyphocaris anonyx
Scina sp.

Sagitta macrocephala
S. zetesios
Eukrohnia fowleri
Nectonemertes miriabilis
Spiratella helicoides
Histioteuthis boneltiana

Table 3.4—4

SPECIES GROUPS AND ASSOCIATED SPECIES OF EASTERN AUSTRALIAN SLOPE ZOOPLANKTON

Section I		Section II		
(Mean temperature [°C] >12 <17.6; mean chlorinity [°/oo] >19.40 <19.60)		(Mean temperature [°C] >17.6 <21.6; mean chlorinity [°/oo] >19.50 <19.65)		
Group A (>12.4°C <17.6°C; >19.41 <19.60°/oo)	Group B (>13.4°C <17.0°C; >19.45 <19.53°/oo)	Group C (>17.6°C <20.8°C; >19.53 <19.59°/oo)	Group D (>19.40°C <20.6°C; >19.54 <19.60°/oo)	Group E (>20.0°C <21.6°C; >19.50 <19.61°/oo)
S. planctonis	E. spinifera	O. cophocerca	D. denticulatum	S. robusta
S. hamata	T. gregaria	O. fusiformis	T. multitentaculata	S. ferox
S. lyra	E. recurva	O. longicauda	M. huxleyi	S. enflata
K. subtilis	N. difficilis		S. magnum	S. bipunctata
			O. rufescens	S. s. pacifica
				S. regularis
				P. draco

Ungrouped Species Occurring with One or More Grouped Species

S. s. tasmanica	I. zonaria	O. intermedia	P. macropus	S. hexaptera
S. decipiens	S. carinatum	O. cornutogastra	R. amboinensis	S. carinatum
		O. parva	C. pinnata	S. s. atlanticum
		F. borealis sargassi	D. gegenbauri	S. neglecta
		F. pellucida	O. cornutogastra	P. macropus
			O. albicans	C. acicula
			S. minima	C. virgula conica
				S. abbreviatum
				S. minima
				D. gegenbauri
				O. albicans

Species from Other Groups Occurring with One or More Grouped Species

S. s. pacifica (E)	S. planctonis (A)	D. denticulatum (D)	S. enflata (E)	D. denticulatum (D)
P. draco (E)	S. lyra (A)	T. multitentaculata (D)	S. s. pacifica (E)	M. huxleyi (D)
S. bipunctata (E)		M. huxleyi (D)	S. bipunctata (E)	S. magnum (D)
S. enflata (E)		S. magnum (D)	S. regularis (E)	S. planctonis (A)
S. magnum (D)		O. rufescens (D)	S. planctonis (A)	S. lyra (A)
T. gregaria (B)			(and each of Group C)	K. subtilis (A)

Probability (P <0.01) of species occurring as species groups or in association with other species.

263

Table 3.4—5
ESTIMATED NUMBER OF SPECIES OF HOLOPLANKTONIC ANIMALS AND MEROPLANKTONIC COELENTERATES

	1935	1967	
Coelenterata			
Antho-, Lepto-, and Scyphomedusae	–	–	–
Meroplankton	535	656	Russell, 1935
Trachy- and Narcomedusae	134	110	Kramp, 1961
Siphonophora			
Calycophorae	72	96	Alvariño after Totton, 1965
Physophorae	34	37	Sears, 1953
Ctenophora	80	?	–
Nemertea	34	?	–
Polychaeta	–	–	–
Tomopteridae	44	10	Tebble, 1962
Chaetognatha	30	55–57	Alvariño, 1967; David, 1963
Crustacea			
Cladocera	7	?	–
Copepoda	754	1200	Fleminger
Amphipoda	–	–	–
Hyperiidea	292	?	–
Gammaridea	44	?	–
Euphausiacea	85	83	Brinton, 1967
Mollusca			
Pteropoda	–	–	–
Thecosomata	51	35	Tesch, 1946, 1948; van der Spoel, 1966; McGowan, 1967
Gymnosomata	41	31	Provot-Fol, 1954
Heteropoda	90	24	Tesch, 1949; McGowan, 1967
Tunicata			
Appendicularia	61	68	*Zoo. Record*
Thaliacea	–	–	–
Doliolidae	12	8–12	Thomson, 1948; Berner, 1967
Salpidae	25	22–25	Yount, Foxton
Pyrosomidae	8	2–10	Thomson, 1948; *Zoo. Record*
Total	2433		
Excluding meroplanktonic Medusae	1898		

(From McGowan, J. A., Oceanic biogeography of the Pacific, in *The Micropalaeontology of Oceans*, Funnell, B. M. and Reidel, W. R., Eds., Cambridge University Press, England, 1971, 10. With permission.)

Table 3.4—6
SPECIES COMPOSITION OF THE FIVE WORLD DISTRIBUTIONAL ZONES
OF PLANKTONIC FORAMINIFERA

Northern and Southern Cold-water Regions

1. Arctic and antarctic zones:
   Globigerina pachyderma (Ehrenberg): Left-coiling variety; right-coiling in subarctic and subantarctic zones.
2. Subarctic and subantarctic zones:
   Globigerina quinqueloba (Natland)
   Globigerina bulloides (d'Orbigny)
   Globigerinita bradyi (Wiesner)
   Globorotalia scitula (Brady)

Transition Zones

3. Northern and south transition zones between cold-water and warm-water regions:
   Globorotalia inflata (d'Orbigny): With mixed occurrences of subpolar and tropical-subtropical species.

Warm-water Region

4. Northern and southern subtropical zones:
   Globigerinoides ruber (d'Orbigny): Pink variety in Atlantic Ocean only.
   Globigerinoides conglobatus (Brady): Autumn species.
   Hastigerina pelagica (d'Orbigny)
   Globigerinita glutinata (Egger)
   Globorotalia truncatulinoides (d'Orbigny)
   Globorotalia hirsuta (d'Orbigny)   Winter species
   Globigerina rubescens (Hofker)   Winter species
   Globigerinella aequilateralis (Brady)   Prefer outer margins of subtropical central water
   Orbulina universa (d'Orbigny)   masses and into transitional zone.
   Globoquadrina dutertrei (d'Orbigny)
   Globigerina falconensis (Blow)
   Globorotalia crassaformis (Galloway and Wissler)

5. Tropical Zone:
   Globigerinoides sacculifer (Brady): Including *Sphaeroidinella dehiscens* (Parker and Jones).
   Globorotalia menardii (d'Orbigny)
   Globorotalia tumida (Brady)
   Pulleniatina obliquiloculata (Parker and Jones)
   Candeina nitida (d'Orbigny)
   Hastigerinella digitata (Rhumbler)
   Globoquadrina conglomerata (Schwager)   Restricted to Indo-Pacific.
   Globigerinella adamsi (Banner and Blow)   Restricted to Indo-Pacific.
   Globoquadrina hexagona (Natland)   Restricted to Indo-Pacific.

The species are listed under the zone where their highest concentrations are observed, but they are not necessarily limited to these areas.

Most species listed under the Subtropical Zones are also common in the tropical waters.

* Usually located in central water masses between 20°N and 40°N, or between 20°S and 40°S latitude.

Table 3.4—7
THE DISTRIBUTION OF CHAETOGNATHS OVER AND SOMEWHAT BEYOND THE CONTINENTAL SHELF OF NORTH CAROLINA

Common	Occasional	Rare

The Inner Shelf—Zone I

Common	Occasional	Rare
S. enflata	K. pacifica	S. bipunctata
S. helenae	P. draco	S. minima
S. tenuis		S. serratodentata

The Outer Shelf—Zone II

Common	Occasional	Rare
S. bipunctata	S. tenuis	K. subtilis
S. enflata		S. hexaptera
S. minima		S. lyra
S. serratodentata		
S. helenae		
K. pacifica		
P. draco		

The Florida Current—Zone III

Common	Occasional	Rare
S. bipunctata		S. minima
S. enflata		
S. serratodentata		
K. pacifica		
P. draco		

(From Pierce, E. L., The Chaetognatha over the continental shelf of North Carolina with attention to their relation to the hydrography of the area, *J. Mar. Res.*, 12(1), 89, 1953. With permission.)

Table 3.4—8
THE WORLDWIDE DISTRIBUTION OF EIGHT SPECIES OF MESO- AND BATHYPELAGIC CHAETOGNATHS AS ELEMENTS OF COSMOPOLITAN, ARCTIC AND ANTARCTIC FAUNAS

Salinity (°/oo) Range, Above in Italics, and Temperature (°C) Range, Below, Are Given for Each Region

Region and Authority	*Eukrohnia fowleri*	*E. hamata*	*Sagitta decipiens*	*S. macrocephala*	*S. marri*	*S. maxima*	*S. planctonis*	*S. zetosios*
Arctic								
Greenland, Davis Strait Kramp, 1917	—	*33.6–34.0* ~0.5	—	—	—	*33.3–34.7* ~0.5–3.0 >34.0	—	—
Umanak Fjord Kramp, 1917	—	*34.1–34.5* 0.5–1.0	—	—	—	—	—	—
Disko Bay Kramp, 1917	—	~0.4–0.9	—	—	—	—	—	—
Near Egedesminde Kramp, 1917	—	*33.75* 0.52	—	—	—	—	—	—
S. Northern Storφ	—	*33.6* ~0.87 0.4	—	—	—	—	—	—
Bredefjord Kramp, 1917	—	*34.4* 3.2	—	—	—	—	—	—
Skovfjord Kramp, 1917	—	*33.9* 1.5	—	—	—	*33.87* 1.56	—	—
Greenland Baffin Bay and Labrador Sea Kramp, 1939	*34.9–35.1* 3.1–5.5	*33.0–34.9* 1.7–5.2	—	—	—	*33.4–35.0* ~1.2–6.0	—	*34.9–35.1* 3.1–5.5
Atlantic								
Gulf of Maine region Bigelow, 1926	—	*32.0–35.0* 1.3–9.0	—	—	—	*32.36–34.9* 1.63–9.0	—	—
Scottish region Fraser, 1952	—	*34.75–35.5*	—	—	—	*34.75–35.5* -1.0–12.0	—	*34.75–35.5* 8.0–9.0
Western Mediterranean Furnestin, 1957a	—	—	*37.47* 15.25 24.30	—	—	—	—	—
Eastern Mediterranean and entrance to Black Sea Furnestin, 1957a	—	—	25.9	—	—	—	—	—
Union of South Africa: (Tafel Bay - Lambert's Bay) Heydorn, 1959	—	*34.85–35.35* 11.92–14.97	—	—	—	—	*35.39* 14.94	—
Florida (off Miami) Owre, 1960	13.4–14.5	7.4	12.8–22.4	—	—	—	—	—

Table 3.4—8 (Continued)
THE WORLDWIDE DISTRIBUTION OF EIGHT SPECIES OF MESO- AND BATHYPELAGIC CHAETOGNATHS AS ELEMENTS OF COSMOPOLITAN, ARCTIC AND ANTARCTIC FAUNAS

Region and Authority	Eukrohnia fowleri	E. hamata	Sagitta decipiens	S. macrocephala	S. marri	S. maxima	S. planctonis	S. zetosios
Pacific								
Chile Fagetti, 1958a	—	12.1–14.2	13.5	—	—	—	—	12.4
British Columbia Lea, 1955		27.0–28.0 8.5–10.0	—	—	—	—	—	
Eastern tropical (off Mexico and Central America) Sund, 1961b	34.65 8.9	34.41–34.82 10.3–11.9	34.40–34.92 8.9–18.0	34.65 8.9	—	34.78 10.8	—	34.65
Eastern Australia-Tasmania Thomson, 1957	—	8.0–16.5	11.0–22.0	8.9	—	—	6.0–24.0	8.9
Kurile-Kamchatka Trench region Tchindonova, 1955	—	33.1–34.5	—	—	—	—	—	
Indian								
Bay of Bengal (off Visakhapatnam) Rao and Ganapati, 1958		—	22.03–34.61 26.5–28.5	—	—		—	—
Antarctic								
Hoces Straits (Drake Passage) Balech, 1962	—	33.77–34.09 2.68–7.16	—	—	—	34.17 5.2	—	—
Antarctic and Subantarctic waters David, 1958b	—	–0.5–7.75	—	—	0.1–1.0	0.3–7.3	—	—
Antarctic-Subantarctic waters Fagetti, 1959	—	0.5–3.5	—	—	—		0.3	—
Antarctic Jameson, 1914	—	–2.7–3.8	—	—	—		—	—
Antarctic Convergence to Subtropical Convergence David, 1958b	—	33.7–34.7	—	—	34.67–34.71	33.4–35.0	—	—

Table 3.4–9
FAUNAL GROUPS OF CHAETOGNATHS
IN THE INDIAN OCEAN

a) Cosmopolitan (common to Atlantic, Indian, and Pacific oceans): *S. lyra, S. enflata, S. hexaptera, S. minima, S. bipunctata, K. subtilis, K. pacifica, P. draco, S. gazellae, S. tasmanica.*

b) Cold-water representants: *S. gazellae, S. tasmanica, E. hamata.*

c) Tropical-equatorial, and restricted to the Indo-Pacific waters: *S. ferox, S. robusta, S. pacifica, S. pulchra, S. neglecta, S. bedoti, S. regularis.*

d) Mesoplanktonic: *S. decipiens, S. planctonis, S. zetesios.*

e) Deep water: *E. hamata* (in low latitudes), *E. fowleri, E. bathypelagica.*

Table 3.4–10
NUMERICALLY IMPORTANT SPECIES OF COPEPODS IN THE
WESTERN ATLANTIC BY REGIONS

Neritic (14 samples) | Gulf Stream (3 samples)

	Freq.	Mean no./m³		Freq.	Mean no./m³
Pseudocalanus minutus	13	559	*Clausocalanus furcatus*	3	27
Centropages typicus	14	450	*Lucicutia flavicornis*	3	9
Oithona similis	13	151	*Oithona plumifera*	3	9
Temora longicornis	8	59	*O. setigera*	3	7
Paracalanus parvus	8	39	*Calocalanus pavo*	2	9
Calanus finmarchicus	11	32	*Farranula gracilis*	3	4
Metridia lucens	12	16	*Mecynocera clausi*	3	2
Candacia armata	9	9			

Slope (15 samples) | Sargasso Sea (11 samples)

	Freq.	Mean no./m³		Freq.	Mean no./m³
			Clausocalanus furcatus	9	7
Centropages typicus	11	76	*Oithona setigera*	11	6
Pseudocalanus minutus	8	16	*Lucicutia flavicornis*	11	4
Oithona similis	6	14	*Ctenocalanus vanus*	6	3
Metridia lucens	11	15	*Farranula gracilis*	6	2
Clausocalanus pergens	6	19	*Mecynocera clausi*	9	2
C. arcuicornis	7	13			
Pleuromamma borealis	8	6			
Oithona atlantica	12	6			

(From Grice, G. D. and Hart, A. D., The abundance, seasonal occurrence, and distribution of the epizooplankton between New York and Bermuda, *Ecol. Monogr.*, 32(4), 297, 1962. With permission.)

Table 3.4—11

ANTARCTIC AND SUBANTARCTIC PELAGIC COPEPODS AND THEIR FAUNAL ASSOCIATIONS

Species	Type	Neritic	Subantarctic	Antarctic	Abyssal
Calanoida					
Calanus australis (Brodsky)	S		Farran, 1929 Vervoort, 1957 Brodsky, 1959		
Calanus propinquus (Brady)	S–B*			Vervoort, 1957	
Calanus tonsus (Brady)	S		Vervoort, 1957		
Calanus similimus (Giesbrecht)	S*		Vervoort, 1957		
Calanoides acutus (Giesbrecht)	V*			Vervoort, 1957	
Megacalanus princeps (Wolfenden)	A				Vervoort, 1957
Bathycalanus bradyi (Wolfenden)	A				Vervoort, 1957
Eucalanus elongatus (Dana)	S–B		Farran, 1929	Hardy and Gunther, 1935	
Eucalanus longiceps (Matthews)	S–B		Vervoort, 1957	Vervoort, 1957	
Rhincalanus gigas (Brady)	V			Vervoort, 1957	
Rhincalanus nasutus (Giesbrecht)				Vervoort, 1957	
Microcalanus pygmaeus (G. O. Sars)	S–B			Vervoort, 1957	
Gaidius affinis (G. O. Sars)	B		Vervoort, 1957		
Gaidius intermedius (Wolfenden)	A				Vervoort, 1957
Gaidius tenuispinus (G. O. Sars)	B–A			Vervoort, 1957	Vervoort, 1957
Gaetanus antarcticus (Wolfenden)	B–A			Vervoort, 1957	Vervoort, 1957
Gaetanus latifrons (G. O. Sars)	A		Vervoort, 1957		Vervoort, 1957
Gaetanus minor (Farran)	B				Vervoort, 1957

S – Surface.
B – Bathypelagic.
A – Abyssal.
V – Show vertical seasonal migration.
* – Species characteristic of Antarctic waters.

Table 3.4–11 (Continued)
ANTARCTIC AND SUBANTARCTIC PELAGIC COPEPODS AND THEIR FAUNAL ASSOCIATIONS

Species	Type	Neritic	Subantarctic	Antarctic	Abyssal
Euchirella latirostris (Farran)	B*		Vervoort, 1957		
Euchirella rostrata (Claus)	B		Vervoort, 1957		
Euchirella rostromagna (Wolfenden)	B*			Vervoort, 1957	
Pseudochirella elongata (Wolfenden)	B*			Vervoort, 1957	
Pseudochirella hirsuta (Wolfenden)	A				Vervoort, 1957
Pseudochirella mawsoni (Vervoort)	B–A		Vervoort, 1957	Vervoort, 1957	Vervoort, 1957
Pseudochirella notacantha (G. O. Sars)	A				Vervoort, 1957 Hardy and Gunther, 1935
Pseudochirella pustulifera (G. O. Sars)	A				
Pseudeuchaeta brevicauda (G. O. Sars)	A				Vervoort, 1957
Undeuchaeta major (Giesbrecht)	B–A		Vervoort, 1957		Vervoort, 1957
Ctenocalanus vanus (Giesbrecht)	B			Vervoort, 1957	Vervoort, 1957
Clausocalanus arcuicornis (Dana)	S		Farran, 1929		
Clausocalanus laticeps (Farran)	S*		Vervoort, 1957		
Farrania frigida (Wolfenden)	A				Vervoort, 1957
Drenanopus pectinatus (Brady)	S	Vervoort, 1957			
Spinocalanus abyssalis (Giesbrecht)	A				Vervoort, 1957
Spinocalanus magnus (Wolfenden)	A				Vervoort, 1957 Farran, 1929
Spinocalanus spinosus (Farran)	A				
Mimocalanus cultrifer (Farran)	A				Vervoort, 1957
Stephus longipes (Giesbrecht)	S			Tanaka, 1960	
Aetideus armatus (Boeck)	B		Vervoort, 1957		

S – Surface.
B – Bathypelagic.
A – Abyssal.
V – Show vertical seasonal migration.
* – Species characteristic of Antarctic waters.

Table 3.4—11 (Continued)

ANTARCTIC AND SUBANTARCTIC PELAGIC COPEPODS AND THEIR FAUNAL ASSOCIATIONS

Species	Type	Neritic	Subantarctic	Antarctic	Abyssal
Euaetideus australis (Vervoort)	B		Vervoort, 1957		
Aetideopsis antarcticus (Wolfenden)	B*				
Aetideopsis minor (Wolfenden)	B*			Farran, 1929	
Chiridius polaris (Wolfenden)	B*				
Undeuchaeta minor (Giesbrecht)	S–B		Farran, 1929		
Euchaeta antarctica (Giesbrecht)	B*			Vervoort, 1957	
Euchaeta austrina (Giesbrecht)	B*			Vervoort, 1957	
Euchaeta biloba (Farran)	S–B		Vervoort, 1957	Vervoort, 1957	
Euchaeta erebi (Farran)	B*			Farran, 1929	
Euchaeta exigua (Wolfenden)	A				Vervoort, 1957
Euchaeta farrani (With)	A				Vervoort, 1957
Euchaeta rasa (Farran)	B–A			Vervoort, 1957	Vervoort, 1957
Euchaeta scotti (Farran)	B–A			Hardy and Gunther, 1935	Hardy and Gunther, 1935
Euchaeta similis (Wolfenden)	B*			Vervoort, 1957	Vervoort, 1957
Valdiviella insignis (Farran)	A			Vervoort, 1957	
Onchocalanus magnus (Wolfenden)	B*			Vervoort, 1957	
Onchocalanus wolfendeni (Vervoort)	B*			Vervoort, 1957	
Cornucalanus robustus (Vervoort)	B–A			Vervoort, 1957	Vervoort, 1957
Cephalophanes frigidus	A			Vervoort, 1957	Vervoort, 1957
Amallophora altera	B*			Vervoort, 1957	
Undinella brevipes (Farran)	A				Vervoort, 1957
Racovitzanus antarcticus (Giesbrecht)	B*			Vervoort, 1957	
Racovitzanus erraticus (Vervoort)	B*			Vervoort, 1957	

S – Surface.
B – Bathypelagic.
A – Abyssal.
V – Show vertical seasonal migration.
* – Species characteristic of Antarctic waters.

Table 3.4–11 (*Continued*)

ANTARCTIC AND SUBANTARCTIC PELAGIC COPEPODS AND THEIR FAUNAL ASSOCIATIONS

Species	Type	Neritic	Subantarctic	Antarctic	Abyssal
Scolecithricella glacialis (Giesbrecht)	S*			Vervoort, 1957	
Scolecithricella dentipes (Vervoort)	B*			Vervoort, 1957	
Scolecithricella emarginata (Farran)	B			Hardy and Gunther, 1936	
Scolecithricella incisa (Farran)	S–B*			Farran, 1929	
Scolecithricella minor (Brady)	B			Hardy and Gunther, 1936	
Scolecithricella ovata (Farran)	B		Vervoort, 1957	Vervoort, 1957	
Scolecithricella polaris (Wolfenden)	B*			Vervoort, 1957	
Scolecithricella robusta (T. Scott)	B			Vervoort, 1957	
Scolecithricella valida (Farran)	A				Vervoort, 1957
Scaphocalanus affinis (G. O. Sars)	B–A			Vervoort, 1957	Vervoort, 1957
Scaphocalanus brevicornis (G. O. Sars)	B			Hardy and Gunther, 1935	
Scaphocalanus echinatus (Farran)	B–A		Farran, 1929		
Scaphocalanus magnus (T. Scott)	A				Vervoort, 1957
Scaphocalanus subbrevicornis (Wolfenden)	B*			Vervoort, 1957	
Temorites brevis (G. O. Sars)	A				Vervoort, 1957
Metridia curticauda (Giesbrecht)	B–A			Vervoort, 1957	Vervoort, 1957

S – Surface.
B – Bathypelagic.
A – Abyssal.
V – Show vertical seasonal migration.
* – Species characteristic of Antarctic waters.

Table 3.4–11 (Continued)

ANTARCTIC AND SUBANTARCTIC PELAGIC COPEPODS AND THEIR FAUNAL ASSOCIATIONS

Species	Type	Neritic	Subantarctic	Antarctic	Abyssal
Metridia gerlachei (Giesbrecht)	S–B*			Vervoort, 1957	
Metridia lucens (Boeck)	S–B		Vervoort, 1957	Vervoort, 1957	
Metridia princeps (Giesbrecht)	B–A				Vervoort, 1957
Pleuromamma borealis (F. Dahl)	B		Farran, 1929		
Pleuromamma gracilis (Claus)	B		Farran, 1929		
Pleuromamma robusta (F. Dahl) f. antarctica (Steuer)	B–A		Vervoort, 1957	Vervoort, 1957	Vervoort, 1957
Lucicutia curta (Farran)	A				Vervoort, 1957
Lucicutia frigida (Wolfenden)	B–A			Vervoort, 1957	Vervoort, 1957
Lucicutia grandis (Giesbrecht)	A				Vervoort, 1957
Lucicutia macrocera (G. O. Sars)	A				Vervoort, 1957
Lucicutia magna (Wolfenden)	A				Farran, 1929
Lucicutia maxima (Steuer)	B			Hardy and Gunther, 1935	
Lucicutia wolfendeni (Sewell)	A				Vervoort, 1957
Disseta palumboi (Giesbrecht)	A				Hardy and Gunther, 1935
Heterorhabdus austrinus (Giesbrecht)	B*			Vervoort, 1957	
Heterorhabdus compactus (G. O. Sars)	A			Hardy and Gunther, 1935	Farran, 1929; Hardy and Gunther, 1935
Heterorhabdus farrani (Brady)	B*			Vervoort, 1957	
Heterorhabdus pustulifer (Farran)	B*			Vervoort, 1957	Vervoort, 1957
Heterostylites major (F. Dahl)	A			Hardy and Gunther, 1935	
Haloptilus fons (Farran)	B				

S – Surface.
B – Bathypelagic.
A – Abyssal.
V – Show vertical seasonal migration.
* – Species characteristic of Antarctic waters.

Table 3.4–11 (Continued)
ANTARCTIC AND SUBANTARCTIC PELAGIC COPEPODS AND THEIR FAUNAL ASSOCIATIONS

Species	Type	Neritic	Subantarctic	Antarctic	Abyssal
Haloptilus ocellatus (Wolfenden)	B*			Vervoort, 1957	
Haloptilus oxycephalus (Giesbrecht)	B			Vervoort, 1957	
Augaptilus glacialis (G. O. Sars)	B			Vervoort, 1957	
Augaptilus megalurus (Giesbrecht)	B			Hardy and Gunther, 1935	
Euaugaptilus laticeps (G. O. Sars)	A				Vervoort, 1957
Euaugaptilus magnus (Wolfenden)	A				Vervoort, 1957
Centraugaptilus rattrayi (T. Scott)	B			Hardy and Gunther, 1935	
Pseudaugaptilus longiremis (G. O. Sars)	A				Vervoort, 1957
Pontoptilus ovalis (G. O. Sars)	A				Vervoort, 1957
Pachyptilus eurygnathus (G. O. Sars)	A				Vervoort, 1957
Arietellus simplex (G. O. Sars)	A				Vervoort, 1957
Phyllopus bidentatus (Brady)	B		Vervoort, 1957	Vervoort, 1957	
Candacia cheirura (Cleve)	S–B		Vervoort, 1957	Vervoort, 1957	
Candacia falcifera (Farran)	B			Vervoort, 1957	
Candacia maxima (Vervoort)	B		Vervoort, 1957	Vervoort, 1957	
Paralabidocera antarctica (I. C. Thompson)	S			Vervoort, 1957	
Cyclopoida					
Pseudocyclopina belgicae (Giesbrecht)	S			Giesbrecht, 1902	

S – Surface.
B – Bathypelagic.
A – Abyssal.
V – Show vertical seasonal migration.
* – Species characteristic of Antarctic waters.

Table 3.4–11 (Continued)
ANTARCTIC AND SUBANTARCTIC PELAGIC COPEPODS AND THEIR FAUNAL ASSOCIATIONS

Species	Type	Neritic	Subantarctic	Antarctic	Abyssal
Mormonilla phasma (Giesbrecht)	B–A			Hardy and Gunther, 1935	Hardy and Gunther, 1935
Oithona frigida (Giesbrecht)	S		Vervoort, 1957	Vervoort, 1957	
Oithona similis (Claus)	S–B		Vervoort, 1957	Vervoort, 1957	
Ratania atlantica (Farran)	S–B			Vervoort, 1957	
Oncaea conifera (Giesbrecht)	S–B			Vervoort, 1957	
Oncaea curvata (Giesbrecht)	B			Vervoort, 1957	
Oncaea mediterranea (Giesbrecht)	S–B		Vervoort, 1957	Vervoort, 1957	
Oncaea notopus (Giesbrecht)	B			Vervoort, 1957	
Oncaea venusta (Philippi)	B			Tanaka, 1960	
Conea rapax (Giesbrecht)	B–A			Hardy and Gunther, 1935	
Lubbockia aculeata (Giesbrecht)	B–A			Vervoort, 1957	Vervoort, 1957

S – Surface.
B – Bathypelagic.
A – Abyssal.
V – Show vertical seasonal migration.
* – Species characteristic of Antarctic waters.

(From Vervoort, W., Biogeography and ecology in Antarctica: Notes on the biogeography and ecology of free-living copepoda, *Monographiae Biologicae*, 15, 394, 1965. With permission.)

Table 3.4–12
THE WORLDWIDE DISTRIBUTION OF EUPHAUSIIDS

The approximate latitudinal range of each species is stated. The Atlantic and Pacific Oceans are divided latitudinally into the region north of 40°N, the eastern (E), central (C), and western (W) areas of the region between 40°N and 40°S, and the region south of 40°S. Species occurring in the Mediterranean (Med.) are detailed separately. The Indian Ocean is divided into the eastern (E), central (C), and western (W) areas of the region north of 40°S, and the region south of 40°S, with species found in the Red Sea (Red) detailed separately.

Species	Latitudinal range	Atlantic N of 40°N	Atlantic 40°N–40°S E	C	W	Med.	Atlantic S of 40°S	Pacific N of 40°N	Pacific 40°N–40°S E	C	W	Pacific S of 40°S	Indian N of 40°S E	C	W	Red	Indian S of 40°S
Bentheuphausia																	
B. amblyops	50°N 50°S	X	X	X	X		X	X	X	X	X	X	X	X	X		
Thysanopoda																	
T. monacantha	40°N–10°S		X		X				X	X	X		X	X	X		
T. cristata	35°N–40°S			X	X					X	X				X		
T. tricuspidata	35°N–30°S		X	X	X					X	X			X	X		
T. aequalis	40°N–40°S		X	X	X					X	X			X	X		
T. subaequalis	40°N–40°S				X	X				X					X		
T. obtusifrons	35°N–35°S		X	X	X		X			X	X				X		
T. pectinata	35°N–35°S		X		X				X	X	X		X	X	X		
T. orientalis	40°N–40°S				X				X	X	X			X	X		
T. microphthalma	40°N–40°S		X	X	X									X	X		
T. acutifrons	{70°N–40°N / 40°S–60°S}	X			X		X	X				X			X		X
T. cornuta	55°N–40°S		X		X			X	X	X	X						
T. ergregia	40°N–50°S		X						X	X	X				X		
T. spinicaudata	30°N–30°S								X	X	X	X					
Meganyctiphanes																	
M. norvegica	70°N–30°N	X				X											

Table 3.4–12 (Continued)
THE WORLDWIDE DISTRIBUTION OF EUPHAUSIIDS

Species	Latitudinal range	Atlantic N of 40°N	Atlantic 40°N–40°S E	Atlantic 40°N–40°S C	Atlantic 40°N–40°S W	Atlantic Med.	Atlantic S of 40°S	Pacific N of 40°N	Pacific 40°N–40°S E	Pacific 40°N–40°S C	Pacific 40°N–40°S W	Pacific S of 40°S	Indian N of 40°S E	Indian N of 40°S C	Indian N of 40°S W	Indian N of 40°S Red	Indian S of 40°S
Nyctiphanes																	
N. couchii	60°N–30°N	X				X											
N. australis	35°S–50°S											X					
N. capensis	30°S–40°S		X														
N. simplex	30°N–20°S				X				X								
Pseudeuphausia																	
P. latifrons	40°N–35°S										X		X	X	X		
P. sinica	30°N–15°N										X					X	
Euphausia																	
E. americana	40°N–10°S		X	X	X												
E. eximia	40°N–30°S		X	X		X			X								
E. krohnii	65°N–0°S	X	X	X	X	X											
E. mutica	40°N–40°S		X	X	X				X	X	X		X	X	X	X	
E. brevis	40°N–40°S		X	X	X	X			X	X	X		X	X	X	X	
E. diomedeae	25°N–25°S								X	X	X		X	X	X	X	
E. recurva	40°N–40°S		X						X								
E. superba	55°S–75°S						X					X					X
E. vallentini	45°S–60°S						X					X					X
E. lucens	35°S–50°S						X					X					X
E. frigida	50°S–65°S						X					X					X
E. pacifica	50°N–35°N							X									
E. nana	35°N–25°N										X						
E. crystallorophias	65°S–75°S						X					X					X
E. tenera	40°N–30°S		X	X	X				X	X	X		X	X	X		
E. similis	40°N–50°S						X		X		X	X	X	X	X		X
E. similis var. *armata*	15°N–50°S		X				X		X		X	X	X				X
E. mucronata	0°–10°S								X								
E. sibogae	0°–20°S										X						

Table 3.4–12 (Continued)

THE WORLDWIDE DISTRIBUTION OF EUPHAUSIIDS

Species	Latitudinal range	Atlantic: N of 40°N	Atlantic 40°N–40°S: E	C	W	Med.	Atlantic: S of 40°S	Pacific: N of 40°N	Pacific 40°N–40°S: E	C	W	Pacific: S of 40°S	Indian N of 40°S: E	C	W	Red	Indian: S of 40°S
E. distinguenda	30°N–20°S								X				X	X	X		
E. lamelligera	25°N–10°S								X				X	X	X	X	
E. gibba	20°S–40°S								X		X						
E. gibboides	40°N–40°S		X	X	X				X	X	X			?	?	?	
E. fallax	35°N–20°S									X	X				?		
E. sanzoi	20°N															X	
E. pseudogibba	30°N–30°S		X	X	X				X	X	X		X	X	X		
E. paragibba	20°N–20°S									X	X		X	X	X		
E. hemigibba	40°N–40°S		X	X	X	X			X	X	X		X	X	X		
E. spinifera	30°S–45°S		X				X					X					X
E. hanseni	25°S–40°S		X														
E. longirostris	40°S–55°S						X					X					X
E. triacantha	50°S–65°S						X					X					X
Tessarabrachion																	
T. oculatum	50°N–35°N							X									
Thysanoëssa																	
T. spinifera	60°N–25°N							X	X								
T. longipes	60°N–45°N							X									
T. inspinata	55°N–35°N							X									
T. inermis	75°N–40°N	X						X									
T. longicaudata	75°N–40°N	X						X									
T. parva	40°N–20°N / 25°S–40°S		X		X				X	X	X		X				
T. gregaria	50°N–10°N / 20°S–50°S	X	X	X	X	X	X	X	X			X	X				
T. vicina	50°S–75°S						X					X					X
T. macrura	50°S–75°S						X					X					X
T. raschii	75°N–40°N	X						X									

Table 3.4–12 (Continued)

THE WORLDWIDE DISTRIBUTION OF EUPHAUSIIDS

Species	Latitudinal range	Atlantic N of 40°N	Atlantic 40°N–40°S E	C	W	Med.	Atlantic S of 40°S	Pacific N of 40°N	Pacific 40°N–40°S E	C	W	Pacific S of 40°S	Indian N of 40°S E	C	W	Red	Indian S of 40°S
Nematoscelis																	
N. difficilis	45°N–20°N	X												X			
N. megalops	60°N–10°N / 20°S–55°S	X				X	X					X	X	X	X		X
N. tenella	40°N–40°S		X	X	X				X	X	X		X		X		
N. microps	40°N–40°S		X	X	X				X	X	X		X	X ?	X		
N. atlantica	40°N–40°S		X	X	X	X					X						
N. lobata	15°N–5°N																
N. gracilis	30°N–30°S		X	X			X		X	X	X		X	X	X		
Nematobrachion																	
N. flexipes	40°N–40°S		X	X	X				X	X	X		X	X	X		
N. sexspinosum	30°N–30°S		X	X	X						X			?	X		
N. boöpis	60°N–55°S	X	X	X	X				X	X	X	X	X	X	X		
Stylocheiron																	
S. carinatum	40°N–40°S		X	X	X				X	X	X		X	X	X	X	
S. affine	40°N–40°S		X	X	X				X	X	X		X	X	X	X	
S. suhmii	50°N–40°S		X						X	X	X		X	X	X	X	
S. microphthalma	35°N–25°S		X			X			X		X		X		X		
S. insulare	10°N–10°S			X	X						X						
S. elongatum	60°N–40°S	X			X			X	X	X	X		X	X	X	X	
S. indicum			X														
S. longicorne	60°N–50°S	X	X	X	X	X			X	X	X		X	X	X		
S. abbreviatum	50°N–40°S		X	X	X	X			X	X	X		X	X	X		
S. maximum	60°N–60°S	X	X	X	X	X			X	X	X	X	X	X	X	X	
S. robustum	30°N–30°S		X	X					X	X							

Figure 3.4–1

South–North Section along 140°W (approx.)

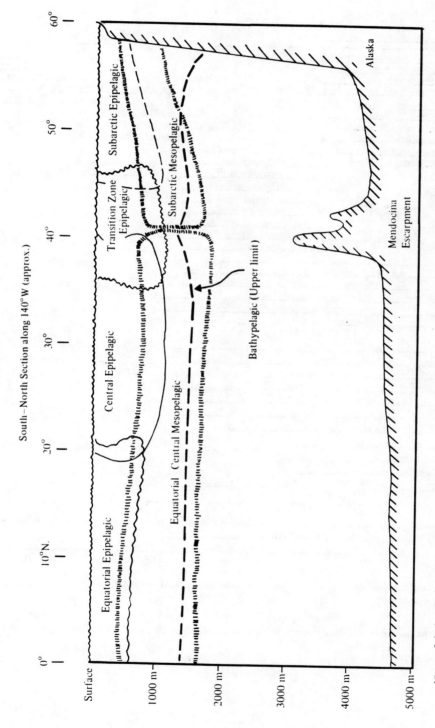

Figure 3.4–1. Bathymetric and latitudinal zonation of associations of euphausiid species in mid-oceanic profile in the North Pacific.

(From Brinton, E., The distribution of Pacific euphausiids, *Bull. Scripps Inst. Oceanogr. Univ. Calif.*, 8, 195, 1962. With permission.)

Figure 3.4–2

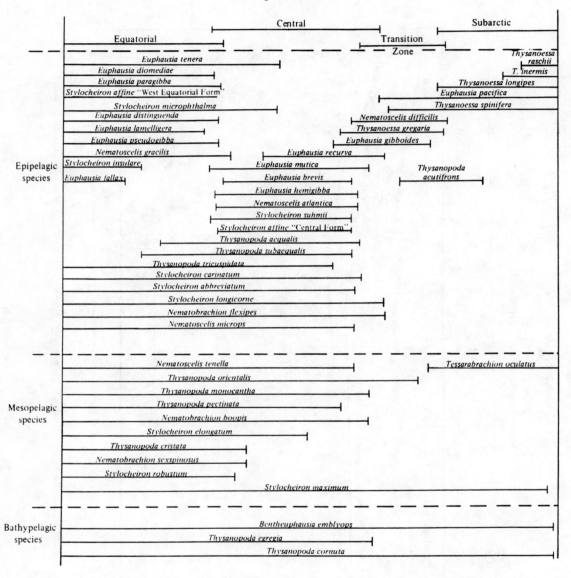

Figure 3.4–2. Species composition of bathymetric euphausiid associations in the North Pacific. Relationships to horizontal zones are also indicated.

(From Brinton, E., The distribution of Pacific euphausiids, *Bull. Scripps Inst. Oceanogr. Univ. Calif.*, 8, 194, 1962. With permission.)

Table 3.4–13

THE EUPHAUSIID FAUNA OF THE HIGH SEAS, AS COMPARED TO THOSE OF LITTORAL AND SUBLITTORAL PROVINCES IN REGARD TO LATITUDINAL EXTENTS

High Seas Euphausiid Groups	Littoral Fauna (Ekman)	Euphausiid Boundary Species
Thysanoessa longipes "Spined," *T. inermis*	Arctic	*Thysanoessa rashchii*
Subarctic group	East Asiatic Temperate, Northwest American Temperate	
Transition-zone group, North	Transition	*T. spinifera*
East equatorial group	Pacific Tropical American	*Nyctiphanes simplex*
West equatorial group	Tropical Indo West Pacific	*E. lamelligera*
Trans-equatorial group		*P. latifrons*
—	—	—
	Peruvian North Chilean	*E. mucronata*
Transition-zone group, South	New Zealand	*N. australis*
Subantarctic group	Antiboreal South American	—
Antarctic group	Antarctic	*E. crystallorophias*

Figure 3.4—3

COMPOSITE DISTRIBUTION PATTERNS OF PACIFIC EUPHAUSIIDS

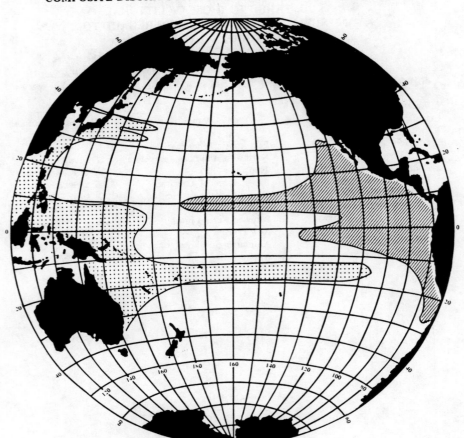

A. The western equatorial (Indo-Australian) group which includes *Euphausia pseudogibba, E. fallax, E. sibogae, Pseudeuphausia latifrons, Nematoscelis lobata, Stylocheiron insulare,* and *S. affine* "Indo-Australian Form." The eastern equatorial group includes *Euphausia lamelligera, E. distinguenda, E. eximia, Stylocheiron affine* "East Equatorial Form," and *Nyctiphanes simplex.*

(From Brinton, E., The distribution of Pacific euphausiids, *Bull. Scripps Inst. Oceanogr. Univ. Calif.,* 8, 212, 1962. With permission.)

Figure 3.4–3 *(Continued)*

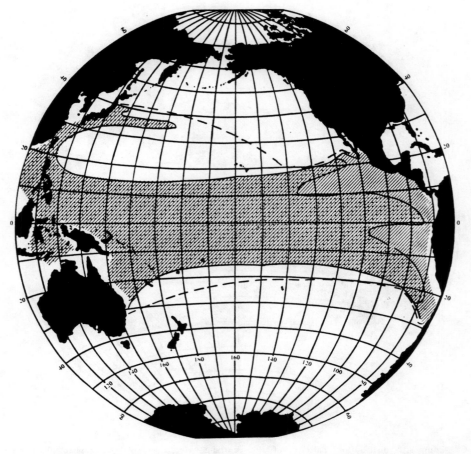

B.   Composite distributions of groups of epipelagic trans-equatorial euphausiid species. Cross-hatched part: *Euphausia diomediae* and *Nematoscelis gracilis*. Strippled part: *Thysanopoda tricuspidata*, *Euphausia paragibba*, and *Stylocheiron microphthalma*. The dashed line indicates the limits of range of an equatorial-west central species, *Euphausia tenera*.

(From Brinton, E., The distribution of Pacific euphausiids, *Bull. Scripps Inst. Oceanogr. Univ. Calif.*, 8, 210, 1962. With permission.)

Figure 3.4–3 *(Continued)*

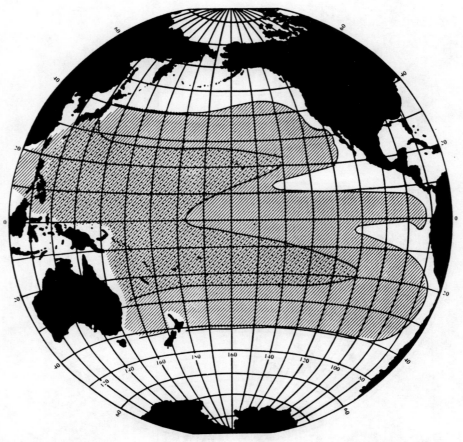

C. Composite distribution patterns of the central-equatorial mesopelagic euphausiid species. Cross-hatched part: *Thysanopoda orientalis, T. monocantha, T. pectinata, Nematoscelis tenella, Nematobrachion boopis,* and *Stylocheiron elongatum.* Stippled part: *Nematobrachion sexspinosus, Thysanopoda cristata,* and *Stylocheiron robustum.*

(From Brinton, E., The distribution of Pacific euphausiids, *Bull. Scripps Inst. Oceanogr. Univ. Calif.,* 8, 209, 1962. With permission.)

Figure 3.4—4

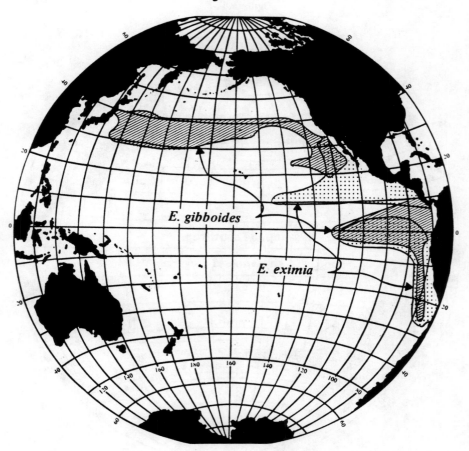

Figure 3.4–4. Distributions of the large transition-zone–equatorial-zone *Euphausia* species *E. gibboides* and *E. eximia*, showing differences in ranges in the North Pacific and similarities in the South Pacific.

(From Brinton, E., The distribution of Pacific euphausiids, *Bull. Scripps Inst. Oceanogr. Univ. Calif.*, 8, 204, 1962. With permission.)

Figure 3.4–5

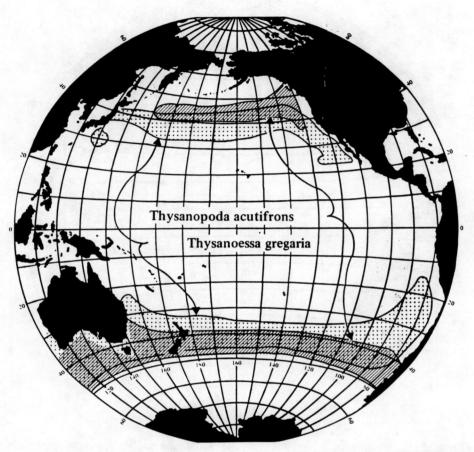

Figure 3.4–5. Distributions of the transition-zone species *Thysanopoda acutifrons* and *Thysanoessa gregaria*. Species of this zone, which lies between subarctic (or subantarctic) and central waters, are all antitropical (bipolar).

(From Brinton, E., The distribution of Pacific euphausiids, *Bull Scripps Inst. Oceanogr. Univ. Calif.*, 8, 203, 1962. With permission.)

3.5 Indicator Species and Their Associated Water Masses

Table 3.5–1
WATER MASSES OF THE OCEANS

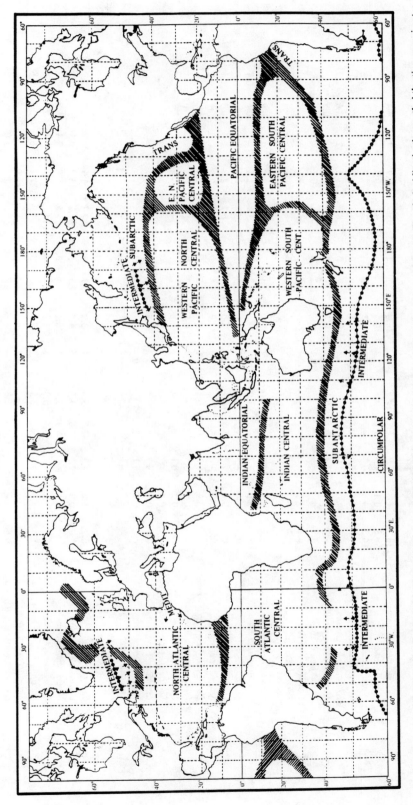

Note: Hatched bands delimit the major water masses; dotted lines indicate convergences where intermediate waters are formed. The dashed line in the north Atlantic approximates the southern boundary of "Gulf Stream water."

Table 3.5—2

INCIDENCES OF OCCURRENCE OF MAJOR PLANKTONIC SPECIES IN SEVEN WATER MASSES OF THE PACIFIC

	138.1	21.8	31.9	8.1	0.9	140.5	28.0	83.7	1.8	105.9	60.7	3.1	4.4	5.2
Calanus helgolandicus	138.1	21.8	31.9	8.1	0.9	140.5	28.0	83.7	1.8	105.9	60.7	3.1	4.4	5.2
Acartia tonsa	59.8	10.1	4.6	1.6	0.1	2.8	0.6	2.0	–	0.1	0.3	–	–	–
Nyctiphanes simplex	16.3	4.6	0.7	0.2	+	–	4.3	–	–	2.6	7.7	+	–	1.2
Sagitta euneritica	7.2	3.6	0.3	0.8	–	14.8	8.6	3.6	0.3	1.3	0.2	0.6	0.6	–
Rhincalanus nasutus	1.3	–	2.0	0.3	+	6.4	3.1	0.9	0.9	8.2	11.5	2.2	3.6	7.8
Heterorhabdus papilliger	–	0.5	0.3	0.1	0.1	1.4	0.4	–	0.1	–	0.2	–	0.5	–
Mysid gen. et sp.	0.8	+	0.6	+	–	0.4	+	0.2	0.1	–	0.1	–	–	–
Candacia bipinnata	0.1	+	–	+	–	0.4	–	–	+	–	–	–	–	–
Paracalanus sp.	0.1	–	–	–	–	0.4	–	–	–	–	–	–	–	–
Aglaura hemistoma	0.1	0.9	14.7	–	4.0	0.7	0.7	2.4	3.1	11.4	0.2	0.8	0.8	3.2
Pleuromamma abdominalis	+	1.0	1.7	0.6	1.0	0.9	2.1	8.7	1.9	5.4	1.1	–	0.5	0.8
Sagitta bierii	0.8	1.4	4.3	1.2	1.2	3.1	3.0	5.9	5.9	15.7	4.3	0.4	–	0.8
Diphyes dispar	0.2	0.8	1.7	1.5	0.1	2.4	2.0	6.7	1.5	6.7	2.8	1.0	0.3	1.4
Muggiaea atlantica	0.1	0.7	–	–	+	2.0	2.0	3.4	0.3	1.1	0.2	0.6	+	0.1
Eucalanus bungii californicus	–	3.4	0.3	0.2	0.1	0.1	0.3	0.5	+	–	0.5	–	–	–
Pleuromamma borealis	+	0.9	0.5	0.2	–	1.4	0.5	0.2	0.3	1.4	0.8	–	–	–
Conchoecia striola	–	0.2	0.1	+	+	–	0.8	+	+	–	–	–	–	–
Sagitta pseudoserratodentata	–	0.1	–	0.2	–	–	0.7	–	–	–	–	–	–	–
Labidocera trispinosa	0.3	0.1	+	+	+	–	0.3	–	0.2	–	0.1	–	–	–
Stylocheiron affine	+	0.1	0.1	0.2	–	0.1	0.1	0.1	0.1	–	–	–	–	–
Euphausia gibboides	–	+	–	–	–	–	0.1	–	–	–	–	–	–	–
Conchoecia alata	–	+	–	–	+	–	+	–	–	–	–	–	–	–
Sagitta enflata	0.2	1.6	4.3	3.7	5.8	5.4	3.9	3.7	4.8	47.2	51.1	20.2	6.3	7.2
Clausocalanus spp.	1.0	0.3	7.2	15.6	4.6	8.6	4.7	7.5	5.8	39.2	15.8	0.1	–	+
Nannocalanus minor	–	–	–	8.4	5.4	1.2	0.1	0.4	0.8	19.5	5.8	8.4	5.8	3.2
Eucalanus attenuatus	–	–	–	0.4	0.9	1.8	1.0	0.1	0.1	14.3	10.3	7.4	2.4	1.6
Oikopleura longicauda	1.4	3.8	3.3	6.1	3.5	13.2	7.2	6.9	3.7	13.9	7.2	6.7	0.2	1.6
Labidocera acutifrons	–	–	–	–	–	0.4	–	+	0.1	1.8	16.6	–	–	–
Euchaeta marina/E. tenuis	0.6	0.1	1.4	0.8	1.3	1.4	–	2.7	–	9.5	1.8	1.0	–	1.0
Eucalanus subtenuis	+	0.2	2.2	0.3	0.6	0.5	–	0.1	0.1	8.5	4.5	3.8	2.9	2.5
Euaetideus bradyi	0.3	0.1	0.9	0.3	0.2	0.8	0.8	0.8	0.4	8.6	2.5	–	0.5	0.4
Candacia curta	–	+	0.2	2.1	0.2	0.5	0.4	1.5	0.6	3.9	1.6	–	0.5	0.2
Desmopterus papilio	–	0.2	+	–	+	0.1	0.1	0.1	–	3.9	0.8	0.3	0.6	0.4

UW – Upwelled Water.
CCW – California Current Water.

Table 3.5–2 (*Continued*)

INCIDENCES OF OCCURRENCE OF MAJOR PLANKTONIC SPECIES IN SEVEN WATER MASSES OF THE PACIFIC

	PCW	TSW											
Limacina inflata	+	0.1	0.9	0.1	+	0.1	1.7	2.3	2.1	2.1	0.2	–	+
Doliolum denticulatum	–	+	0.4	0.1	0.2	0.4	0.1	0.8	0.2	3.4	1.2	19.4	5.4	0.2
Salpa democratica	–	–	–	–	0.1	0.3	+	1.8	2.2	1.1	0.2	1.4	–	1.0
Temora discaudata	–	+	0.3	0.1	+	0.2	+	–	+	1.0·	1.8	1.7	1.0	0.4
Conchoecia magna	+	0.1	0.1	+	+	0.1	0.2	0.2	0.1	1.6	0.5	+	+	0.1
Fritillaria formica	+	0.2	0.2	+	0.3	0.3	0.8	–	0.5	1.5	0.1	+	–	–
Candacia aethiopica	–	+	–	0.1	–	0.1	–	0.1	–	1.5	0.1	–	–
Conchoecia giesbrechti	–	0.1	+	+	0.1	0.1	0.3	0.1	0.2	1.4	0.2	–	–	0.1
Liriope tetraphylla	–	–	–	–	0.1	0.1	0.3	0.2	0.1	0.3	1.4	–	–
Euchaeta wolfendeni	–	–	1.6	–	–	+	–	–	–	–	–	–
Sagitta pacifica	+	+	0.6	1.1	0.1	–	–	–	–	–	0.2	0.8	0.6
Copila mirabilis	–	–	–	–	–	–	–	0.2	0.3	–	1.1	0.6	0.2
Sapphirina spp.	+	+	+	+	0.1	0.2	–	0.1	0.1	1.0	0.1	1.1	1.6	0.2
Euphausia eximia	+	0.1	0.1	0.2	0.1	0.2	0.3	0.2	–	0.8	1.6	0.4	1.1
Amphogona apicata	–	–	–	–	–	0.1	0.1	–	0.7	1.6	0.4	–	1.1
Eudoxoides spiralis	+	+	+	–	0.1	0.1	0.1	0.1	0.1	0.5	–	–	–
Pleuromamma quadrungulata	–	0.2	0.3	0.1	+	0.2	0.1	0.1	0.2	0.4	–	–	–

PCW – Central Pacific Water.
TSW – Tropical Surface Water.

(From Longhurst, A. R., Diversity and trophic structure of zooplankton communities in the California Current, *Deep-Sea Res.*, 14, 402, © 1967 Pergamon Press. With permission.)

Table 3.5–3
WATER MASS PREFERENCES OF
UNGROUPED SPECIES IN
THE PACIFIC

Central Tasman and
Southwest Tasman Water Masses

Tunicata:	*"Ihlea magalhanica"*
	Pegea confederata
	Pyrosoma atlanticum
	Iasis zonaria
	Oikopleura parva
	Fritillaria borealis
Chaetognatha:	*Sagitta serratodentata tasmanica*
Euphausiacea:	*Euphausia similis*
	E. similis var. *armata*
Hyperiidae:	*Brachyscelus crustulum*

South Equatorial and
Coral Sea Water Masses

Tunicata:	*Cyclosalpa pinnata*
	Brooksia rostrata
	Rittierella amboinensis
	Oikopleura intermedia
	O. cornutogastra
	Fritillaria formica
Chaetognatha:	*Sagitta neglecta*
Euphausiacea:	*Nematoscelis microps*
	Stylocheiron abbreviatum
Hyperiidae:	*Phrosina semilunata*
	Primno macropus
	Anchylomera blossevillei
Sergestidae:	*Lucifer hanseni*
	L. typus
Pteropoda:	*Creseis acicula*
	C. virgula conica

Table 3.5–4
SPECIES CHARACTERISTIC OF ARCTIC OR BOREAL WATER
FOUND AT LOWER LATITUDES IN THE ATLANTIC

Calanus hyperboreus	*Dimophyes arctica*	*Spiratella helicina*
Metridia longa	*Sagitta maxima*	*Sergestes arcticus*
Pareuchaeta norvegica	*Eukrohnia hamata*	
Pareuchaeta barbata		

Table 3.5—5
SPECIES OF THE LUSITANIAN STREAM IN THE EASTERN NORTH ATLANTIC

Siphonophora
Rosacea plicata
R. cymbiformis
Nectopyramis diomedeae
N. thetis
Bassia bassensis
Vogtia (all species)
Hippopodius hippopus
Muggiaea spp.
Eudoxoides spiralis
Chuniphyes multidentata
Lensia – all species except *L. conoidea*
Stephanomia bijuga
Velella velella

Medusae
Rhopalonema velatum
Nausithoe punctata
Pelagia noctiluca

Chaetognatha
Sagitta lyra
S. serratodentata atlantica
S. bipunctata
Krohnitta subtilis

Polychaeta
Travisiopsis lobifera
Lagisca hubrechti
Sagitella kowalewskii

Mollusca
Euclio polita
Janthina britannica

Crustacea
Nematoscelis megalops
Stylocheiron spp.
Vibilia spp.
Phronima spp.
Sapphirina spp.
Phyllosoma larvae

Thaliacea
Cyclosalpa spp.
Ritteriella spp.
Thalia democratica
Thetys vagina
Iasis zonaria
Ihlea asymmetrica
Salpa maxima
Doliolina mulleri
Doliolum nationalis

Table 3.5—6
OCEANIC SPECIES FOUND IN THE MIXED WATER OF THE SCOTTISH SHELF IN THE EASTERN NORTH ATLANTIC

Frequent	Intermediate	Rare
Coelenterata		
Dimophyes arctica	*Chelophyes appendiculata*	*Chuniphyes multidentata*
Lensia conoidea	*Lensia fowleri*	*Vogtia* spp.
Sulculeolaria biloba	*Hippopodius hippopus*	*Nectopyramis* spp.
Physophora hydrostatica	*Arachnactes* larvae	*Rosacea* spp.
Agalma elegans	*Staurophora mertonsii*	*Nausithoe* spp.
Laodicea undulata	*Pelagia noctiluca*	*Pantachogon haeckeli*
Cosmetira pilosella		*Halicreas* spp.
Phialidium hemisphericum		*Periphylla periphylla*
		Rhopalonema velatum
Chaetognatha		
Sagitta serratodentata	*Sagitta maxima*	*Sagitta serratodentata*
f. *tasmanica*	*Sagitta lyra*	f. *atlantia*
	Eukrohnia hamata	*Sagitta hexaptera*
		Sagitta zetesios
Polychaeta and Nemertea		
Tomopteris septentrionalis	*Nectonemertes mirabilis*	*Lagisca hubrechti*
		Vanadis formosa

Table 3.5—6 (*Continued*)
**OCEANIC SPECIES FOUND IN THE MIXED WATER
OF THE SCOTTISH SHELF IN THE
EASTERN NORTH ATLANTIC**

Frequent	Intermediate	Rare
	Mollusca	
Clione limacina	*Clio pyramidata*	*Diacria trispinosa*
	Clio cuspidata	*Spiratella helcoides*
	Pneumodermopsis ciliata	*Spiratella helicina*
	Tracheloteuthis riseii	
	Taonidium pfefferi	
	Copepoda	
Rhincalanus nasutus	*Euchaeta hebes*	Other species of
Eucalanus elongatus	*Gaetanus pileatus*	*Euchaeta*
Pareuchaeta norvegica	*Gaidius tenuispinus*	*Pareuchaeta* and
Pleuromamma robusta	*Euchirella curticaudata*	*Euchirella*
Euchirella rostrata	*Metridia longa*	Most other bathypelagic species
Calanus hyperboreus	*Phaenna spinifera*	
Oithona spinirostris		
	Other Crustacea	
Thysanoessa longicaudata	*Euphausia krohni*	*Nematoscelis megalops*
Sergestes arcticus	*Stylocheiron longicorne*	*Thysanopoda acutifrons*
	Brachyscelus crusculum	*Nematobrachion boopis*
	Munnopsis murrayi	*Stylocheiron elongatum*
		Vibilia spp.
		Phronima sedentaria
		Scina spp.
		Ammalopeneus elegans
	Fish Larvae	
Maurolicus mulleri	*Bathylagus* spp.	*Nansenia groenlandica*
Myctophum glaciale	*Fierasfer* spp.	*Stomias boa*
Gadus poutassou	*Paralepis coregonoides*	*Argyropelecus hemigymnus*

Figure 3.5–1

ASSOCIATIONS OF MAJOR SPECIES AND SPECIES GROUPS OF EUPHAUSIIDS
WITH WATER MASSES OF THE PACIFIC

Water mass designations are given in A; corresponding distributions of species are shown in B to S.

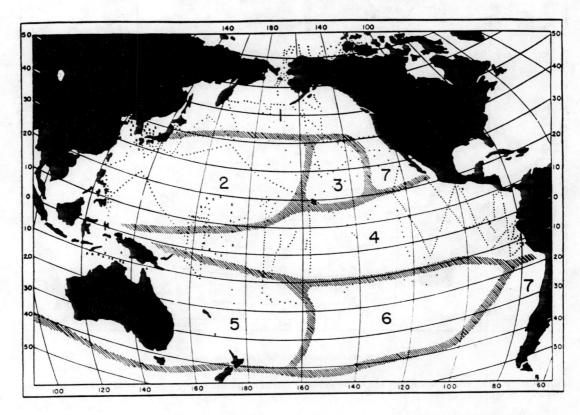

A. Water masses of the Pacific (100–500 m) according to Sverdrup, Johnson, and Fleming (1942). (1) Pacific Subarctic Water, (2) Western North Pacific Central Water, (3) Eastern North Pacific Central Water, (4) Pacific Equatorial Water, (5) Western South Pacific Central Water, (6) Eastern South Pacific Central Water, (7) Transition Water.

(From Bieri, R., The distribution of the planktonic Chaetognatha in the Pacific and their relationship to the water masses, *Limnol. Oceanogr.*, 4(1), 4, 1959.)

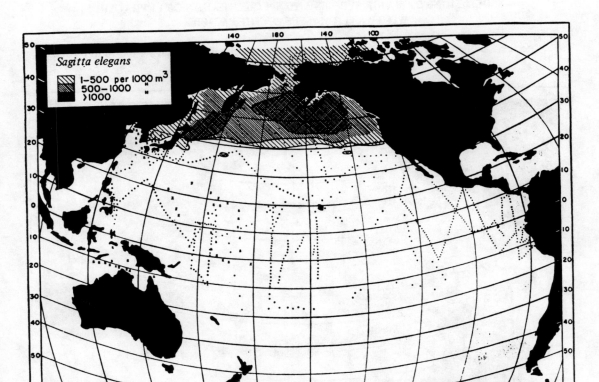

B. Distribution of *Sagitta elegans* in the Pacific and regions to the north.

(From Bieri, R., The distribution of the planktonic Chaetognatha in the Pacific and their relationship to the water masses, *Limnol. Oceanogr.*, 4(1), 4, 1959.)

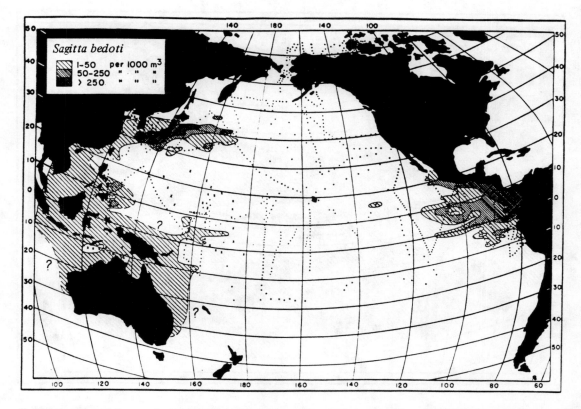

C. Known extent of the two *Sagitta bedoti* populations in the Pacific. The distribution of this species is reminiscent of the equatorial-west-central species such as *Sagitta robusta*, but it is apparently unable to cross the oceanic equatorial region.

(From Bieri, R., The distribution of the planktonic Chaetognatha in the Pacific and their relationship to the water masses, *Limnol. Oceanogr.*, 4(1), 11, 1959.)

Figure 3.5—1 (*Continued*)

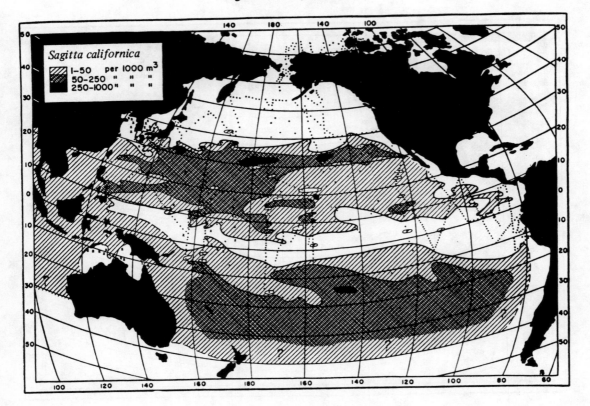

D. Distribution of *Sagitta californica* in the Pacific.

(From Bieri, R., The distribution of the planktonic Chaetognatha in the Pacific and their relationship to the water masses, *Limnol. Oceanogr.*, 4(1), 16, 1959.)

Figure 3.5–1 (*Continued*)

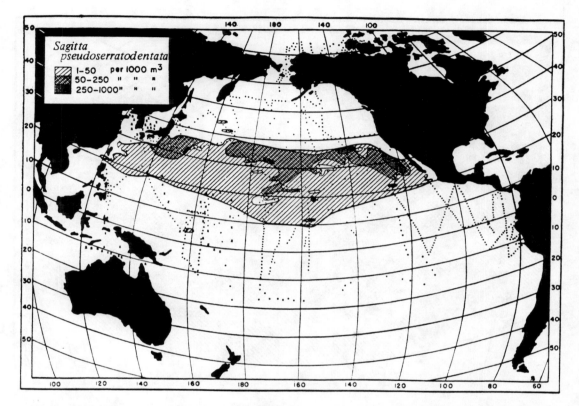

E. Distribution of *Sagitta pseudoserratodentata* in the Pacific. The limits of this species approximate fairly closely the limits of the Pacific Central Water.

(From Bieri, R., The distribution of the planktonic Chaetognatha in the Pacific and their relationship to the water masses, *Limnol. Oceanogr.*, 4(1), 10, 1959.)

Figure 3.5—1 *(Continued)*

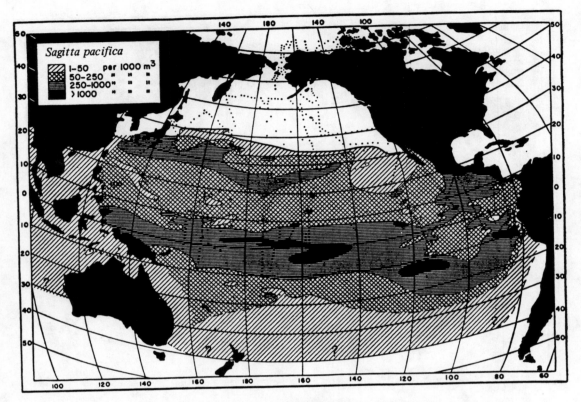

F. Occurrence of *Sagitta pacifica* (*serratodentata* group) in the Pacific.

(From Bieri, R., The distribution of the planktonic Chaetognatha in the Pacific and their relationship to the water masses, *Limnol. Oceanogr.*, 4(1), 11, 1959.)

Figure 3.5–1 (*Continued*)

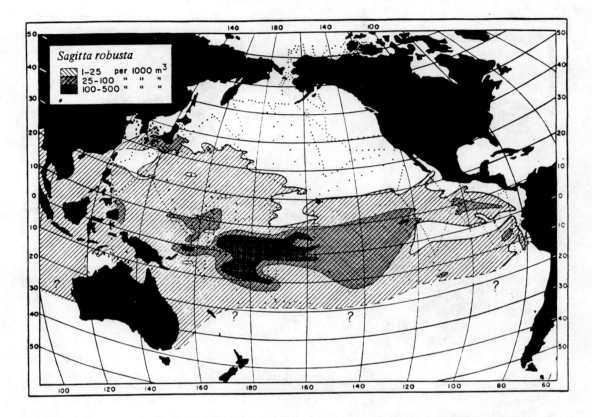

G. Occurrence of *Sagitta robusta* in the Pacific. This species is absent from the Eastern North Pacific Central Water.

(From Bieri, R., The distribution of the planktonic Chaetognatha in the Pacific and their relationship to the water masses, *Limnol. Oceanogr.*, 4(1), 13, 1959.)

Figure 3.5—1 *(Continued)*

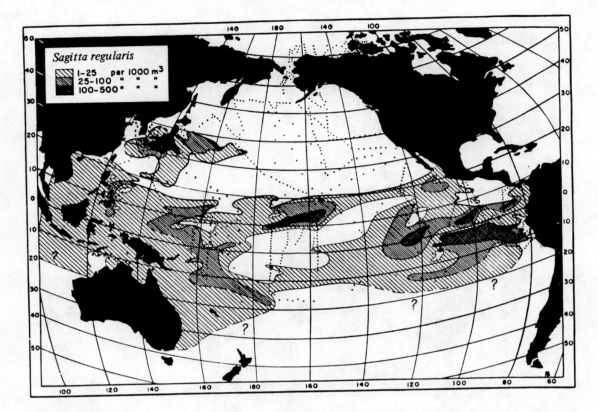

H. Distribution of *Sagitta regularis* in the Pacific. The distribution of this species is similar to that of *Sagitta robusta*.

(From Bieri, R., The distribution of the planktonic Chaetognatha in the Pacific and their relationship to the water masses, *Limnol. Oceanogr.*, 4(1), 13, 1959.)

Figure 3.5 −1 (*Continued*)

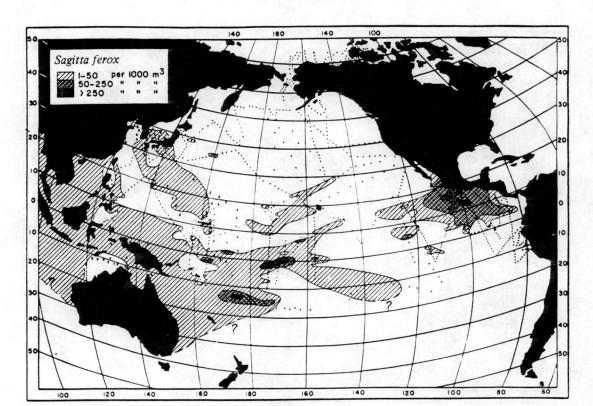

I. Distribution of *Sagitta ferox* in the Pacific. This species is similar in its distribution to *Krohnitta pacifica*.

(From Bieri, R., The distribution of the planktonic Chaetognatha in the Pacific and their relationship to the water masses, *Limnol. Oceanogr.*, 4(1), 15, 1959.)

Figure 3.5 −1 *(Continued)*

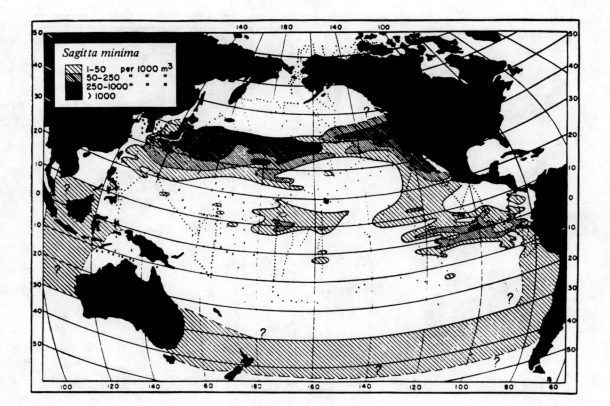

J. Distribution of *Sagitta minima* in the Pacific. This species is most common in the regions of mixing of water masses.

(From Bieri, R., The distribution of the planktonic Chaetognatha in the Pacific and their relationship to the water masses, *Limnol. Oceanogr.*, 4(1), 19, 1959.)

Figure 3.5–1 *(Continued)*

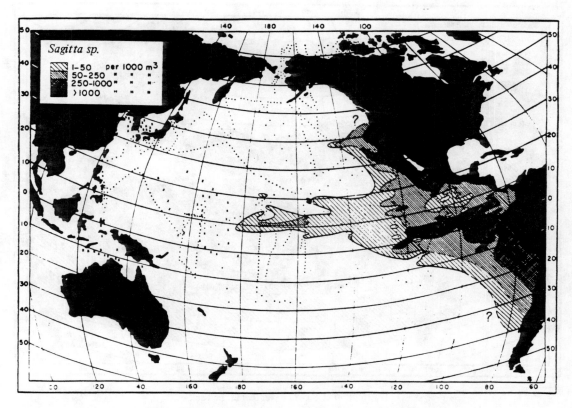

K. Distribution of *Sagitta* sp. (*serratodentata* group) in the Pacific.

(From Bieri, R., The distribution of the planktonic Chaetognatha in the Pacific and their relationship to the water masses, *Limnol. Oceanogr.,* 4(1), 10, 1959.)

Figure 3.5–1 (*Continued*)

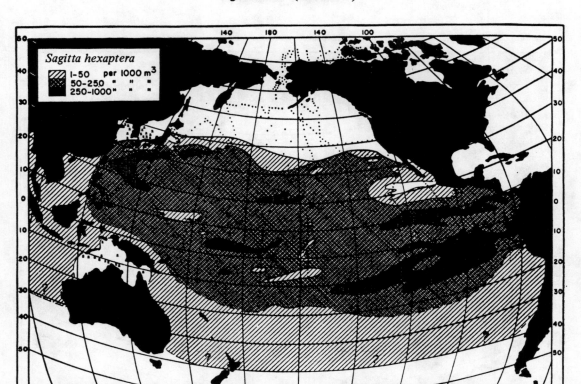

L. Distribution of *Sagitta hexaptera* in the Pacific.

(From Bieri, R., The distribution of the planktonic Chaetognatha in the Pacific and their relationship to the water masses, *Limnol. Oceanogr.*, 4(1), 19, 1959.)

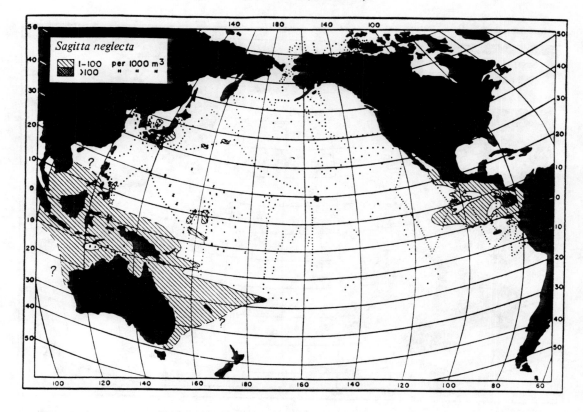

M. Known extent of *Sagitta neglecta* in the Pacific. It has essentially the same distribution as *Sagitta bedoti*, but is not so abundant as that species.

(From Bieri, R., The distribution of the planktonic Chaetognatha in the Pacific and their relationship to the water masses, *Limnol. Oceanogr.*, 4(1), 17, 1959.)

Figure 3.5 −1 *(Continued)*

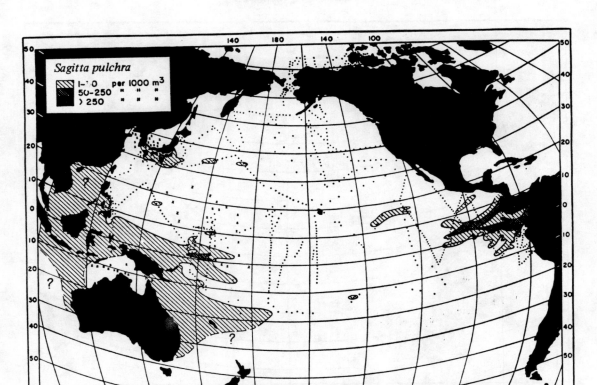

N. Known extent of *Sagitta pulchra* in the Pacific. It is similar in its distribution to *Sagitta bedoti* and *S. neglecta*.

(From Bieri, R., The distribution of the planktonic Chaetognatha in the Pacific and their relationship to the water masses, *Limnol. Oceanogr.*, 4(1), 17, 1959.)

Figure 3.5–1 *(Continued)*

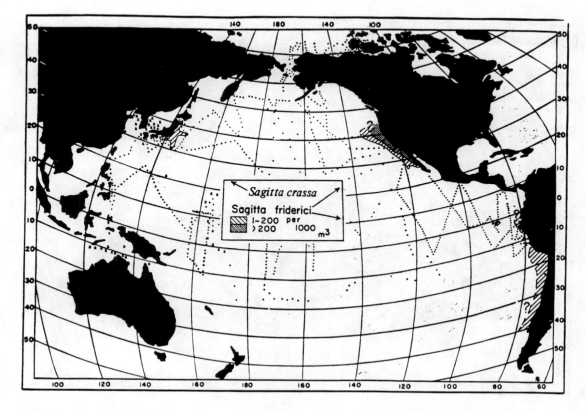

O. Known extent of the two *Sagitta friderici* populations in the Pacific. This species dominates the California and Peru Currents. *Sagitta crassa* off Japan is a closely related species.

(From Bieri, R., The distribution of the planktonic Chaetognatha in the Pacific and their relationship to the water masses, *Limnol. Oceanogr.*, 4(1), 8, 1959.)

Figure 3.5–1 *(Continued)*

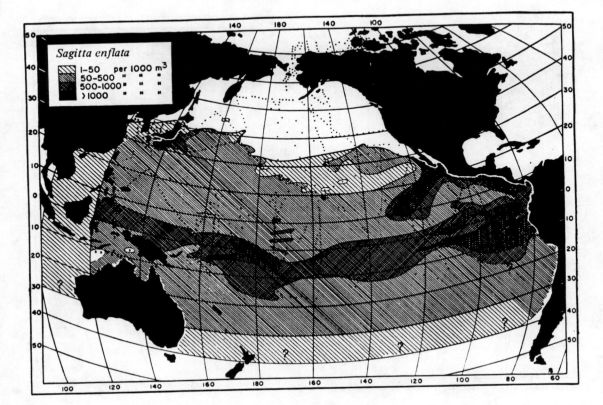

P. Occurrence of *Sagitta enflata* in the Pacific.

(From Bieri, R., The distribution of the planktonic Chaetognatha in the Pacific and their relationship to the water masses, *Limnol. Oceanogr.*, 4(1),16, 1959.)

Figure 3.5–1 (*Continued*)

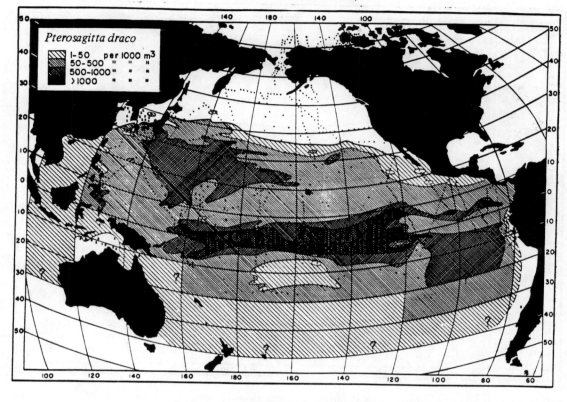

Q. Occurrence of *Pterosagitta draco* in the Pacific.

(From Bieri, R., The distribution of the planktonic Chaetognatha in the Pacific and their relationship to the water masses, *Limnol. Oceanogr.,* 4(1), 8, 1959.)

Figure 3.5—1 (*Continued*)

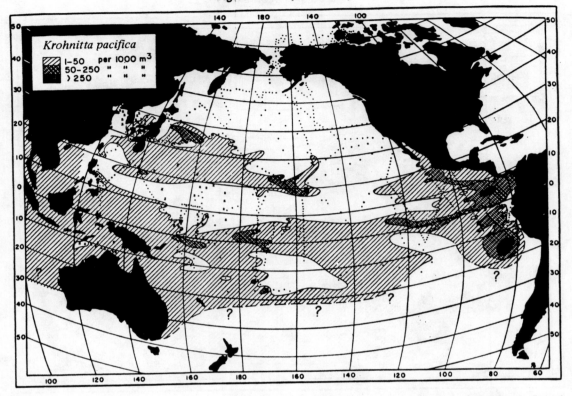

R. Occurrence of *Krohnitta pacifica* in the Pacific. Note that it extends slightly into the Eastern North Pacific Central Water northeast of Hawaii.

(From Bieri, R., The distribution of the planktonic Chaetognatha in the Pacific and their relationship to the water masses, *Limnol. Oceanogr.*, 4(1), 15, 1959.)

Figure 3.5–1 *(Continued)*

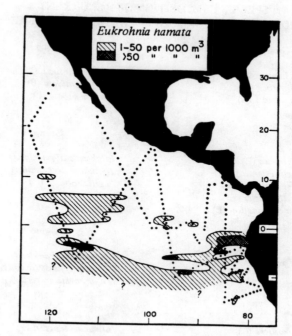

S. Concentration of *Eukrohnia hamata* in the upper 300 meters of water in the eastern equatorial Pacific.

(From Bieri, R., The distribution of the planktonic chaetognatha in the Pacific and their relationship to the water masses, *Limnol. Oceanogr.*, 4(1), 12, 1959. With permission.)

Table 3.5−7
SPECIES COMPOSITION OF RECURRENT
ZOOPLANKTON GROUPS IN THE NORTH PACIFIC*

Group I

Euphausia hemigibba (E)
Euphausia mutica (E)
Euphausia recurva (E)

Euphausia tenera (E)
Nematoscelis atlantica (E)
Nematoscelis microps (E)
Nematoscelis tenella (E)
Stylocheiron carinatum (E)

Stylocheiron submii (E)
Pterosagitta draco (C)
Sagitta enflata (C)

Sagitta hexaptera (C)
Sagitta pacifica (C)

Creseis virgula (P)
Limacina bulimoides (P)

Limacina inflata (P)
Associated:
  Euphausia brevis (E)
  Hyalocylix striata (P)
  Limacina trochiformis (P)

Group II

Euphausia pacifica (E)
Thysanoessa longipes (E)
Tessarabrachion oculatus (E)
Sagitta elegans (C)
Limacina helicina (P)
Associated:
  Thysanoessa inermis (E)
  Sagitta scrippsae (C)

Group III

Cavolinia inflexa (P)
Clio pyramidata (P)
Styliola subula (P)

Group IV

Euphausia diomediae (E)
Nematoscelis gracilis (E)
Sagitta robusta (C)

Group V

Euphausia gibboides (E)
Nematoscelis difficilis (E)
Thysanoessa gregaria (E)

Group VI

Limacina lesueuri (P)
Sagitta pseudoserratodentata (C)

Group VII

Atlanta lesueuri (H)
Atlanta turriculata (H)

No affinities

Euphausia paragibba (E)
Sagitta ferox (C)
Cavolinia longirostris (P)
Carinaria japonica (H)

Oxygyrus keraudreni (H)
Protatlanta souleveti (H)

Pterosoma planum (H)
Pterotrachea hippocamous (H)
Pterotrachea minuta (H)

*Major taxon identifications are: (C) chaetognath; (E) euphausiid; (H) heteropod; and (P) pteropod.

(From Fager, E. W. and McGowan, J. A., Zooplankton species groups in the north Pacific, *Science*, 140(3536), 455, © 1963, American Association for the Advancement of Science. With permission.)

Figure 3.5—2

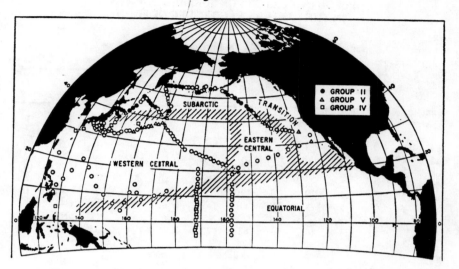

Figure 3.5—2   Distribution of zooplankton Groups II, IV, and V from Table 3.5—6 and their relationships to major water masses. Open circles are sampling stations where no members of the groups were found.

(From Fager, E. W. and McGowan, J. A., Zooplankton species groups in the north Pacific, *Science*, 140(3536), 456, © 1963, American Association for the Advancement of Science. With permission.)

Figure 3.5—3

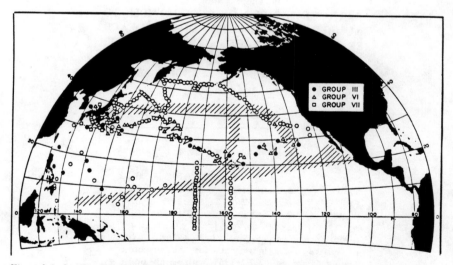

Figure 3.5—3   Distribution of zooplankton Groups III, VI, and VII from Table 3.5—6 and their relationships to major water masses. Open circles are sampling stations where no members of the groups were found.

(From Fager, E. W. and McGowan, J. A., Zooplankton species groups in the north Pacific, *Science*, 140(3536), 457, © 1963, American Association for the Advancement of Science. With permission.)

Table 3.5—8
COPEPOD INDICATORS OF SURFACE CURRENTS
IN THE EASTERN PACIFIC OFF OREGON

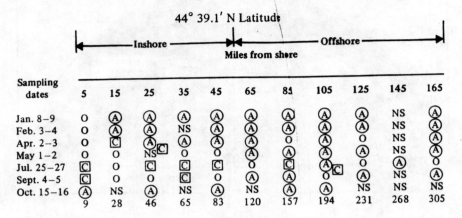

44° 39.1′ N Latitude

Inshore ← Miles from shore → Offshore

Sampling dates	5	15	25	35	45	65	85	105	125	145	165
Jan. 8—9	O	Ⓐ	Ⓐ	Ⓐ	Ⓐ	Ⓐ	Ⓐ	Ⓐ	Ⓐ	NS	Ⓐ
Feb. 3—4	O	Ⓐ	Ⓐ	NS	Ⓐ	Ⓐ	Ⓐ	Ⓐ	O	NS	Ⓐ
Apr. 2—3	O	Ⓒ	Ⓐ	Ⓐ Ⓒ	O	Ⓐ	O	Ⓐ	Ⓐ	NS	Ⓐ
May 1—2	O	O	NS Ⓒ	Ⓐ	Ⓐ	Ⓐ	Ⓐ	Ⓐ	O	NS	Ⓐ
Jul. 25—27	Ⓒ	O	Ⓒ	Ⓒ	Ⓒ	O	Ⓒ	Ⓐ Ⓒ	O	Ⓐ	O
Sept. 4—5	Ⓒ	O	O	Ⓒ	O	Ⓐ	Ⓐ	O	Ⓐ	NS	Ⓐ
Oct. 15—16	Ⓐ	NS	Ⓐ	NS	Ⓐ	NS	Ⓐ	Ⓐ	NS	NS	NS
	9	28	46	65	83	120	157	194	231	268	305

Kilometers from shore

Ⓐ — *Acartia danae.*
Ⓒ — *Centropages memurrichia*
Ⓐ Ⓒ — Both species present.
O — Neither species present.
NS — No sample taken.

(From Cross, F. A. and Small, L. F., Copepod indicators of surface water movements off the Oregon coast, *Limnol. Oceanogr.,* 12, 62, 1967. With permission.)

Table 3.5—9
RATIO OF NUMBER OF OCCURRENCES TO NUMBER OF SAMPLES DURING EACH SEASON FOR THE COPEPODS AND SAMPLING TRANSECT OF TABLE 3.5—7 TO SHOW INFLUENCE OF SEASONS DURING 1962

	Inshore		Offshore	
	A. danae	*C. memurrichi*	*A. danae*	*C. memurrichi*
Spring	3:9	2:9	8:10	0:10
Summer	0:5	4:5	2:6	2:6
Fall	3:8	2:8	6:7	0:7
Winter	7:9	0:9	10:10	0:10

(From Cross, F. A. and Small, L. F., Copepod indicators of surface water movements off the Oregon coast, *Limnol. Oceanogr.,* 12, 62, 1967. With permission.)

Figure 3.5—4

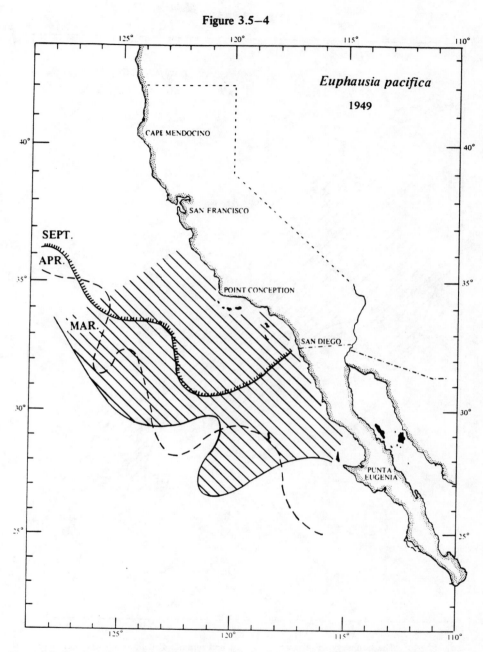

Figure 3.5—4.   The boundary and the winter, spring, and summer distributions of *Euphausia pacifica* in the California Current, 1949.

(From Brinton, E., The distribution of Pacific euphausiids, *Bull. Scripps Inst. Oceanogr. Univ. Calif.*, 8, 227, 1962. With permission.)

Figure 3.5—5

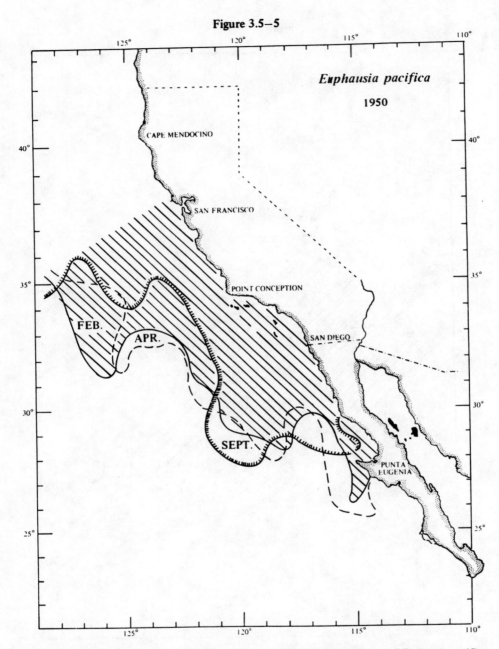

Figure 3.5—5.   The boundary and the winter, spring, and summer distributions of *Euphausia pacifica* in the California Current, 1950.

(From Brinton, E., The distribution of Pacific euphausiids, *Bull. Scripps Inst. Oceanogr. Univ. Calif.*, 8, 228, 1962. With permission.)

Figure 3.5—6

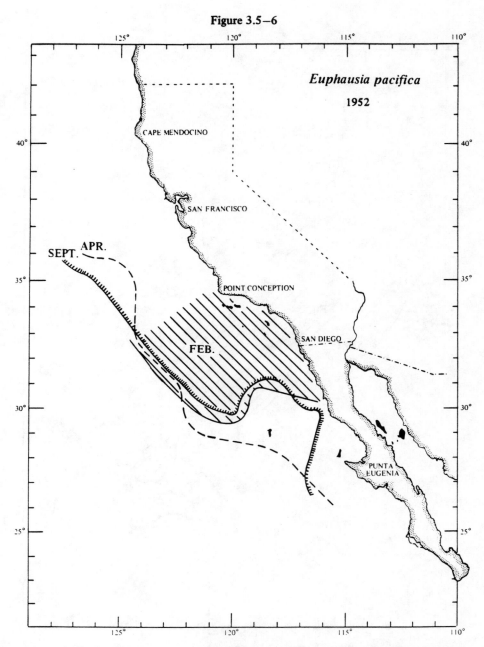

Figure 3.5—6. The boundary and the winter, spring, and summer distributions of *Euphausia pacifica* in the California Current, 1952.

(From Brinton, E., The distribution of Pacific euphausiids, *Bull. Scripps Inst. Oceanogr. Univ. Calif.*, 8, 229, 1962. With permission.)

Figure 3.5–7

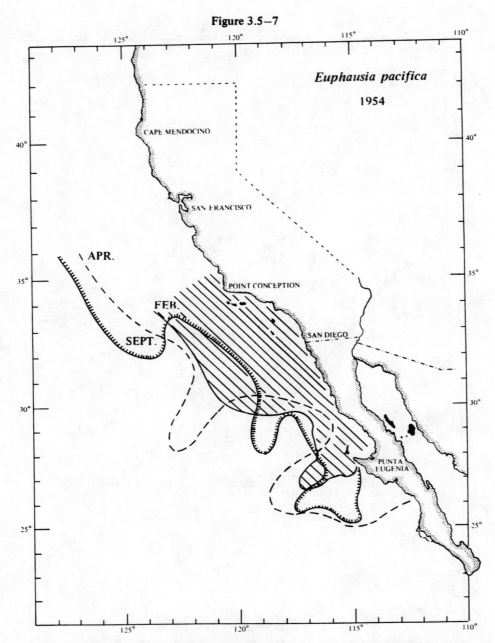

Figure 3.5 –7. The boundary and the winter, spring, and summer distributions of *Euphausia pacifica* in the California Current, 1954.

(From Brinton, E., The distribution of Pacific euphausiids, *Bull. Scripps Inst. Oceanogr. Univ. Calif.*, 8, 230, 1962. With permission.)

Figure 3.5–8

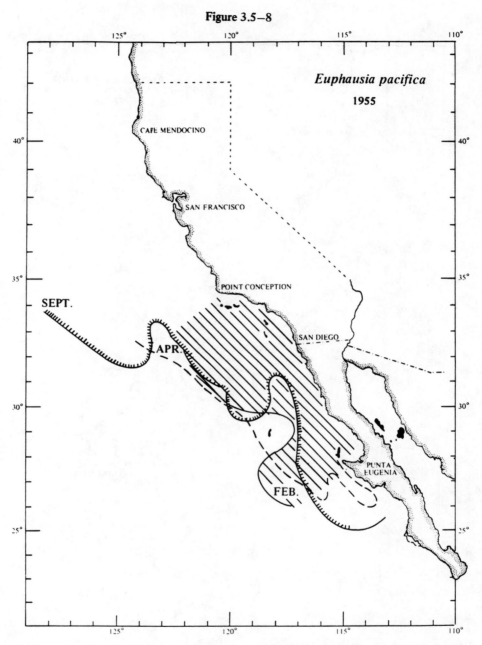

Figure 3.5–8. The boundary and the winter, spring, and summer distributions of *Euphausia pacifica* in the California Current, 1955.

(From Brinton, E., The distribution of Pacific euphausiids, *Bull. Scripps Inst. Oceanogr. Univ. Calif.*, 8, 231, 1962. With permission.)

Figure 3.5—9

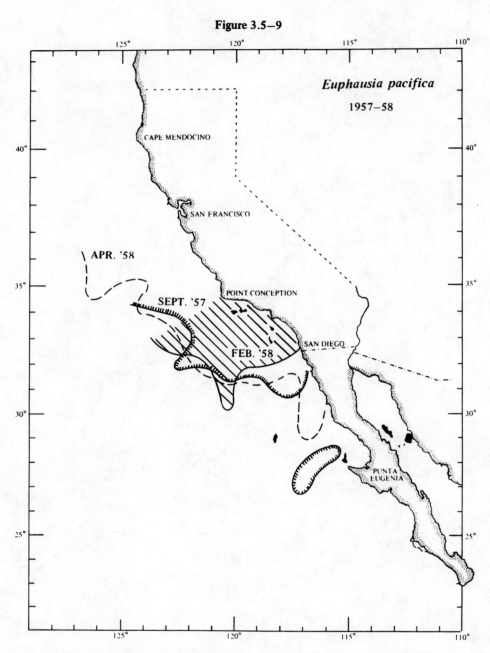

Figure 3.5–9. The boundary and the winter, spring, and summer distributions of *Euphausia pacifica* in the California Current, 1957–1958.

(From Brinton, E., The distribution of Pacific euphausiids, *Bull. Scripps Inst. Oceanogr. Univ. Calif.*, 8, 232, 1962. With permission.)

Figure 3.5–10

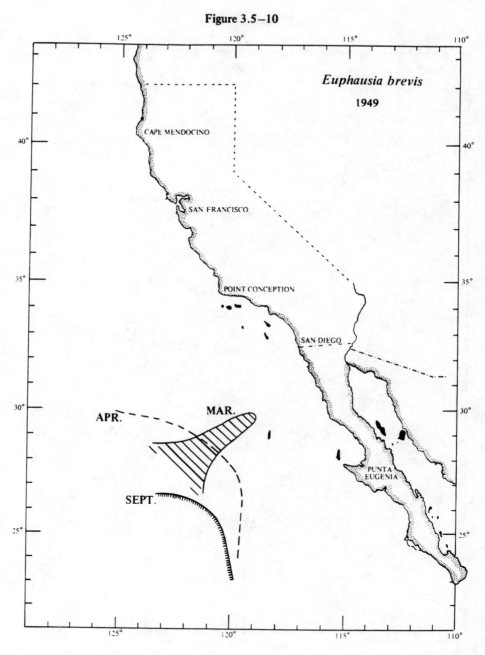

Figure 3.5–10. The boundary and the winter, spring, and summer distributions of *Euphausia brevis* in the California Current, 1949.

(From Brinton, E., The distribution of Pacific euphausiids, *Bull. Scripps Inst. Oceanogr. Univ. Calif.*, 8, 233, 1962. With permission.)

Figure 3.5–11

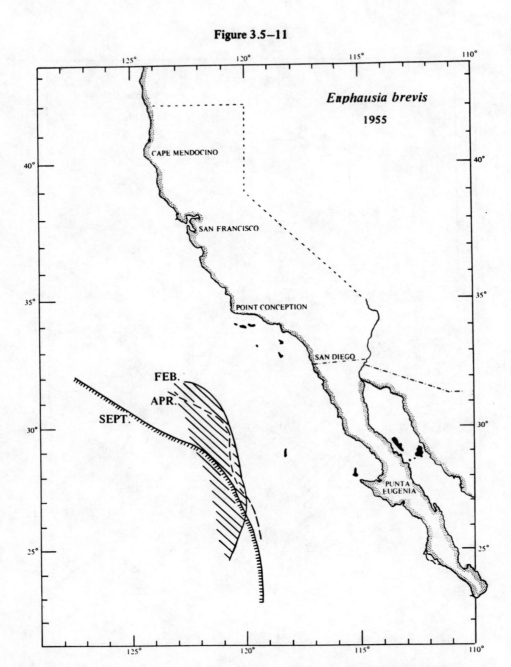

Figure 3.5–11. The boundary and the winter, spring, and summer distributions of *Euphausia brevis* in the California Current, 1955.

(From Brinton, E., The distribution of Pacific euphausiids, *Bull. Scripps Inst. Oceanogr. Univ. Calif.*, 8, 234, 1962. With permission.)

Figure 3.5–12

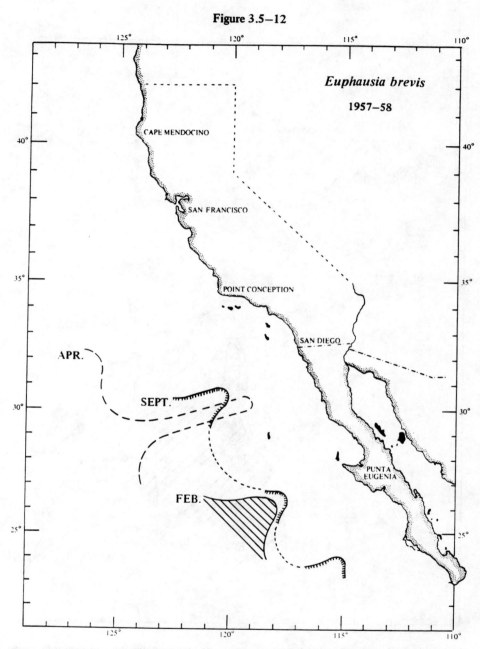

Figure 3.5–12. The boundary and the winter, spring, and summer distributions of *Euphausia brevis* in the California Current, 1957–1958.

(From Brinton, E., The distribution of Pacific euphausiids, *Bull. Scripps Inst. Oceanogr. Univ. Calif.*, 8, 235, 1962. With permission.)

Figure 3.5–13

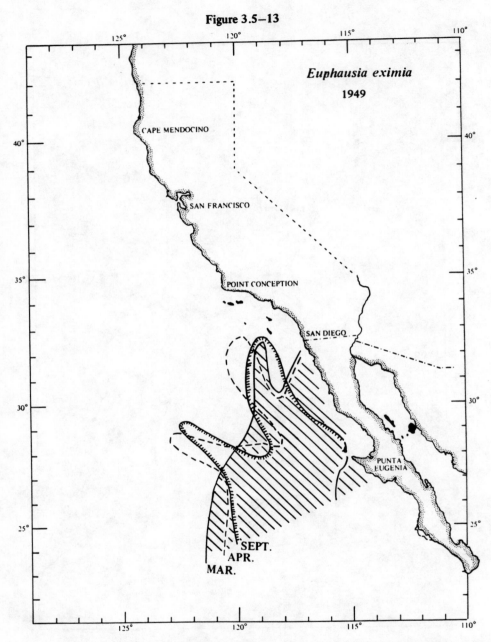

Figure 3.5–13. The boundary and the winter, spring, and summer distributions of *Euphausia eximia* in the California Current, 1949.

(From Brinton, E., The distribution of Pacific euphausiids, *Bull. Scripps Inst. Oceanogr. Univ. Calif.*, 8, 236, 1962. With permission.)

Figure 3.5–14

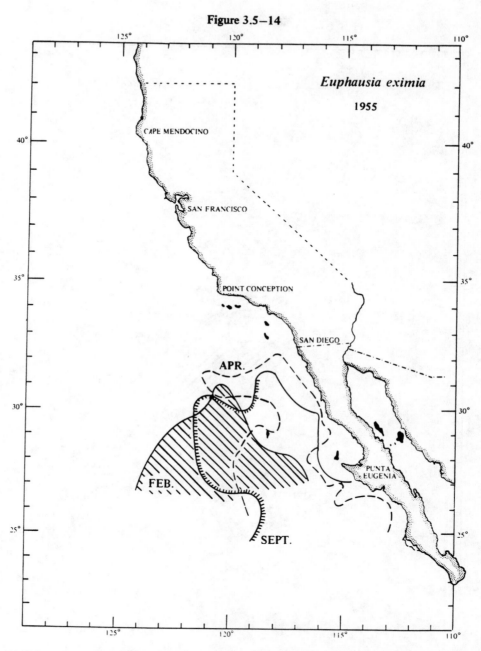

Figure 3.5–14. The boundary and the winter, spring, and summer distributions of *Euphausia eximia* in the California Current, 1955.

(From Brinton, E., The distribution of Pacific euphausiids, *Bull. Scripps Inst. Oceanogr. Univ. Calif.*, 8, 237, 1962. With permission.)

Figure 3.5–15

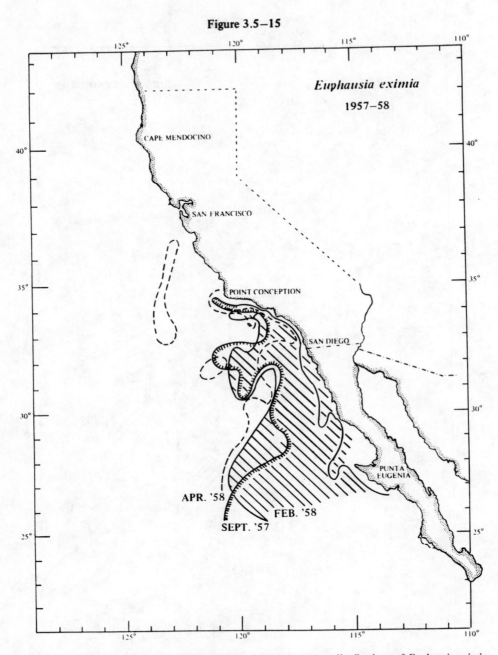

Figure 3.5–15. The boundary and the winter, spring, and summer distributions of *Euphausia eximia* in the California Current, 1957–1958.

(From Brinton, E., The distribution of Pacific euphausiids, *Bull. Scripps Inst. Oceanogr. Univ. Calif.,* 8, 238, 1962. With permission.)

Figure 3.5−16

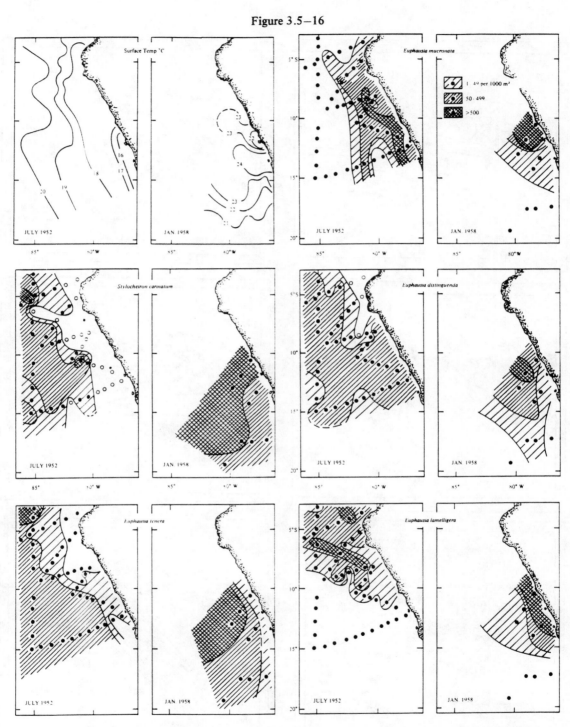

Figure 3.5−16.   Distributions of surface temperature and euphausiid species in the northern part of the Peru Current, 1952−1958.

(From Brinton. E., The distribution of Pacific euphausiids, *Bull. Scripps Inst. Oceanogr. Univ. Calif.*, 8, 239, 1962. With permission.)

Figure 3.5—17

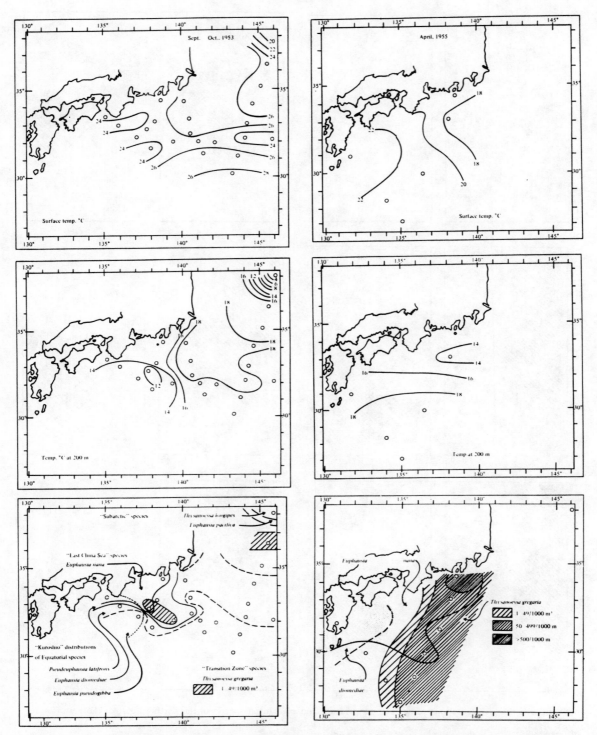

Figure 3.5—17.   Distributions of temperature and euphausiid species in the southeastern waters of Japan, 1953—1955.

(From Brinton, E., The distribution of Pacific euphausiids, *Bull. Scripps Inst. Oceanogr. Univ. Calif.,* 8, 218, 1962. With permission.)

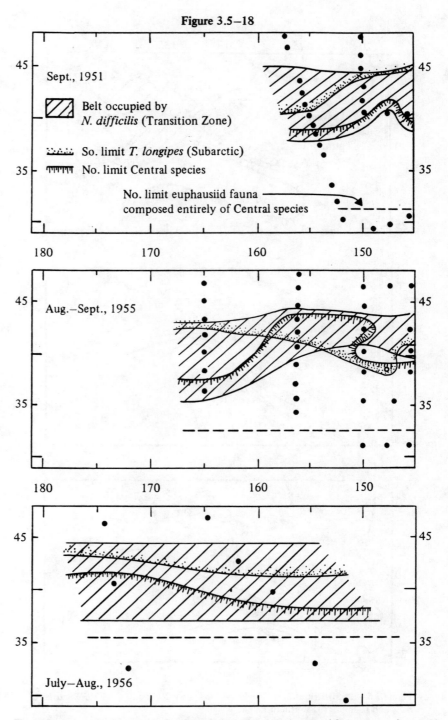

Figure 3.5—18

Sept., 1951

⬛ Belt occupied by *N. difficilis* (Transition Zone)

⬝⬝⬝⬝ So. limit *T. longipes* (Subarctic)

⊤⊤⊤⊤ No. limit Central species

No. limit euphausiid fauna ─────
composed entirely of Central species

Aug.–Sept., 1955

July–Aug., 1956

Figure 3.5—18.   Positions of the boundaries of subarctic, transition-zone, and central euphausiids in the mid-North Pacific, 1951–1956.

(From Brinton, E., The distribution of Pacific euphausiids, *Bull. Scripps Inst. Oceanogr. Univ. Calif.*, 8, 221, 1962. With permission.)

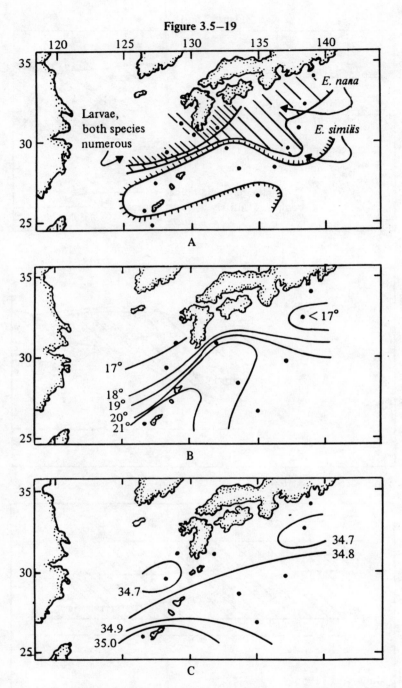

Figure 3.5–19

Figure 3.5–19. Distribution *a*. The East China Sea – Kuroshio species *Euphausia nana* and *E. similis* off southern Japan.

(From Brinton, E., The distribution of Pacific euphausiids, *Bull. Scripps Inst. Oceanogr. Univ. Calif.*, 8, 220, 1962. With permission.)

Figure 3.5—20

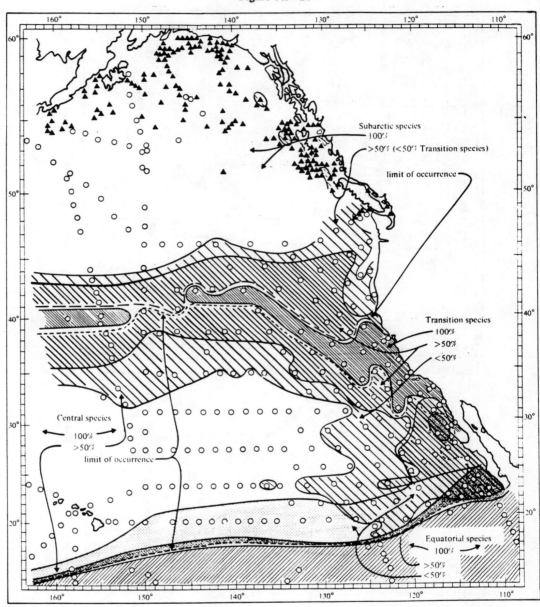

Figure 3.5—20. Associations of euphausiid species in the northeastern Pacific. The amount of a faunal group at a station is indicated as the percentage of the total euphausiid count (all species) made up by the summed counts of the species in that fauna.

(From Brinton, E., The distribution of Pacific euphausiids, *Bull. Scripps Inst. Oceanogr. Univ. Calif.* 8, 224, 1962. With permission.)

Figure 3.5—21

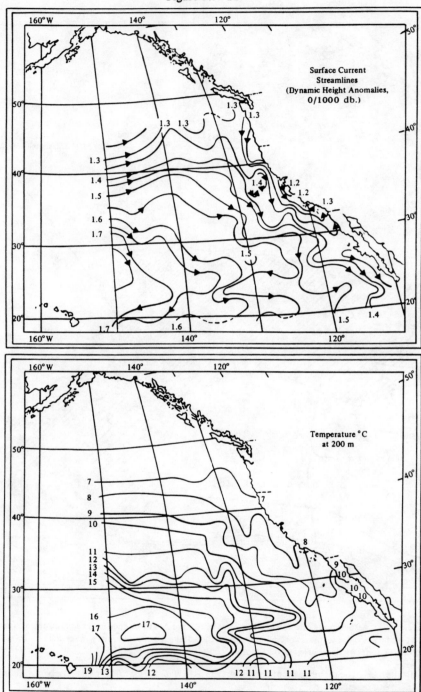

Figure 3.5—21. Surface currents and temperatures at a depth of 200 m in the northeastern Pacific.

(From Brinton, E., The distribution of Pacific euphausiids, *Bull. Scripps Inst. Oceanogr. Univ. Calif.*, 8, 225, 1962. With permission.)

3.6 Length, Height, Weight, and Sample Volume Relationships in Major Groups

Table 3.6–1
RELATIONSHIPS OF LENGTH TO WEIGHT IN MAJOR ZOOPLANKTON GROUPS OF THE CALIFORNIA CURRENT REGION[a]

Ocular[b] μm units[b]	No/g	Obcular[b] μm units[b]	No/g
Amphipods		**Cladocerans**	
10	1023	3	11,731
20	376	6	3921
30	159		
40	101	**Copepods**	
50	64		
60	48	5	5814
70	29	10	1873
80	26	15	698
90	14	20	254
100	10	25	123
110	7	30	70
		35	50
Chaetognaths		40	31
20	6994	**Crustacean larvae**	
30	3808	(a) Anomuran Zoea	
40	2168		
50	1522	3	752
60	762	5	254
70	614	10	162
80	425	15	146
90	330	20	128
100	253	25	111
110	200	30	100
120	142		
130	110	(b) Brachyuran Zoea	
140	75		
150	61	4	3403
160	44	7	2762
170	40	10	949
180	34	13	500
190	31	15	200
200	27	25	65
210	23		
220	18	(c) Megalopa	
230	16		
240	14	15	500
250	13	20	235
260	11	30	105
270	9	40	68
280	6	50	39
290	5	60	23

[a] This table is based upon one gram samples of each size category of a functional group sorted from a variety of plankton collections characterized by pronounced diversity in species and in developmental stages.

[b] One ocular μm unit = 0.167 mm at 7X magnification.

Table 3.6–1 (*Continued*)
RELATIONSHIPS OF LENGTH TO WEIGHT IN MAJOR
ZOOPLANKTON GROUPS OF THE CALIFORNIA CURRENT REGION[a]

Ocular[b] μm units[b]	No/g	Ocular[b] μm units[b]	No/g
Ctenophores		**Euphausiids**	
Usually weighed			
		5	11,111
Decapods		10	6243
(a) Galatheids		15	4200
		20	1426
30	286	25	930
40	125	30	778
50	36	35	451
60	28	40	308
70	18	50	175
80	15	55	125
90	10	60	99
		65	89
(b) Hoplocarids		70	60
		80	45
8	1368	90	31
20	775	100	21
30	253	110	17
40	217	120	13
50	157	130	10
60	83	140	8
70	71	150	7
80	50	160	6
90	40	170	5
		210	4
(c) Natantians		240	3
		270	2
5	7143	300	1
15	1744	330	1
25	765	360	1
35	539	390	1
45	379		
55	147	**Heteropods**	
65	54		
75	42	5	7000
85	29	10	2147
95	20	15	941
105	18	20	525
115	15	30	375
125	13	40	222
135	11	50	185
145	8	60	158
155	7	70	105
165	6	80	83
175	5	90	61
185	4	100	48
195	3		
205	2		
215	1.5	(Larger forms	
250	1	usually weighed)	

[a] This table is based upon one gram samples of each size category of a functional group sorted from a variety of plankton collections characterized by pronounced diversity in species and in developmental stages.

[b] One ocular μm unit = 0.167 mm at 7X magnification.

Table 3.6–1 (*Continued*)
RELATIONSHIPS OF LENGTH TO WEIGHT IN MAJOR
ZOOPLANKTON GROUPS OF THE CALIFORNIA CURRENT REGIONS[a]

Ocular[b] μm units[b]	No/g	Ocular[b] μm units[b]	No/g
Larvaceans		**Ostracods (cont.)**	
10–15	13,200	15	501
15–20	7173	20	341
20–25	5200	25	96
25–30	3848	30	87
30–35	2564		
35–40	2197	**Pteropods**	
40–50	1923		
50–60	405	5	3520
60–70	267	10	1444
		15	910
Medusae		20	327
Usually weighed		25	185
		30	140
Mysids		35	122
		40	100
15	5400	45	62
25	3420	50	42
35	2500	55	33
45	1075		
55	625	**Radiolarians**	
65	300		
75	154	4	5106
85	133	5	3998
95	48	6	2431
110	15		
		Thaliaceans	
Ostracods		Usually weighed	
5	3465	**Siphonophores**	
10	1496	Usually weighed	

[a] This table is based upon one gram samples of each size category of a functional group sorted from a variety of plankton collections characterized by pronounced diversity in species and in developmental stages.
[b] One ocular μm unit = 0.167 mm at 7X magnification.

Table 3.6—2
THE AMOUNT OF INTERSTITIAL LIQUID IN
DRAINED "WET" PLANKTON SAMPLES

Total volume of drained "wet" plankton (ml)	Volume of interstitial liquid (ml)	Volume of organisms only (ml)	% interstitial liquid
35	12.0	23.0	34
20	5.5	14.5	28
10	4.0	6.0	40
32	11.0	21.0	34
32	5.5	26.5	17
11	5.0	6.0	45
23	7.5	15.5	33
12	5.0	7.0	42
24	8.0	16.0	33
12	4.5	7.5	38
10	4.0	6.0	40
27	9.0	18.0	33

Table 3.6—3
PERCENTAGE DECREASE IN VOLUME OF PRESERVED PLANKTON SAMPLES FROM
ORIGINAL LIVE VOLUME, GIVEN FOR SELECTED TIME INTERVALS

Original volume (ml)	Final volume (ml)	% decrease from original live plankton volume					
		Imm. after pres.	1 day after pres.	10 days after pres.	1 month after pres.	1 year after pres.	2 years after pres.
13	11	92.3	92.3	84.6	84.6	84.6	84.6
12	10	83.3	83.3	83.3	83.3	83.3	83.3
(44)	35	(87)	(87)	(85)	(80)	(80)	(80)
50	32	84.0	80.0	70.0	68.0	64.0	64.0
48	24	70.8	52.1	52.1	52.1	50.0	50.0
25	12	76.0	60.0	52.0	52.0	48.0	48.0
84	32	75.0	45.2	41.6	39.3	38.1	38.1
57	20	82.4	52.6	42.1	42.1	35.1	35.1
77	27	62.4	49.4	40.3	39.0	36.4	35.1
42	10	61.9	38.1	33.3	30.1	23.8	23.8
119	23	47.3	31.8	20.2	20.2	17.8	17.8
93	12	64.4	39.7	22.6	18.3	14.0	12.9

3.7 Sampling Equipment and Comparative Efficiencies

Table 3.7–1

SOME CONTEMPORARY PLANKTON SAMPLING NETS AND THEIR CHARACTERISTICS

	Mouth area, m^3	Form	Mesh width, mm	Porosity	Open area ratio	Reference
Low-speed Nets (< 3 Knots)						
Coarse gauze (> 0.4 mm)						
Bongo net	0.38	Cone	0.51	0.51	6.8	McGowan and Brown, 1966
FAO-Larval tuna	0.79	Cyl-cone[1]	0.51	0.51	4.8	Matsumoto, 1966
WP-3 (Interim)	1.00	Cyl-cone	1.00	0.58	3.7	Fraser, 1966
CalCOFI Standard	0.79	Cyl-cone	0.55	0.36	3.2	Smith et al.
Tropical Juday–large	1.00	Red-cone[2]	0.45	0.40	3.1	Bogorov, 1959
Medium gauze (0.2–0.4 mm)						
CalCOFI Anchovy Egg	0.20	Cyl-cone	0.33	0.46	7.8	Smith et al.
Australian Clarke-Bumpus	0.012	Cone	0.27	0.44	5.3	Tranter, 1965
Indian Ocean Standard	1.00	Cyl-cone	0.33	0.46	4.3	Currie, 1963
NORPAC net	0.16	Cone	0.35	0.46	3.7	Motoda and Osawa, 1964
ICITA		Cone	0.28	0.42	3.1	Jossi, 1966
Marunaka	0.28	Cone	0.33	0.45	2.4	Nakai, 1962
Hensen Egg	0.42	Red-cone	0.30	0.44	2.1	Künné, 1933
Marutoku A	0.16	Cone	0.33	0.45	1.7	Nakai, 1962
Fine gauze (< 0.2 mm)						
WP-2	0.25	Cyl-cone	0.20	0.45	6.0	Fraser, 1966
Tropical Juday-Reg	0.50	Red-cone	0.17	0.32	4.2	Bogorov, 1959
Kitahara	0.05	Red-cone	0.11	0.32	4.2	Nakai, 1962
Flowmeter	0.38	Cyl-cone	0.17	0.32	3.2	Currie and Foxton, 1957
Bé net MPS	0.25	Pyramid	0.20	0.45	2.7	Bé, 1962a
Marutoku A	0.16	Cone	0.11	0.32	1.2	Nakai, 1962

[1] Cylinder-cone.
[2] Reduction cone-cone.

Table 3.7–1 (Continued)

SOME CONTEMPORARY PLANKTON SAMPLING NETS AND THEIR CHARACTERISTICS

	Mouth area, m³	Form	Mesh width, mm	Porosity	Open area ratio	Reference
Mixed gauzes						
International Standard	0.20	Cone	0.23 0.08 10.00	0.36 0.20	2.6	Ostenfeld and Jespersen, 1924
N70	0.37	Cone	0.37 0.17	0.34 0.32	2.4	Foxton, 1956
High-speed Nets (> 3 Knots)						
Coarse gauze (> 0.4 mm)						
Miller high speed I	0.0081	Red-cone[2]	0.95	0.57	28.0	Miller, 1961
Jet net	0.0110	Red-cone-Enc[3]	0.44	0.48	27.0	Clarke, 1964
Miller high speed II	0.0081	Red-cone	0.53	0.52	26.0	Miller, 1961
Medium gauze (0.2–0.4 mm)						
Miller high speed II	0.0081	Red-cone	0.26	0.44	22.0	Miller, 1961
Gulf III modified	0.0320	Red-cone-Enc	0.38	0.44	13.1	Bridger, 1958
Hardy recorder	0.00016	Red-disc-Enc	0.22	0.37	11.8	Glover, 1953
Isaacs high speed	0.0005	Red-cyl-Enc	0.24	0.30	11.2	Ahlstrom et al., 1958
Catcher	0.0410	Red-cone-Enc	0.46	0.55	11.0	Bary et al., 1958
Hardy sampler	0.00029	Red-cone-Enc	0.22	0.37	8.1	Glover, 1953
Gulf III	0.1300	Red-cone-Enc	0.38	0.44	3.2	Gehringer, 1952
Small Hardy indicator	0.00013	Red-disc-Enc	0.22	0.37	2.3	Glover, 1953
Standard Hardy indicator	0.0011	Red-disc-Enc	0.22	0.37	1.4	Glover, 1953
Fine gauze (< 0.2 mm)						
Jashnov high speed	0.25	Pyramid	0.17	0.32	1.9	Jashnov, 1961

[3] Encased gauze section.

Table 3.7–2
INITIAL FILTRATION PERFORMANCE OF SOME CONTEMPORARY SAMPLERS

Towing velocity: 100 cm/sec

Plankton sampler	Filtration efficiency (F)	Mesh velocity (v') (cm/sec)	Approach velocity (v) (cm/sec)	Dynamic pressure at v cm/sec ($\frac{1}{2} q v^2$) (g/cm²)	Filtration pressure (K, $\frac{1}{2} q v^2$) (g/cm²)	Towing velocity which will give an approach velocity of: (cm/sec)			
						v = 5	v = 10	v = 15	v = 20
Bongo	0.96	14	7	0.027	0.097	69	138	207	276
FAO larval tuna	0.97	20	10	0.054	0.176	49	97	146	194
WP-3 interim	0.98	26	15	0.120	0.239	33	65	98	130
CalCOFI standard	0.91	29	10	0.054	0.383	48	97	146	194
CalCOFI anchovy egg	0.95	12	6	0.016	0.089	90	179	269	358
Australian Clarke-Bumpus	0.88	17	7	0.027	0.158	52	104	156	208
Indian Ocean standard	0.96	22	10	0.053	0.244	49	98	147	196
NORPAC	0.96	26	12	0.072	0.309	42	84	126	168
ICITA	0.95	31	13	0.084	0.456	39	78	117	156
Marunaka	0.88	37	16	0.138	0.553	30	61	91	121
Marutoku A (medium)	0.80	47	21	0.228	0.817	24	47	71	94
WP-2	0.94	16	7	0.025	0.161	71	142	213	284
Flowmeter	0.90	28	9	0.041	0.476	56	112	167	223
Bé multiple	0.88	33	15	0.110	0.534	34	68	102	136
Marutoku A (fine)	0.59	49	16	0.125	1.237	32	64	96	128
Jashnov	0.78	41	13	0.089	0.871	38	76	113	151

Table 3.7—3
BOLTING CLOTH NUMBERS AND
THEIR RESPECTIVE MESH APERTURES

No. of cloth	Mesh opening in microns	Mesh opening in in.	Open area %	No. of cloth	Mesh opening in microns	Mesh opening in in.	Open area %
1800	1800	.0709	61%	280	280	.0110	45%
1558	1558	.0614	60%	275	275	.0108	45%
1340	1340	.0528	59%	263	263	.0104	45%
1179	1179	.0464	59%	253	253	.0100	44%
1050	1050	.0413	59%	243	243	.0096	45%
947	947	.0373	58%	233	233	.0092	44%
860	860	.0339	57%	223	223	.0088	43%
782	782	.0308	56%	210	210	.0083	43%
760	760	.0299	56%	202	202	.0080	47%
706	706	.0278	53%	183	183	.0072	46%
656	656	.0258	53%	165	165	.0065	45%
630	630	.0248	52%	153	153	.0060	45%
602	602	.0237	51%	130	130	.0051	43%
571	571	.0225	50%	116	116	.0046	43%
526	526	.0207	50%	110	110	.0043	42%
505	505	.0199	50%	102	102	.0040	39%
471	471	.0185	50%	93	93	.0037	38%
452	452	.0178	48%	86	86	.0034	36%
423	423	.0167	47%	80	80	.0031	35%
405	405	.0159	47%	73	73	.0029	34%
390	390	.0154	47%	64	64	.0025	30%
363	363	.0143	47%	53	53	.0021	21%
351	351	.0138	46%	44	44	.0017	17%
333	333	.0131	46%	35	35	.0014	16%
316	316	.0124	46%	28	28	.0011	18%
308	308	.0121	45%	25	25	.0010	17%
295	295	.0116	45%	20	20	.0008	16%

Table 3.7—4
AVERAGE APERTURE SIZE OF STANDARD GRADE DUFOUR BOLTING SILK

Silk no.	Meshes/ in.	Size of aperture (mm)	Silk no.	Meshes/ in.	Size of aperture (mm)
0000	18	1.364	10	109	0.158
000	23	1.024	11	116	0.145
00	29	0.752	12	125	0.119
0	38	0.569	13	129	0.112
1	48	0.417	14	139	0.099
2	54	0.366	15	150	0.094
3	58	0.333	16	157	0.086
4	62	0.318	17	163	0.081
5	66	0.282	18	166	0.079
6	74	0.239	19	169	0.077
7	82	0.224	20	173	0.076
8	86	0.203	21	178	0.069
9	97	0.168	25	200	0.064

(From Sverdrup, H. V., Johnson, M. W., and Fleming, R. H., Observations and collections at sea, *The Oceans, Their Physics, Chemistry, and General Biology,* © 1942, renewed 1970, 378. By permission of Prentice-Hall, Inc., Englewood Cliffs, New Jersey.)

Table 3.7—5
SOME CHARACTERISTICS OF 200 μ BOLTING MATERIAL USED IN PLANKTON NETS

Gauze	Mesh openings of sample (μ)	Sifting surface (%)	Diameter of filaments (μ) Warp	Woof	Manu-facturer*	Whether tested
Nytal DIN 1171 30–200 μ	190–220	34.5	140	140	A	No
HD 200 μ	180–200	41.5	110	110	A	No
7xxx 200 μ	200–215	43	110	100	A	Yes
7 200 μ	190–215	47.5	$80 + 2 \times 60$	80	A	Yes
200 μ	200–215	47.5	90	90	A	No
7P 200 μ	190–210	55	70	70	A	Yes
Estal Mono P.E. 200 μ	205–215	44.5	108	108	A	Yes
Monodur 200	200–250	40	127	127	B	Yes
Nitex 202	195–210	45	$88 + 2 \times 64$	88	C	Yes

* Manufacturers: *A*: Schweizer Seidengasenfabrik A.G., 9425 Thal, St. Gallen, Switzerland; *B*: Vereinigte Seiden webereien A.G., 415 Krefeld, Federal Republic of Germany; *C*: Swiss origin, dealer Tobler, Ernst & Traber, 71 Murray Street, New York, N.Y.

Table 3.7—6
THE COMBINED EFFECTS OF MESH SIZE AND POROSITY ON FILTRATION EFFICIENCY

Mesh width (mm)	Porosity	Open area ratio	Reduction in efficiency %
1.17	0.60	5.3	0
0.66	0.54	4.8	0
0.33	0.48	4.2	0
0.27	0.44	3.9	3
0.14	0.36	3.2	4
0.08	0.31	2.7	8
0.06	0.26	2.3	12

Table 3.7—7
DECREASE IN MESH WIDTH OF SILK ON IMMERSION IN WATER

Material		Mean mesh width, mm	Standard error
New silk	Dry	0.312	± 0.0034
	Wet	0.261	± 0.0029
Used silk	Dry	0.236	± 0.0040
	Wet	0.221	± 0.0073

Table 3.7—8
CHEMICAL RESISTANCE OF SOME MATERIALS USED IN PLANKTON NET GAUZE

Chemical	Nylon	Perlon	Polyester	Silk
Concentrated acids	Medium	Medium	Good	–
Dilute acids	Good	Good	Good	Good
Alkalis	Good	Good	Medium	Poor
Alcohols	Good	Good	–	–
Oxidizing agents	Medium	Medium	Good	–
Bleaching agents	Medium	Medium	Good	–
Sunlight	Medium	Medium	Good	Medium
Organic solvents	Good	Good	Good	Good
Soap	Good	Good	Good	Good
Petrol	Good	Good	–	–
Formic acid	Poor	Poor	–	–
Phenol	Poor	Poor	–	–

Section 4

Phytoplankton Populations

4.1 Biomass as Related to Area, Depth and Season

Figure 4.1—1   PHYTOPLANKTON PRODUCTION

See special foldout section at back of the book, p. 384.

(From FAO Department of Fisheries, *Atlas of the Living Resources of the Seas,* Food and Agricultural Organization of the United Nations, Rome, 1972.)

Figure 4.1–1

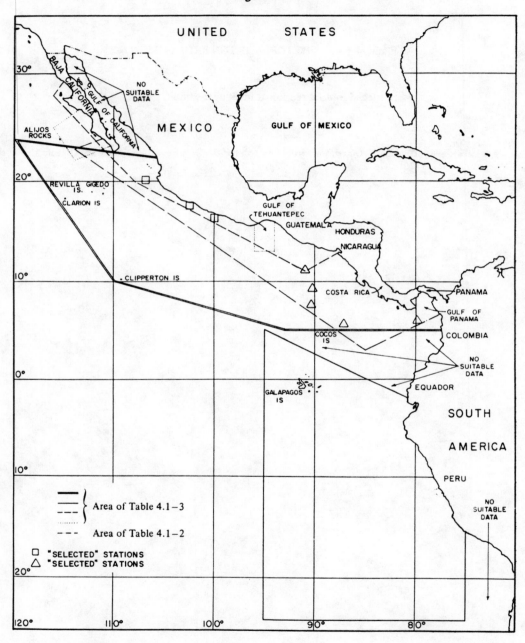

Figure 4.1–1. Areas of groups of station-pairs and stations listed in Tables 4.1–6 and 4.1–7, and positions of selected stations from some groups for which data are given.

(From Blackburn, M., Relationships between standing crops at three successive trophic levels in the eastern tropical Pacific, *Pac. Sci.*, 20, 41, 1966.)

Table 4.1–1

VERTICAL AND SEASONAL DISTRIBUTIONS OF CHLOROPHYLL *a* AND PRIMARY PRODUCTIVITY AT SELECTED STATIONS OFF THE OREGON COAST

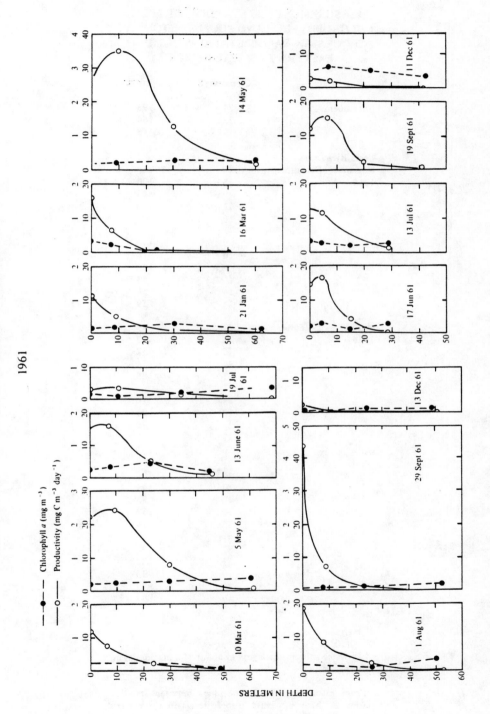

(From Anderson, G. C., The seasonal and geographic distribution of primary productivity off the Washington and Oregon coasts, *Limnol. Oceanogr.*, 9, 294, 1964. With permission.)

Table 4.1—2
MEASUREMENTS OF STANDING CROPS OF CHLOROPHYLL a[1] AND ZOOPLANKTON[2] AT OCEANOGRAPHIC STATIONS IN THE EASTERN TROPICAL PACIFIC[3]

Date	Chlorophyll	Zooplankton
November, 1955	24	193
	22	164
	25	143
Costa Rica Dome,	20	84
November—December,	22	250
1959	8.8	120
	9.6	43
	12	100
	10	190
	9.4	140
	20	70
	8.7	120
	11	54
	14	56
	12	59
November—December,	24	49
1956	25	37
	43	32
	44	54
	51	85
	74	314
	60	125
	24	95
	27	250
	36	135
	45	166
	70	104
	32	95
	33	139
	32	114
	27	96
	59	233
	30	33
	33	77
	49	47

[1] Mg/m^2.

[2] Ml/10^3 m^3.

[3] In water columns or layers to about 100 m and 300 m, respectively.

(From Blackburn, M., Relationships between standing crops at three successive trophic levels in the eastern tropical Pacific, *Pac. Sci.*, 20, 40, 1966. With permission.)

Table 4.1–3

MEASUREMENTS OF STANDING CROPS OF CHLOROPHYLL a[1],
ZOOPLANKTON[2], CARNIVOROUS MICRONEKTON[2] AND COPEPODS[2] AT OCEANOGRAPHIC STATIONS
IN THE EASTERN TROPICAL PACIFIC[3]

Noon chlorophyll	Noon zooplankton	Night zooplankton	G.M. zooplankton	Night micronekton	Noon copepods	Night copepods	G.M. copepods
April 23–June 20, 1958							
9.7	23	24	23	2.6	0.36	1.51	0.73
6.4	18	21	19	1.0	0.77	0.96	0.86
14	10	20	14	3.3	0.51	0.89	0.67
10	13	31	20	2.8	0.64	1.50	0.98
16	30	50	39	6.7	2.46	1.42	1.87
13	79	57	67	1.0	1.89	4.52	2.92
12	49	67	57	4.4	1.67	1.94	1.80
15	72	100	85	8.5	3.22	5.45	4.19
12	85	84	84	4.0	7.00	6.75	6.87
14	86	56	69	10.0	1.74	7.30	3.56
22	63	87	74	9.5	3.87	4.25	4.06
46	100	163	131	13.9	7.85	15.06	10.87
39	270	299	284	19.1	30.96	11.90	19.20
36	85	101	93	6.5	3.63	1.82	2.57
81	74	79	76	4.1	4.57	3.97	4.26
41	113	148	129	7.5	5.28	3.05	4.01
29	30	104	91	5.9	2.28	6.16	3.74
45	90	160	120	10.0	2.23	7.05	3.96
39	130	150	140	7.9	6.63	5.49	6.03
50	114	182	144	16.7	3.66	6.88	5.02
120	117	206	155	15.5	4.16	10.30	6.55
49	74	111	91	24.5	0.85	6.12	2.28
27	108	84	95	15.4	3.74	5.81	4.66
26	127	143	135	12.7	5.77	4.26	4.96
11	48	48	48	16.7	2.51	4.72	3.43
23	107	122	114	13.8	6.31	7.38	6.82
37	44	52	48	14.5	2.61	4.68	3.49
April 30–May 27, 1960							
11	37	77	53	1.1	1.08	5.63	2.47
7.4	22	32	27	2.2	1.28	1.32	1.30
6.0	56	50	53	5.5	2.72	3.70	3.17
7.4	38	59	47	5.9	2.50	1.65	2.03
4.8	61	63	62	4.0	1.45	2.82	2.02
6.0	42	41	41	2.8	2.94	0.61	1.34
9.3	68	57	62	6.8	1.42	2.68	1.95
19	53	95	71	1.4	1.01	7.85	2.82
19	53	115	78	5.5	1.01	1.44	1.21

[1] Mg/m².
[2] Ml/10³ m³.
[3] In water columns or layers to about 100 m, 300 m, 90 m, and 300 m, respectively, with exceptions noted. G. M. is geometric mean.

Table 4.1–3 (*Continued*)
MEASUREMENTS OF STANDING CROPS OF CHLOROPHYLL *a*,[1]
ZOOPLANKTON,[2] CARNIVOROUS MICRONEKTON[2] AND COPEPODS[2] AT OCEANOGRAPHIC STATIONS IN THE EASTERN TROPICAL PACIFIC[3]

September 15–December 14, 1960

12		150		3.2
18		170		8.0
11		150		10.8
15		120		11.1
8.8		80		6.6
3.0		20		2.3
12		60		4.4
16		80		7.4
34		200		1.4
5.7		50		0.6
3.0		50		1.2
9.2		140		2.0
7.7		40		2.7
2.4		40		1.4
2.8		10		1.6
5.0		30		4.8
4.2		50		6.1
1.8		80		5.7

January 15–February 25, 1959

24	24	47	34	12.9
18	9	18	13	0.0
41	18	10	13	0.0
22	37	88	57	12.1
33	111	95	103	6.6
29	98	43	65	8.1
21	197	396	279	6.6
24	166	239	199	11.9
105	196	175	185	12.1
25	91	71	80	5.5
30	158	77	110	9.7
107	164	130	146	5.9
36	60	121	85	6.2
31	54	33	42	1.4

[1] Mg/m^2.

[2] $Ml/10^3 \ m^3$.

[3] In water columns or layers to about 100 m, 300 m, 90 m, and 300 m, respectively, with exceptions noted. G. M. is geometric mean.

Table 4.1–3 (*Continued*)

MEASUREMENTS OF STANDING CROPS OF CHLOROPHYLL *a*,[1] ZOOPLANKTON,[2] CARNIVOROUS MICRONEKTON[2] AND COPEPODS[2] AT OCEANOGRAPHIC STATIONS IN THE EASTERN TROPICAL PACIFIC[3]

August 13–September 22, 1959

6.6	4.3
7.5	4.6
16	1.4
12	3.2
23	31.4
9.7	8.0
7.6	5.1
6.8	6.1
12	12.7
4.4	0.5
11	8.0
12	9.6
10	8.0

[1] Mg/m^2.

[2] $Ml/10^3 \ m^3$.

[3] In water columns or layers to about 100 m, 300 m, 90 m, and 300 m, respectively, with exceptions noted. G. M. is geometric mean.

(From Blackburn, M., Relationships between standing crops at three successive trophic levels in the eastern tropical Pacific, *Pac. Sci.*, 20, 38, 1966. With permission.)

Table 4.1–4
TOTAL AND PARTIAL CORRELATION COEFFICIENTS AMONG LOGGED STANDING CROP DATA FOR STATION PAIRS IN TABLES 4.1–6 AND 4.1–7

n	r_{cz}	r_{cm}	r_{zm}	$r_{cz.m}$	$r_{cm.z}$	$r_{zm.c}$
36	+0.702[2]	+0.615[2]	+0.626[2]	+0.515[2]	+0.345[1]	+0.317
11	+0.966[2]	+0.860[2]	+0.840[2]	+0.880[2]	+0.348	+0.068
18	+0.718[2]	+0.314	+0.365	+0.682[2]	+0.078	+0.207
14	+0.376	+0.137	+0.781[2]	+0.435	−0.271	+0.794[1]
13	–	+0.592[1]	–	–	–	–
35	+0.124	–	–	–	–	–

n	r_{ch}	r_{cm}	r_{hm}	$r_{ch.m}$	$r_{cm.h}$	$r_{hm.c}$
36	+0.613[2]	+0.615[2]	+0.602[2]	+0.386[1]	+0.390[1]	+0.361[1]
8	+0.929[2]	+0.874[2]	+0.849[2]	+0.728	+0.440	+0.202

Note: The letters c, z, h, and m indicate chlorophyll *a* zooplankton, copepods, and carnivorous micronekton, respectively.

[1,2] Indicate significance at 5% and 1% levels of probability; n is number of station-pairs, or stations.

(From Blackburn, M., Relationships between standing crops at three successive trophic levels in the eastern tropical Pacific, *Pac. Sci.*, 20, 44, 1966. With permission.)

4.2 Taxon Diversity as a Function of Depth and Area

Table 4.2–1
DISTRIBUTION OF CELLULAR CARBON BETWEEN
PHYTOPLANKTON GROUPS IN WATER SAMPLES FROM VARIOUS DEPTHS
OFF THE COAST OF SOUTHERN CALIFORNIA

	Depth, m				
	0	50	100	200	600
Organic C (μg/liter) in:					
Diatoms	2.1	5.5	0.2	0	0
Dinoflagellates	7.1	6.5	1.6	0.2	0
Coccolithophorids	1.7	9.2	0.6	0	0.1
Monads	1.2	3.2	0.9	0.5	0.4
Phytoplankton (total)	12.1	24.4	3.3	0.7	0.5
Biomass (μg C/liter) based on:					
Direct examination	13	25	3.8	1.1	0.5
ATP	24	22	5.6	1.9	0.5
Chlorophyll	14	22	5.0	0.2	–
DNA	250	200	95	50	25
Total particulate organic carbon (μg/liter)	45	30	9.0	9.0	13

Note: Also shown for comparative purposes are the total particulate organic carbon value and the biomass estimates based on ATP, DNA, and chlorophyll. ATP, DNA, and chlorophyll values (in μg) have been converted to organic carbon (in μg) by multiplying by the factors 250, 50, and 100, respectively. These factors are based on laboratory studies of phytoplankton cultures.

Table 4.2–2
OCCURRENCE OF BACTERIA IN VARIOUS ESTUARINE ENVIRONMENTS

(Percentage of Species)

	Water			
Species	Bottom	1 m from bottom	Surface	Sea grass community
Bacillus subtilis	45	39.5	22	10–25
B. megaterium	18	7	5.5	0
B. sphaericus	0	7.5	0	0
Corynebacterium globiforme	0	7.0	0	0
C. flavum	0	0	0	10
C. miltinum	0	0	0	5
Actinomyces spp.	18	0	5.5	10–25
Staphylococcus candidus	0	8	8	0
S. roseus	0	8	0	0
Mycoplana dimorpha	19	23	54	40
M. citrea	0	0	0	5
Sarcina lutea	0	0	5.5	0
Pigmented strains	20	38	27.5	50
Ratio g pos./g neg. strains	1.9	2.0	0.7	0.9

Table 4.2–3
FLORA OF DIATOMS FOUND IN
THE PLANKTON-COLORED LAYER AT
THE BOTTOM OF ARCTIC SEA ICE
OFF BARROW, ALASKA

Amphiprora kryophila
Gomphonema exiguum v. arctica
Navicula algida
N. crucigeroides
N. directa

N. gracilis v. inaequalis
N. kjellmanii
N. obtusa
N. transitans

N. transitans v. derasa
N. transitans v. erosa
N. trigoncephla

N. valida
Nitzschia lavuensis

Pinnularia quadratarea
P. quadratarea v. biconstracta
P. quadratarea v. capitata
P. quadratarea v. constricta
P. quadratarea v. stuxbergii

P. semiinflata
P. semiinflata v. decipiens
Pleurosigma stuxbergii
P. stuxbergii v. rhomboides
Stenoneis inconspiqua

Table 4.2–4
COUNTS OF PHYTOPLANKTON AND PROTISTANS (AS INFUSORIA) FROM THE INDICATED VOLUMES OF SEA WATER TAKEN FROM VERTICAL SERIAL SAMPLES IN SEVERAL NORWEGIAN FJORDS

Drøbak, March 8, 1916

Depth, m	0	5	10	20	30	10	50
Temperature, °C	0.0	1.46	1.99	2.19	4.41	5.24	5.56
Salinity, °/$_{oo}$	30.08	31.25	31.58	32.27	33.31	33.89	34.08
Density, σ_t	24.17	25.03	25.26	25.74	26.43	26.78	26.91
O_2, cc/l	–	7.46	7.59	7.24	6.72	6.61	6.53
O_2, °/$_o$	–	93.0	96.5	94.5	92.0	92.5	92.0
No. cc centrifuged	100	160	100	100	100	100	100

Diatoms

l, sum of all species	2725	920	860	470	100	–	130
Cerataulina Bergonii	–	38	30	–	–	–	–
Chaetoceras constrictum	–	88	40	–	–	–	–
debile	–	–	80	--	–	–	–
decipiens	170	–	–	–	–	–	–
laciniosum	–	181	160	–	–	–	–
Gyrosigma fasciola	–	–	–	–	10	–	–
Lauderia glacialis	960	150	130	110	70	–	–
Navicula sp.	–	6	–	–	10	–	–
Rhizosolenia setigera	5	–	–	–	–	–	–
Skeletonema costatum	50	–	–	–	–	–	130
Thalassiosira decipiens	20	6	70	–	–	–	–
gravida	780	438	350	360	10	–	–
Thalassiothrix nitzschioides	10	–	–	–	–	–	–
Tropidoneis Lepidoptera	–	13	–	–	–	–	–

Flagellata

Eutreptia Lanowii	740	63	30	–	–	20	10

Peridiniales

Ceratium longipes	–	6	–	–	–	–	–
tripos	–	6	–	–	–	–	–
Diplopsalis lenticula	–	–	10	–	–	–	–
Gymnodinium Lohmanni	30	56	–	10	–	10	–
sp.	–	25	–	–	–	–	–
Peridinium achromaticum	–	13	20	–	–	–	–
conicum	–	6	–	–	–	–	–
parallelum	–	6	–	–	–	–	–
pellucidum	–	6	–	–	–	–	–
roseum	–	–	10	–	–	–	–
sp.	–	19	–	–	–	–	10
Pouchetia sp.	–	–	–	–	–	–	–
Torodinium robustum	80	19	20	–	–	–	–

Table 4.2–4 (*Continued*)

COUNTS OF PHYTOPLANKTON AND PROTISTANS (AS INFUSORIA) FROM THE INDICATED VOLUMES OF SEA WATER TAKEN FROM VERTICAL SERIAL SAMPLES IN SEVERAL NORWEGIAN FJORDS

Infusoria

Laboea conica	10	19	–	–	–	–	–
crassula	10	231	90	20	10	–	–
strobila	–	31	–	–	–	–	–
vestita	–	77	–	–	–	–	–
sp.	–	–	10	–	–	–	–
Lohmaniella oviformis	170	231	–	–	–	–	–
sp.	–	–	30	10	–	–	–
Mesodinium, small	–	75	260	–	–	–	–
bigger	20	513	60	20	30	–	10
Ptychocylis urnula	–	6	–	–	–	–	–
Tintinnopsis sp.	–	13	–	10	–	–	–
Infusoria indeterminata	–	6	10	–	–	–	–

Drøbak, March 15, 1916

Depth, m	0	2	5	10	30
Temperature, °C	1.0	5.57	5.69	6.09	6.25
Salinity, °/$_{\circ\circ}$	31.33	34.15	34.21	34.56	34.80
Density, σ_t	25.11	26.95	26.98	27.21	27.39
O_2, cc/l	7.44	5.91	5.97	5.70	5.62
O_2, °/$_\circ$	92.5	83.0	84.5	81.5	81.0
No. cc centrifuged	100	100	100	100	100

Diatoms

All species, sum	4030	1140	480	10	100
Biddulphia aurita	10	–	–	–	–
Chaetoceras decipiens	50	20	–	–	70[1]
diademia	–	260	–	–	–
Coscinodiscus radiatus	–	10	–	–	–
Lauderia glacialis	1150	10	130	10[1]	10
Licmophora sp.	20	10	–	–	–
Navicula sp.	20	20	20	–	10
Nitzschia sp.	–	–	–	–	10
Rhabdonema arcuatum	10	–	–	–	–
Skeletonema costatum	–	240	–	–	–
Thalassiosira decipiens	250	–	10	–	–
gravida	2200	470	300	–	–
Nordenskioldii	320	–	–	–	–
Thalassiothrix nitzschioides	–	–	20	–	–

Flagellata

Eutreptia Lanowii	200	10	–	–	–

[1] Dead cells.

Table 4.2—4 (*Continued*)
COUNTS OF PHYTOPLANKTON AND PROTISTANS (AS INFUSORIA) FROM THE INDICATED VOLUMES OF SEA WATER TAKEN FROM VERTICAL SERIAL SAMPLES IN SEVERAL NORWEGIAN FJORDS

Cilioflagellata

Ceratium furca	10	20	–	–	–
Dinophysis acuta	–	10	–	–	–
Gymnodinium Lohmanni	30	20	10	20	–
sp.	–	10	–	10	–
Peridinium pallidum	–	–	10	–	–
sp.	–	–	10	–	–
Pouchetia sp.	–	10	–	–	–
Torodinium robustum	60	10	10	–	–

Infusoria

Cyttarocylis denticulata	–	–	10	–	–
Laboea crassula	10	10	–	–	–
sp.	–	–	–	–	20
Lohmanniella oviformis	40	90	70	30	–
sp.					
Mesodinium, small	–	10	40	–	–
bigger	50	150	30	10	–
Strombidium spinosum	–	10	70	–	–
Tintinnus acuminatus	–	–	–	–	10
Infusoria indeterminata	140	40	–	–	–

Drøbak, March 15, 1916

Depth, m	0	2	5	10	20	30
Temperature, °C	1.0	1.02	4.68	5.79	6.14	6.11
Salinity, °/oo	30.69	30.91	33.41	34.22	34.71	34.73
Density, σ_t	24.60	24.79	26.48	26.98	27.32	27.35
O_2, cc/l	7.95	7.77	6.34	5.87	5.73	5.90
O_2, °/o	98.0	95.5	87.0	83.0	82.0	84.5
P_H	7.92	7.95	7.96	7.98	7.98	8.02
No. cc centrifuged	25	25	25	100	50	100

Diatoms

All species, sum	26,040	34,480	16,680	6060	60	240
Biddulphia aurita	120	680	–	90	–	–
sinensis	–	–	–	–	20	–
Chaetoceras decipiens	1000	640	–	–	–	–
debile	280	1640	–	–	–	40
teres	120	–	–	–	–	–
Lauderia glacialis	9520	12,840	6720	1960	40	30
Licmophora sp.	10	40	–	20	–	–
Melosira Borreri	–	–	–	380	–	80
Navicula sp.	–	40	80	20	–	–
Nitzschia closterium	–	–	–	–	–	10
sp.	–	40	–	–	–	–
Pleurosigma sp.	–	40	–	–	–	–

Table 4.2—4 (*Continued*)

COUNTS OF PHYTOPLANKTON AND PROTISTANS (AS INFUSORIA) FROM THE INDICATED VOLUMES OF SEA WATER TAKEN FROM VERTICAL SERIAL SAMPLES IN SEVERAL NORWEGIAN FJORDS

Rhizosolenia semispina	–	–	40	–	–	–
Skeletonema costatum	–	–	–	200	–	–
Thalassiosira decipiens	–	–	–	20	–	–
gravida	12,720	15,960	6520	1590	–	80
Nordenskiöldii	2240	2320	3320	1780	–	–
Thalassiothrix nitzschioides	–	240	–	–	–	–

Flagellata

Eutreptia Lanowii	1760	1600	40	10	–	–

Peridiniales

Ceratium furca	–	–	–	20	–	–
fusus	–	–	–	10	–	–
Dinophysis norvegica	20	–	–	10	–	–
Gonyaulax sp.	40	–	40	–	–	–
Gymnodinium Lohmanni	120	–	40	50	20	–
sp.	–	40	–	–	–	–
Peridinium achromaticum	–	40	–	–	–	–
ovatum	–	40	–	10	–	–
Pouchetia sp.	80	40	40	20	–	–
Prorocentrum micans	–	–	–	10	–	–
Torodinium robustum	40	80	–	–	–	–

Infusoria

Laboea crassula	80	240	–	70	–	–
strobila	40	40	–	20	–	–
vestita	80	80	280	–	–	–
Lohmanniella oviformis	80	–	–	–	20	10
Mesodinium	–	240	280	100	20	–
Strombidium reflexum	80	160	40	10	–	10
Infusoria indeterminata	–	–	–	10	–	10

Drøbak, March 24, 1916

Depth, m	0	2	5	10	20	30	40	50
Temperature, °C	1.7	1.43	1.38	1.38	1.62	3.04	5.38	6.04
Salinity, °/oo	30.06	30.10	30.12	30.23	30.45	31.93	34.12	34.71
Density, σ_t	24.06	24.11	24.13	24.21	24.38	25.47	26.95	27.34
O_2, cc/l	8.36	8.31	8.35	8.26	7.98	6.92	6.02	5.88
O_2, °/o	104.0	103.0	103.5	102.5	99.5	90.5	84.5	84.0
No. cc centrifuged	10	20	10	10	25	25	50	50

Diatoms

All species, sum	347,200	371,450	476,700	393,600	272,180	94,820	75,600	200
Asterionella japonica	200	–	–	–	–	–	–	–
Biddulphia aurita	37,600	43,200	41,000	58,700	33,040	12,520	900	–
Cerataulina Bergonii	–	–	–	–	80	–	–	–

Table 4.2–4 *(Continued)*

COUNTS OF PHYTOPLANKTON AND PROTISTANS (AS INFUSORIA) FROM THE INDICATED VOLUMES OF SEA WATER TAKEN FROM VERTICAL SERIAL SAMPLES IN SEVERAL NORWEGIAN FJORDS

Chaetoceras boreale	2200	3400	2200	2200	1480	280	–	–
constrictum	9700	9400	14,000	14,600	3120	1760	–	–
compressum	7000	3800	13,600	22,500	6520	440	–	–
curvisetum	2800	3500	7400	3600	100	600	100	–
debile	45,800	43,400	58,300	41,200	22,920	8920	300	–
decipiens	1000	1600	2700	900	720	680	–	–
diadema	7000	11,500	18,900	11,700	4760	720	–	–
breve	200	200	200	–	–	–	–	–
laciniosum	9800	6800	10,300	9300	3960	1760	–	–
scolopendra	3000	9500	7500	5300	600	320	–	–
simile	700	–	200	–	–	–	–	–
teres	800	600	600	2500	880	–	–	–
Detonula cystifera	2000	400	500	600	–	–		
Lauderia glacialis	6000	6800	8500	9800	8160	6040	1260	80
Leptocylindrus danicus	800	500	1400	500	1200	–	20	–
minimus	600	200	500	1500	720	840	–	–
Navicula sp.	1800	900	1600	2600	680	–	140	40
Nitzschia seriata	–	300	200	200	280	–	–	–
delicatissima	–	–	400	–	–	–	–	–
Rhizosolenia semispina	2700	900	200	1900	1460	620	–	–
setigera	100	300	100	300	160	160	–	–
Thalassiosira decipiens	–	200	–	600	–	–	40	–
gravida	26,900	20,100	19,500	22,700	20,080	11,480	1300	80
Nordenskiöldii	94,100	112,300	135,500	165,100	137,840	39,040	3420	–
sp.	–	400	200	–	–	–	–	–
Thalassiothrix longissima	–	50	100	–	–	–	–	–
nitzschioides	400	400	–	500	80	160	–	–
Coscinodiscus radiatus	–	–	–	–	–	40	–	–
Skeletonema costatum	86,000	90,800	129,300	24,800	12,440	8440	80	–

Flagellata

Eutreptia Lanowii	3400	3700	4100	5100	3760	1040	40	–

Peridiniales

Dinophysis acuminata	–	–	–	100	–	–	40	20
Gymnodinium Lohmanni	200	50	400	500	320	–	–	–
Peridinium achromaticum	–	–	–	–	–	40	–	–
ovatum	–	–	–	–	–	–	20	–
Prorocentrum micans	–	–	–	–	–	40	–	–
Torodinium robustum	100	–	100	–	–	–	–	–

Infusoria

Laboea conica	100	700	600	–	–	–	–	–
crassula	–	200	–	–	–	–	–	–
strobila	–	50	–	100	120	40	–	–
vestita	3000	5400	6500	3500	720	80	20	–
Lohmanniella oviformis	100	200	200	240	40	–	–	–
Mesodinium	900	1900	1800	1100	480	480	20	80
Strombidium sp.	100	200	100	600	40	–	–	–

Table 4.2—4 (*Continued*)

COUNTS OF PHYTOPLANKTON AND PROTISTANS (AS INFUSORIA) FROM THE INDICATED VOLUMES OF SEA WATER TAKEN FROM VERTICAL SERIAL SAMPLES IN SEVERAL NORWEGIAN FJORDS

Drøbak, April, 1916

Depth, m	0	5	10	20	30	50
Temperature, °C	1.4	0.52	0.55	0.55	4.50	5.98
Salinity, °/oo	24.90	25.05	25.12	25.42	33.27	34.85
Density, σ_t	19.95	20.10	20.16	20.40	26.38	27.45
O_2, cc/l	8.84	8.84	8.87	8.80	6.33	6.03
O_2, °/o	105.5	103	103.5	103	86.5	86
No. cc centrifuged	10	10	10	10	25	50

Diatoms

	0	5	10	20	30	50
All species, sum	142,750	342,000	431,750	258,750	46,640	1290
Biddulphia aurita	2900	18,700	32,700	25,200	3340	–
sinensis	–	–	–	–	–	20
Chaetoceras boreale	700	1100	1100	2300	–	–
breve	500	500	200	400	–	–
constrictum	700	3000	2600	1300	320	–
compressum	500	1800	1400	4300	680	–
curvisetum	1300	5100	1600	4300	1160	–
debile	17,400	35,400	29,000	32,400	4280	–
decipiens	1000	600	800	–	–	–
diadema	4000	4900	6100	2600	480	–
laciniosum	1100	2700	3500	2100	320	–
scolopendra	2000	4400	1500	5400	–	–
simile	800	200	800	–	–	–
teres	700	800	1800	1400	80	–
Detonula cystifera	600	700	100	400	–	–
Eucampia groenlandica	–	–	–	100	–	–
Lauderia glacialis	–	100	200	900	7120	340
Leptocylindrus danicus	–	–	–	400	–	–
Navicula sp.	–	900	500	2000	40	20
Nitzschia seriata	800	–	–	200	–	–
Rhizosolinia semispina	1550	3050	3250	3550	280	10
setigera	100	100	700	500	80	–
Skeletonema costatum	99,200	188,800	231,000	94,100	4360	80
Thalassiosira decipiens	–	–	–	–	1600	–
gravida	300	2100	1800	5900	12,840	580
Nordenskiöldii	4800	64,200	108,700	67,700	9520	200
sp.	200	700	600	–	–	–
Thalassiothrix longissima	–	–	100	–	40	–
nitzschioides	1600	2200	800	1300	–	40

Flagellata

	0	5	10	20	30	50
Eutreptia Lanowii	–	1500	500	500	200	40

Table 4.2–4 (*Continued*)

COUNTS OF PHYTOPLANKTON AND PROTISTANS (AS INFUSORIA) FROM THE INDICATED VOLUMES OF SEA WATER TAKEN FROM VERTICAL SERIAL SAMPLES IN SEVERAL NORWEGIAN FJORDS

Peridiniales

Dinophysis norvegica	–	–	100	–	–	–
Diplopsalis lenticula	–	–	100	–	–	–
Glenodinium bipes	–	200	100	–	–	–
Gonyaulax triacantha	–	–	100	–	–	–
Gymnodinium Lohmanni	200	–	100	–	–	40
Peridinium pellucidum	–	200	–	300	–	–
Steinii	–	–	–	100	–	–
sp.	100	200	700	–	–	–
Prorocentrum micans	–	–	–	–	–	40
Protoceratium reticulatum	–	–	100	–	–	–

Infusoria

Laboea conica	100	1200	200	–	–	–
crassula	–	400	300	–	–	–
strobila	–	–	100	200	–	–
vestita	100	6000	3600	500	40	–
Lohmanniella oviformis	–	–	100	–	–	–
Mesodinium	–	600	–	100	–	–
Strombidium sp.	–	–	–	200	–	–

Nesodden, August 28, 1917

Depth about 50 m

Depth, m	1	5	10	20	30	40	50
Temperature, °C	–	–	–	15.13	5.41	–	–
Salinity, °/°°	20.43	20.46	20.49	30.56	32.04	32.22	33.15
Density, σ_t	–	–	–	22.55	25.20	–	–
O_2, cc/l	6.47	6.49	6.37	4.57	4.09	4.01	3.18
O_2, °/°	–	–	–	76	57	–	–
No. cc centrifuged	100(10)	100(10)	100	100	50	50	50

Diatoms

Cerataulina Bergonii	20	60	180	–	–	–	–
Chaetoceras curvisetum	320	260	300	–	–	–	100
Coscinodiscus radiatus	40	140	90	–	–	–	–
Lauderia glacialis	–	–	–	–	20	–	–
Skeletonema costatum	6000	2580	17,380	60	1340	–	–
Rhizosolenia fragilissima	8240	7660	10,460	30	–	–	–

Peridiniales

Ceratium furca	2620	2030	1340	–	–	–	–
fusus	3160	2130	1870	–	–	–	–
macroceros	10	–	–	–	–	–	–
tripos	2450	1960	1050	–	–	–	–
Dinophysis acuminata	300	160	180	10	60	20	–
acuta	380	180	120	–	20	–	–
rotundata	180	20	100	–	–	–	–
Gonyaulax polyedra	17,280	6680	3340	10	–	–	–
spinifera	60	140	60	–	–	–	–

COUNTS OF PHYTOPLANKTON AND PROTISTANS (AS INFUSORIA) FROM THE INDICATED VOLUMES OF SEA WATER TAKEN FROM VERTICAL SERIAL SAMPLES IN SEVERAL NORWEGIAN FJORDS

Peridinium achromaticum	–	20	–	–	–	–	–
conicum	–	–	30	–	–	–	–
depressum	–	20	–	30	–	–	–
divergens	140	240	120	40	60	–	–
pallidum	40	40	40	–	–	–	–
pellucidum	20	–	–	–	–	–	–
pyriforme	–	20	–	–	–	–	–
Steinii	880	420	340	30	–	–	–
Prorocentrum micans	35,100	26,500	24,480	220	280	180	–
Torodinium robustum	–	–	–	70	20	–	–

Infusoria

Amphorella subulata	180	80	300	–	–	–	–
Cyttarocylis denticulata	–	40	–	–	–	–	–
Laboea conica	–	80	20	–	–	–	–
delicatissima	–	–	40	10	–	–	–
emergens	120	240	340	10	–	20	–
strobila	60	40	40	–	–	–	–
vestita	60	380	40	–	–	–	–
Lohmanniella oviformis	100	80	160	10	20	40	–
Mesodinium	35,500	47,500	13,320	320	580	–	–
Tintennopsis sp.	120	80	20	–	–	–	–
Infusoria indeterminata	460	440	140	–	–	–	–
Nauplii of Copepoda	100	210	200	30	–	–	–

Steilene, August 28, 1917

Depth 53 m

Depth, m	1	5	10	20	30	40	50
Temperature, °C	17.59	17.58	17.71	8.19	5.51	5.42	–
Salinity, $^{\circ}/_{\circ\circ}$	20.24	20.23	20.58	29.59	32.08	32.76	33.05
Density, σ_t	14.14	14.14	14.39	23.03	25.32	25.87	–
O_2, cc/l	6.39	6.36	6.30	5.04	4.82	4.31	3.92
O_2, $^{\circ}/_{\circ}$	104	104	103	73	67	60	–
No. cc centrifuged	100	50	50	50	50	50	50

Diatoms

Chaetoceras curvisetum	2920	160	–	–	–	–	–
Coscinodiscus radiatus	180	170	120	40	40	40	20
Rhizosolenia fragilissima	100	–	–	–	–	–	–

Peridiniales

Ceratium bucephalum	20	20	–	–	–	–	–
furca	70	60	–	–	–	–	–
fusus	480	460	20	–	–	–	–
macroceras	30	10	–	–	–	–	–
tripos	800	730	200	–	–	–	–

Table 4.2–4 (*Continued*)

COUNTS OF PHYTOPLANKTON AND PROTISTANS (AS INFUSORIA) FROM THE INDICATED VOLUMES OF SEA WATER TAKEN FROM VERTICAL SERIAL SAMPLES IN SEVERAL NORWEGIAN FJORDS

Dinophysis acuminata	80	100	–	20	–	20	–
acuta	80	180	–	–	–	20	–
norvegica	20	–	–	–	–	–	–
rotundata	40	20	–	–	–	–	–
Gonyaulax polyedra	60	100	20	–	–	–	20
Peridinium conicum	–	40	60	–	–	–	–
depressum	20	–	–	20	–	–	–
divergens	210	200	40	–	–	–	–
Steinii	–	60	20	–	–	20	–
Prorocentrum micans	4540	4300	2180	60	20	20	40
Torodinium robustum	–	–	–	80	–	–	–

Infusoria

Laboea conica	940	720	140	–	–	–	–
delicatissima	280	60	–	–	–	–	–
emergens	260	240	520	–	–	–	–
strobila	80	80	20	–	–	–	–
Lohmanniella oviformis	80	160	–	–	–	20	–
Infusoria indeterminata	500	380	240	–	–	–	–
Mesodinium	32,000	25,580	5020	380	40	40	20
Nauplii of Copepoda	80	80	40	20	–	–	–

Drøbak, August 28–29, 1917

	St. 4, August 29				St. 3, August 28			
Depth, m	1	5	10	20	20	30	40	50
Temperature, °C	17.0	16.91	16.92	14.07	12.99	12.03	–	6.54
Salinity, °/oo	20.16	20.35	20.57	24.38	28.24	28.80	30.12	32.16
Density, σ_t	14.21	14.38	14.54	18.02	21.20	21.82	–	25.27
O_2, cc/l	6.27	6.22	6.18	5.78	5.57	5.60	5.48	6.01
O_2, °/o	101	100	100	91	88	87	–	85
No. cc centrifuged	100	100(10)	50	50	50	50	50	50
Halosphaera viridis	–	20	–	–	–	40	20	–

Diatoms

Chaetoceras curvisetum	3780	520	2340	1600	–	–	120	–
Coscinodiscus radiatus	190	240	160	80	60	–	–	–
Thalassiothrix nitzschioides	–	–	–	–	–	–	–	220
Cerataulina Bergonii	–	–	–	–	–	–	–	60

Peridiniales

Ceratium bucephalum	60	–	10	10	–	20	–	–
furca	140	90	200	30	–	–	–	–
fusus	1350	810	580	200	–	20	–	–
longipes	10	–	–	–	–	20	–	–
macroceros	30	30	70	20	20	–	–	–
tripos	1320	1160	870	340	–	20	20	–

Table 4.2–4 (*Continued*)

COUNTS OF PHYTOPLANKTON AND PROTISTANS (AS INFUSORIA) FROM THE INDICATED VOLUMES OF SEA WATER TAKEN FROM VERTICAL SERIAL SAMPLES IN SEVERAL NORWEGIAN FJORDS

	St. 4, August 29				St. 3, August 28			
Dinophysis acuminata	80	220	120	–	–	–	–	–
acuta	200	280	220	100	–	–	–	–
rotundata	20	–	–	20	–	–	–	–
Gonyaulax polyedra	140	180	220	80	–	–	–	–
spinifera	–	20	–	–				
Peridinium conicum	–	–	–	20	–	–	–	–
depressum	–	–	30	–	–	–	–	–
divergens	640	740	590	100	–	–	–	–
pallidum	–	20	20	–	–	–	–	40
Steinii	–	–	–	–	–	–	–	80
Prorocentrum micans	9240	6600	6200	3600	260	360	40	80
Protoceratium reticulatum	20	–	20	–	–	–	–	–
Torodinium robustum	–	–	–	–	80	–	–	–

Infusoria

Laboea conica	660	560	1120	240	60	20	–	–
delicatissima	–	–	100	20	–	–	–	–
emergens	100	260	360	60	40	40	40	100
strobila	50	180	100	–	–	–	–	–
Lohmanniella oviformis	180	40	20	20	60	20	–	60
Mesodinium	72,500	77,100	52,200	7420	260	140	–	–
Tintinnopsis campanula	20	–	–	–	–	–	–	–
Infusoria indeterminata	460	360	340	240	20	20	–	–
Nauplii of Copepoda	80	100	140	60	100	80	–	–

Oslofjord, W. of Mölen, September 5, 1917

Depth about 200 m

Depth, m	1	5	10	20	30	40	50	75
Temperature, °C	16.95	17.09	17.24	15.92	14.56	13.26	12.88	5.82
Salinity, °/oo	14.43	20.20	23.87	29.84	31.15	31.87	32.64	33.08
Density, σ_t	9.87	14.22	16.98	21.82	23.11	23.96	24.61	26.08
O_2, cc/l	6.11	6.10	5.36	5.54	5.63	5.72	5.70	6.24
O_2, °/o	95	99	89	93	93	93	93	88
P_H	8.12	8.16	8.08	8.08	8.08	8.06	8.06	8.04
Oxidizability: No. cc $\frac{n}{100}$ KMnO$_4$ used/l	81.0	53.6	36.5	25.7	13.6	24.2	–	17.8
No. cc O_2 used/l	4.52	2.99	2.04	1.44	0.76	1.35	–	0.99
No. cc centrifuged	100	50	100	50	50	50	50	50
Halosphaera viridis	–	–	30	40	20	20	–	–

Diatoms

Cerataulina Bergonii	20	–	–	–	–	–	–	–
Chaetoceras curvisetum	200	220	–	–	–	–	–	–
Coscinodiscus radiatus	140	320	20	20	–	20	20	20
Leptocylindrus danicus	60	100	–	–	–	–	–	–
Melosira Borreri	320	–	–	–	–	–	–	–
Rhizosolenia fragilissima	140	–	–	–	–	–	–	–

Table 4.2—4 (*Continued*)
COUNTS OF PHYTOPLANKTON AND PROTISTANS (AS INFUSORIA) FROM THE INDICATED VOLUMES OF SEA WATER TAKEN FROM VERTICAL SERIAL SAMPLES IN SEVERAL NORWEGIAN FJORDS

Peridiniales

Ceratium bucephalum	–	80	50	–	–	–	–	–
furca	70	–	–	–	–	–	–	–
fusus	570	80	50	–	–	–	–	–
longipes	–	–	10	–	–	–	–	–
macroceros	10	–	10	–	–	–	–	–
tripos	210	20	20	–	–	–	–	–
Dinophysis acuminata	40	–	20	–	–	40	–	–
acuta	410	–	–	–	–	20	–	–
norvegica	20	–	–	–	–	20	–	–
rotundata	–	–	–	–	–	20	–	–
Gonyaulax polyedra	60	–	–	–	–	–	–	–
Gymnodinium sp.	40	20	20	–	–	–	–	–
Peridinium conicum	20	–	–	–	–	–	–	–
divergens	330	20	–	–	–	20	20	20
pallidum	20	–	–	–	–	–	–	–
Steinii	20	20	20	–	–	20	–	20
Prorocentrum micans	2560	840	840	40	–	40	60	100

Infusoria

Laboea conica	160	180	–	–	–	–	–	–
emergens	–	220	80	–	–	20	40	–
strobila	20	–	–	–	–	–	–	40
Lohmanniella oviformis	200	20	80	–	–	20	20	40
Leprotintinnus sp.	20	–	–	–	–	–	–	–
Mesodinium	120	500	240	20	60	140	40	140
Tintinnopsis campanula	20	40	–	–	–	–	–	–
sp.	–	80	40	20	20	80	40	–
Infusoria indeterminata	200	140	60	–	–	–	–	–
Nauplii of Copepoda	90	60	20	40	–	–	20	–

Svelvikfjord, by Knivsfjeld, September 5, 1917

Depth 16 m

Depth, m	1	5	10	15
Temperature, °C	16.05	17.56	18.68	16.05
Salinity, °/$_{oo}$	7.16	19.77	24.14	27.10
Density, σ_t	4.49	13.80	16.87	19.70
O$_2$, cc/l	6.02	5.84	5.23	4.91
O$_2$, °/$_o$	88.5	95	89	81
P$_H$	7.68	8.08	8.06	8.04
Oxidizability: No. cc $\frac{n}{100}$ KMnO$_4$ used/l	100.0	65.0	36.2	35.0
No. cc O$_2$ used/l	5.59	3.63	2.02	1.96
No. cc centrifuged	100	50	50	50
Halosphaera viridis	–	–	20	–

Table 4.2–4 (*Continued*)

COUNTS OF PHYTOPLANKTON AND PROTISTANS (AS INFUSORIA) FROM THE INDICATED VOLUMES OF SEA WATER TAKEN FROM VERTICAL SERIAL SAMPLES IN SEVERAL NORWEGIAN FJORDS

Diatoms

Chaetoceras curvisetum	–	600	–	–
Coscinodiscus radiatus	40	320	40	20
Melosira distans	–	20	20	–
Nitzschia sigma	–	–	–	80
Pleurosigma sp.	–	–	20	–
Tabellaria flocculosa	–	40	–	–

Peridiniales

Ceratium bucephalum	–	40	20	–
fusus	20	100	–	20
birundinella	–	20	–	–
longipes	20	–	–	–
Dinophysis acuminata	–	20	–	–
rotundata	–	60	–	–
Gymnodinium sp.	–	220	–	–
Peridinium conicum	–	20	–	–
divergens	–	60	–	20
pallidum	–	40	–	–
Steinii	–	–	20	–
sp. (brown)	20	–	–	–
Prorocentrum micans	40	20	–	–

Infusoria

Laboea emergens	–	80	–	–
sp.	–	80	–	–
Lohmanniella oviformis	40	160	20	40
Tintinnopsis beroidea	–	140	20	–
campanula	–	80	–	–
sp.	–	460	140	–
Infusoria indeterminata	–	20	20	–
Nauplii of Copepoda	–	60	60	–

Drammensfjord, by Hernaestangen, September 5, 1917

Depth, m	1	5	10	15	20	30
Temperature, °C	15.10	15.02	16.10	12.36	7.18	4.65
Salinity, °/oo	0.21	0.25	11.40	20.32	26.06	29.25
Density, σ_t	0.55	0.53	7.72	15.22	20.40	23.19
O_2, cc/l	7.00	6.94	4.68	4.51	3.11	0.77
O_2, °/o	97	96	71	67	43	10
P_H	7.58	7.58	7.49	7.59	7.40	7.40
Oxidizability:						
No. cc $\frac{n}{100}$ KMnO₄ used/l	100.0	118.0	78.8	48.0	40.0	33.2
No. cc O_2 used/l	5.60	6.60	4.40	2.68	2.23	1.85
No. cc centrifuged	50	50	50	50	50	50

Table 4.2—4 (*Continued*)

COUNTS OF PHYTOPLANKTON AND PROTISTANS (AS INFUSORIA) FROM THE INDICATED VOLUMES OF SEA WATER TAKEN FROM VERTICAL SERIAL SAMPLES IN SEVERAL NORWEGIAN FJORDS

Diatoms

Coscinodiscus radiatus	20	–	20	–	–	–
Cyclotella sp.	–	20	–	–	–	–
Navicula sp.	–	60	–	–	–	–
Tabellaria flocculosa	–	200	–	–	–	–

Peridiniales

Ceratium fusus	–	–	–	–	40	–
hirundinella	60	40	–	–	–	–
Dinophysis acuminata	–	–	60	–	–	–
acuta	–	–	40	–	–	–
Gymnodinium sp.	–	–	80	20	20	–
Peridinium depressum	–	–	–	–	20	–
sp. (brown)	–	40	100	–	–	–

Infusoria

Laboea emergens	220	–	–	–	–	–
Lohmanniella oviformis	810	280	120	20	–	60
Tintinnopsis sp.	40	–	240	–	–	–
Infusoria indeterminata	20	–	40	–	–	–
Nauplii of Copepoda	–	–	–	20	–	–

Drøbak, September 6, 1917

Depth, m	1	5	10	20	30	40
Temperature, °C	16.75	16.36	15.30	9.88	8.06	7.56
Salinity, °/$_{oo}$	19.64	20.21	26.59	29.14	30.63	31.26
Density, σ_t	13.87	14.39	19.49	22.43	23.86	24.42
O_2, cc/l	6.22	5.96	5.27	4.94	4.87	5.04
O_2, °/$_o$	100	95	86	74	71	73
P_H	8.19	8.13	7.99	7.87	7.79	7.79
Oxidizability:						
No. cc $\frac{n}{100}$ KMnO$_4$ used/l	58.0	48.4	29.4	24.7	21.2	23.2
No. cc O_2 used/l	3.23	2.70	1.64	1.38	1.18	1.29
No. cc centrifuged	100	100	50	50	50	50
Halosphaera viridis	–	10	60	–	–	–
Distephanus speculum	20	–	–	–	–	–

Diatoms

Chaetoceras curvisetum	100	40	40	180	–	–
Cerataulina Bergonii	–	–	–	–	20	–
Coscinodiscus radiatus	1040	920	20	40	40	20
Rhizosolenia alata	–	60	–	–	–	–

Table 4.2–4 (*Continued*)

COUNTS OF PHYTOPLANKTON AND PROTISTANS (AS INFUSORIA) FROM THE INDICATED VOLUMES OF SEA WATER TAKEN FROM VERTICAL SERIAL SAMPLES IN SEVERAL NORWEGIAN FJORDS

Peridiniales

Ceratium furca	270	70	–	–	–	20
fusus	1230	880	60	–	–	–
longipes	–	–	20	–	–	–
macroceros	–	–	20	-	–	–
tripos	980	420	–	–	–	–·
Dinophysis acuminata	1180	20	–	–	–	–
acuta	1020	380	100	–	–	–
rotundata	100	40	–	–	–	–
Gonyaulax polyedra	380	80	–	–	–	–
spinifera	30	–	–	–	–	–
Gymnodinium sp.	–	–	40	–	–	–
Peridinium depressum	–	–	–	–	–	20
divergens	1250	220	–	–	–	20
pallidum	130	40	20	–	–	–
pellucidum	60	–	–	–	–	–
pyriforme	–	20	–	–	–	–
Steinii	120	70	40	–	–	–
Prorocentrum micans	5540	2340	600	–	–	–
Torodinium robustum	–	–	80	40	40	20

Infusoria

Amphorella subulata	70	–	–	–	–	–
Cyttarocylis annulata	40	–	–	–	–	–
Claparedei	20	–	–	–	–	–
Laboea conica	400	1700	60	–	–	–
delicatissima	–	80	–	–	–	–
emergens	320	1220	40	40	–	–
strobila	50	40	20	–	–	–
Lohmanniella oviformis	300	300	40	20	120	100
Mesodinium	6880	4760	80	100	200	940
Infusoria indeterminata	1080	620	60	–	–	–
Nauplii of Copepoda	70	60	20	80	20	..

Drøbak, September 22, 1917

Depth, m	1	5	10	20	30	40	50	70
Temperature, °C	14.2	14.2	14.0	12.9	12.22	11.96	9.78	8.11
Salinity, °/$_{oo}$	20.97	21.41	26.21	29.73	31.22	31.75	31.825	32.12
Density, σ_t	15.38	15.73	19.44	22.36	23.64	24.09	24.54	25.02
O_2, cc/l	6.16	6.12	5.70	5.13	5.20	5.50	5.45	5.64
O_2, °/$_o$	95	95	91	82	82	87	82	82
P_H	8.13	8.13	8.10	8.01	7.98	7.98	7.94	7.92
Oxidizability: No. cc $\frac{n}{100}$ KMnO$_4$ used/l	36.5	34.1	17.1	10.5	10.7	10.2	8.2	10.8
No. cc O_2 used/l	2.04	1.91	0.96	0.59	0.60	0.59	0.45	0.61
No. cc centrifuged	50	100	50	50	50	50	50	50
Distephanus speculum	140	100	–	–	20	..	..	–

Table 4.2—4 (*Continued*)

COUNTS OF PHYTOPLANKTON AND PROTISTANS (AS INFUSORIA) FROM THE INDICATED VOLUMES OF SEA WATER TAKEN FROM VERTICAL SERIAL SAMPLES IN SEVERAL NORWEGIAN FJORDS

Diatoms

Chaetoceras curvisetum	1060	560	100	–	–	–	–	–
Coscinodiscus radiatus	3920	2910	1590	760	260	100	40	40
Guinardia flaccida	–	–	–	20	–	–	–	–
Leptocylindrus danicus	40	360	240	–	–	–	–	–
Rhizosolenia fragilissima	800	340	790	–	–	–	–	–
Skeletonema costatum	–	--	60	–	–	–	–	–

Peridiniales

Ceratium furca	4340	1880	410	–	–	–	–	–
fusus	5620	3970	370	60	20	–	–	–
longipes	–	10	10	–	–	–	–	–
tripos	960	2260	120	–	–	–	–	–
Dinophysis acuminata	80	20	--	–	–	–	–	–
acuta	1660	1760	140	–	20	–	–	–
rotundata	20	–	–	–	–	–	–	–
Gonyaulax polyedra	320	100	20	–	–	–	–	–
spinifera	40	20	–	–	–	–	–	–
Gymnodinium Lohmanni	–	80	60	20	20	20	–	–
Peridinium achromaticum	–	–	20	–	–	–	–	–
divergens	280	100	140	–	–	–	–	–
pallidum	–	–	–	20	–	–	–	–
Steinii	80	180	40	20	–	–	–	–
Prorocentrum micans	16,600	11,180	3360	480	20	–	20	–
Torodinium robustum	–	--	60	–	–	–	–	–

Infusoria

Laboea conica	580	440	80	–	–	–	–	–
emergens	–	80	160	80	20	--	–	–
strobila	100	40	–	–	–	–	–	–
vestita	1340	80	20	–	–	–	–	–
Lohmanniella oviformis	80	100	40	–	–	–	–	–
spiralis	–	–	140	–	–	–	–	–
Amphorella subulata	–	60	–	–	–	–	–	–
Infusoria indeterminata	60	160	–	–	–	–	–	–
Mesodinium	1660	220	60	60	20	–	–	–
Nauplii of Copepoda	40	20	20	–	–	–	–	–

4.3 Chemical Composition

Table 4.3—1
ELEMENTARY COMPOSITION OF SOME
AQUATIC PLANTS IN PERCENT ASH-FREE DRY
WEIGHT ± STANDARD DEVIATION

Organism and reference	No. analyses	C	H	O	N	P
Diatoms	18	50.54	10.21	28.83	7.0	1.55
Brandt and Raben, 1920		±3.90	±1.90	±7.10	±3.7	±0.88
Ketchum and Redfield, 1949						
Vinogradov, 1939						
Peridineans	7	48.12	7.50	33.85	10.40	0.80
Brandt and Raben, 1920		±1.90	±1.70	±3.90	±1.40	±.36
Chlorophyceae	18	54.55	7.54	31.21	7.7	2.94
Ketchum and Redfield, 1949		±1.30	±0.50	±4.35	±0.56	±0.83

(From Ryther, J. H., The measurement of primary production, *Limnol. Oceanogr.*, 1(2), 73, 1956. With permission.)

Table 4.3—2
RELATIVE AMOUNTS OF AMINO ACIDS IN
MEMBERS OF THE OCEANIC FOOD CHAIN

	G amino acid/16 g amino acid N		
	Phytoplankton	*Calanus*	Cod[1]
Glutamic acid	13.7	14.5	16.6
Aspartic acid	11.7	11.5	10.6
Lysine	10.4	8.9	10.3
Glycine	9.4	8.6	5.1
Leucine	9.3	9.3	9.3
Alanine	9.1	9.5	7.2
Valine	7.9	7.3	5.8
Serine	6.8	5.2	5.4
Arginine	6.8	7.8	6.7
Threonine	6.5	5.7	5.3
Phenylalanine	5.9	4.3	4.7
Isoleucine	5.7	4.9	4.9
Proline	5.3	4.5	4.2
Tyrosine	3.7	4.7	4.0
Histidine	2.1	1.9	3.5

[1] Data from Connell and Howgate, 1959.

Table 4.3–3

RELATIVE MOLAR QUANTITIES OF AMINO ACIDS IN PLANKTON FROM PLYMOUTH SOUND, 1960, AND IN A SPECIES OF CULTURED ALGAE

	1960 Calanus			1960 Phytoplankton				Chlorella vulgaris*	1961 Calanus			1961 Phytoplankton				
	April/May	June/July	August/Sept	Feb/March	April/May	June/July	August/Sept		April	May	June	April	May	June	July	August
Taurine	0.12	0.10	0.20	–	–	–	–	–	0.19	0.20	0.25	–	–	–	–	–
Aspartic acid	1.22	1.15	1.22	1.17	1.43	1.20	1.02	1.05	1.10	1.25	1.21	1.30	1.19	0.71	1.03	0.95
Threonine	0.68	0.77	0.62	0.86	0.89	0.64	0.83	0.48	0.60	0.59	0.69	0.80	0.68	0.71	0.67	0.64
Serine	0.74	1.06	0.67	1.17	1.43	0.68	1.12	0.54	0.68	0.53	0.64	1.36	1.15	1.17	1.21	1.00
Glutamic acid	1.45	1.33	1.41	1.75	1.43	1.50	1.04	1.28	1.58	1.36	1.49	1.10	1.12	0.71	1.04	0.95
Proline	0.59	0.60	0.45	0.78	0.67	0.68	0.64	1.18	0.55	0.44	0.56	0.68	0.51	0.39	0.59	0.52
Glycine	1.61	1.63	1.63	2.08	2.05	1.80	1.51	1.02	1.51	1.54	1.62	1.90	1.85	1.73	2.04	1.86
Alanine	1.49	1.46	1.57	1.17	1.43	1.48	1.24	1.26	1.48	1.34	1.34	1.10	1.13	0.92	1.13	1.27
Valine	0.79	0.91	0.76	0.61	0.89	0.86	0.94	0.90	0.93	0.97	0.84	0.90	0.97	0.96	0.78	0.80
Isoleucine	0.53	0.56	0.50	0.48	0.59	0.61	0.67	0.57	0.60	0.62	0.56	0.55	0.65	0.75	0.57	0.61
Tyrosine	0.44	0.32	0.35	0.17	0.14	0.35	0.38	0.46	0.38	0.40	0.39	0.26	0.32	0.32	0.26	0.21
Phenylalanine	0.41	0.40	0.32	0.49	0.54	0.39	0.62	0.46	0.37	0.44	0.40	0.33	0.47	0.45	0.43	0.53
Lysine	0.90	0.80	0.88	0.86	0.96	0.95	1.06	0.84	0.99	0.94	0.87	0.50	0.77	0.77	0.83	0.62
Histidine	0.18	0.14	0.18	0.14	0.15	0.17	0.13	0.18	0.22	0.17	0.15	0.17	0.05	0.17	0.12	0.04
Arginine	0.66	0.60	0.62	0.47	0.36	0.74	0.56	0.65	0.71	0.62	0.79	0.52	0.59	0.86	0.54	0.52

Note: Leucine = 1.00.

* Recalculated from the data of Fowden, 1952.

Table 4.3—4
GROSS COMPOSITION OF PHYTOPLANKTON AND A COPEPOD FROM EARLY SPRING TO FALL, 1957, INSHORE NORTH ATLANTIC

	Phytoplankton					Calanus		
	Particulate matter μg dry wt/l	Particulate organic matter μg dry wt/l	N content of particulate matter μg/100 μg dry wt	Protein (= N × 6.25) content of particulate matter μg/100 μg dry wt	Amino acid content of particulate matter μg/100 μg dry wt	Mg N/100 mg dry wt	Protein (= N × 6.25) mg/100 mg dry wt	Mg amino acids/100 mg dry wt
February/March	–	–	0.66	4.12	3.0	9.8	61.3	–
April/May	2,140	870	0.96	6.00	4.2	12.3	76.9	35
June/July	2,040	870	1.41	8.81	8.0	9.6	60.0	20
August/September	1,770	960	0.73	4.58	2.0	11.2	70.0	26

Table 4.3—5
AVERAGE CONCENTRATIONS OF CARBON IN
PHYTOPLANKTON, HERBIVORES, AND CARNIVORES

On the Continental Shelf of the North Atlantic
1956—1958

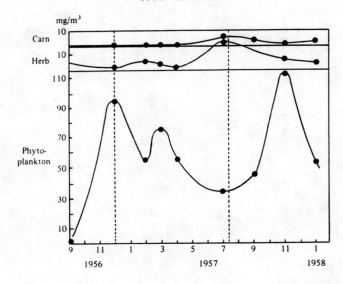

Table 4.3—6
AVERAGE CONCENTRATIONS OF PHOSPHORUS IN
THREE TROPHIC LEVELS AND DISSOLVED PHOSPHORUS IN
SEA WATER

On the Continental Shelf of the North Atlantic
1956—1958

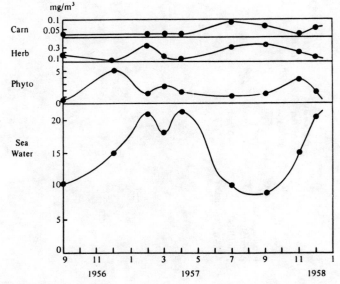

Table 4.3–7
AVERAGE CONCENTRATIONS OF NITROGEN IN
THREE TROPHIC LEVELS AND AVAILABLE NITROGEN IN
SEA WATER*

On the Continental Shelf of the North Atlantic
1956–1958

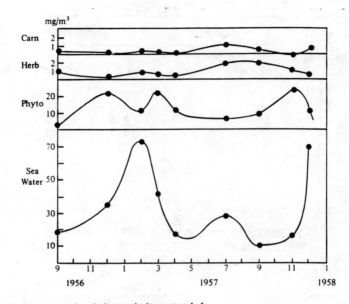

*As ammonia, nitrite, and nitrate totaled.

Table 4.3—8
CONTRIBUTIONS OF ZOOPLANKTON AND PHYTOPLANKTON
TO TOTAL PARTICULATE PHOSPHORUS IN SEA WATER

On the Continental Shelf of the North Atlantic
1957

| | | Avg particulate phosphorus | | | | | |
| | | Phytoplankton | | Zooplankton | | Detritus (by difference) | |
Date	Sea water mg P/m³	mg P/m³	% part P	mg P/m³	% part P	mg P/m³	% part P
February, 1957	3.6	2.94	82	0.04	0.01	0.62	17.2
April, 1957	4.1	1.79	44	0.07	0.02	2.24	54.6
July, 1957	3.4	1.09	32	0.38	11.2	1.93	56.7
September, 1957	3.2	1.32	41	0.39	12.2	1.49	46.5

Table 4.3—9
VARIATIONS IN THE RELATIVE COMPOSITION
AND CALCULATED PHOTOSYNTHETIC QUOTIENT[1]
IN *CHLORELLA PYRENOIDOSA*

| | Calculated PQ nitrogen source | | |
Relative composition by atoms[2] C-H-O-N	Not consid.	NH_4^+	NO_3^-
100–165–61–16	1.11	0.99	1.22
100–162–45–16	1.18		
100–171–39–23	1.23		
100–160–38–16	1.21		
100–173–43–8	1.21		
100–178–36–16	1.22		
100–168–37–12	1.26		
100–172–33–6	1.28		
100–168–28–4	1.31		
100–177–27–2.2	1.33		
100–179–20–1.3	1.34		
100–180–19–1.7	1.35	1.34	1.37

[1] $O_2/ - CO_2$.
[2] C = 100.

(From Ryther, J. H., The measurement of primary production, *Limnol. Oceanogr.*, 1(2), 75, 1956. With permission.)

Table 4.3—10
THE RELATIVE COMPOSITION BY ATOMS OF
C,[1] H, AND O, AND THE CALCULATED
PHOTOSYNTHETIC QUOTIENT[2] IN
VARIOUS PHYTOPLANKTERS

Organism	No. analyses	C-H-O	PQ
Diatoms	8	100—251—48	1.38
Peridians	7	100—184—54	1.16
Stichococcus bacillaria	6	100—160—46	1.17
Chlorella pyrenoidosa	8	100—168—41	1.21
Chlorella vulgaris	1	100—168—44	1.20
Scenedesmus obliquus 1	1	100—180—41	1.25
Scenedesmus obliquus 2	1	100—180—39	1.25
Scenedesmus basilensis	1	100—184—39	1.27
Nitzschia closterium	1	100—164—42	1.20

[1] C = 100.
[2] $O_2 / - CO_2$.

(From Ryther, J. H., The measurement of primary production, *Limnol. Oceanogr.*, 1 (2), 75, 1956. With permission.)

Table 4.3—11
TOTAL CARBON, CARBONATE CARBON, AND ORGANIC CONCENTRATION
IN PHYTOPLANKTONIC MARINE ORGANISMS

Phytoplankton, net collections		Total carbon, % dry wt	Carbonate carbon, % dry wt	Organic carbon	
				% dry wt	% a-f dry wt
12a	Phytoplankton (diatoms)	12.0	—	12.0	40.0
33a	Phytoplankton (diatoms)	7.1	—	7.1	35.3
36b	*Sargassum* sp.	31.5	—	31.5	39.3
	Average	16.8	—	38.2	

Note: All figures given are % of dry wt, excepting last column, which is ash-free dry wt. Where no results are shown, inorganic carbonate was not detectable.

Table 4.3—12
VITAMIN CONTENT OF ALGAE

Species	Vitamin	Quantities/100g dry matter	Ref.
Chlorella pyrenoidosa	thiamine	1—4.1 mg	1
	riboflavin	3.6—8 mg	
	nicotinic ac.	12—24 mg	
	pyridoxine	2.3 mg	
	pantothenic ac.	0.8—2.0 mg	
	biotin	14.8 µg	
	choline	300 mg	
	B_{12} *(E.g.)**	2.2—10 µg	
Chlorella vulgaris	B_{12} *(E.c)**	6.3 µg	2
Chlorella ellipsoidea	B_{12} *(E.g.)**	4.2—8.9 µg	3

Note: See also Kanazawa, A., *Mem. Fac. Fish. Kagoshima Univ.*, 10, 38, 1962.

*B_{12} as assayed with microorganism: *E.g.* = *Euglena gracilis*, *E.c.* = *E. coli*.

(From Provasoli, L., Organic regulation of phytoplankton fertility, *The Sea*, Vol. 2, Hill, M. N., Ed., Interscience, New York, 1963, 182. With permission.)

REFERENCES

1. Combs, C. F., *Science*, 116, 453, 1952.
2. Brown, F., Cuthbertson, W. F. J., and Fogg, G. E., *Nature*, 177, 188, 1955.
3. Hashimoto, Y., *J. Vitaminol.*, 1, 49, 1954.

Table 4.3–13
PRODUCTION OF CARBOHYDRATES IN MARINE ALGAE

Species	Carbohydrate produced (mg/l)	
	Before maximal growth	Highest value (stationary phase)
CHLOROPHYTES		
Dunaliella euchlora	3.1	9.0
Chlorella sp. (No. 580, Indiana Univ. Cult. Coll.)	–	9.0
Chlamydomonas sp. ("Y" R. Lewin)	2.1	10.6
Chlorococcum sp.	–	27.0
Pyramimonas inconstans	2.8	5.4
DIATOMS		
Cyclotella sp.	–	1.5
Nitzschia brevirostris	–	25.6
Melosira sp.	–	60.0
CHRYSOMONADS and CRYPTOMONADS		
Isochrysis galbana	–	25.0
Monochrysis lutheri	1.7	15.7
Prymnesium parvum	5.8–15.9	123.0
Rhodomonas sp.	1.9	8.8
DINOFLAGELLATES		
Amphidinium carteri	–	>5.0
Prorocentrum sp.[1]	–	~20.0
Katodinium dorsalisulcum[2]	–	0.6–2.4 g/l

(From Guillard, R. R. L. and Wangersky, P. J., the production of extracellular carbohydrates by some marine flagellates, *Limnol. Oceanogr.*, 3, 449, 1958. With permission.)

REFERENCES

1. **Collier, A.**, *Limnol. Oceanogr.*, 3, 33, 1958.
2. **McLaughlin, J. J. A., Zahl, P. A., Novak, Z., Marchisotto, J., and Prager, J.**, *Ann. N. Y. Acad. Sci.*, 90, 856, 1960.

Figure 2.1–1 WORLD CATCH (1968) AND ESTIMATED POTENTIALS BY REGIONS

Letters Refer To Suffix Designations Of Regional Maps In Sections 2.2, 2.2, 2.3, 2.5 And Corresponding Areas In The Composite Map Of Figure 2.6–1

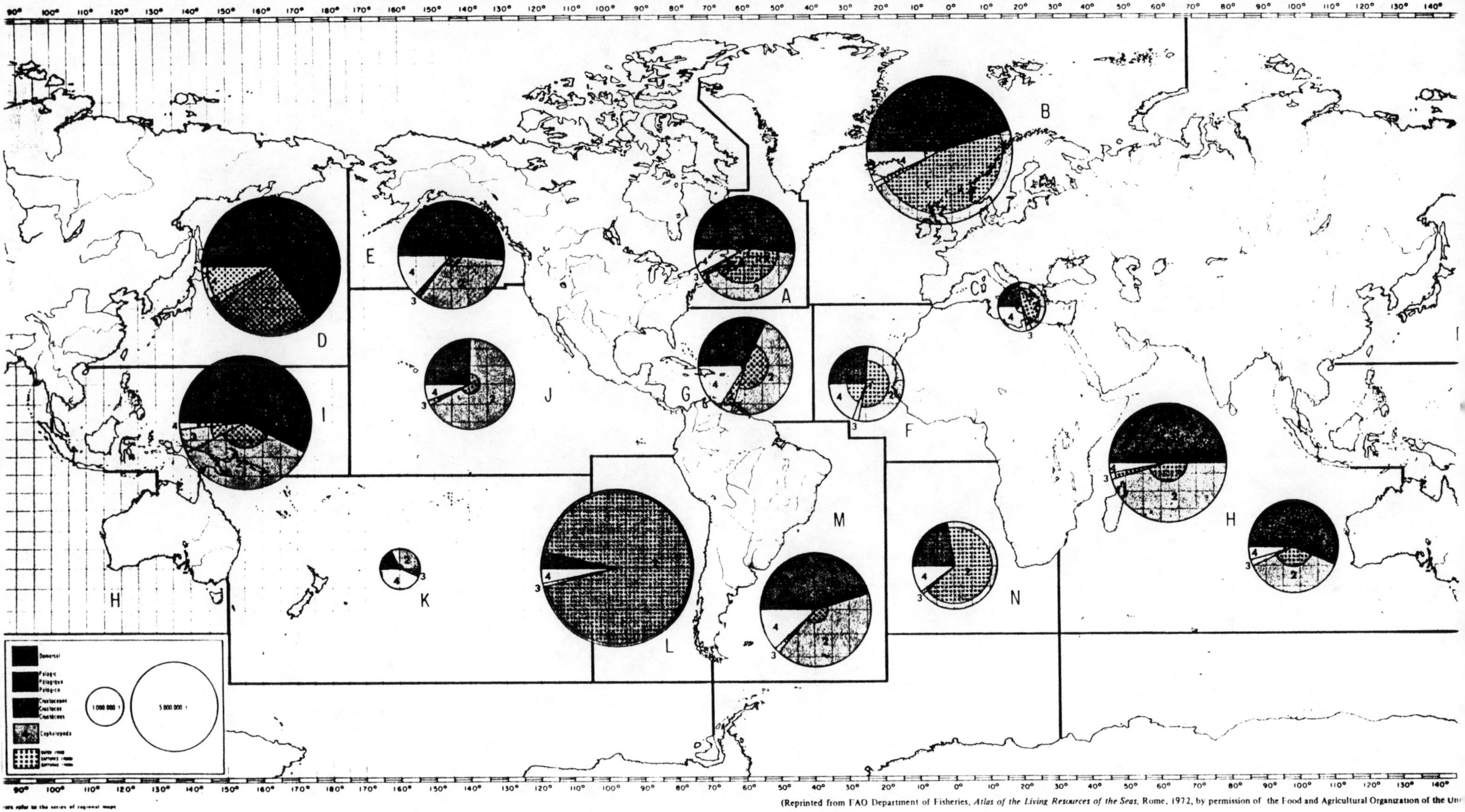

(Reprinted from FAO Department of Fisheries, Atlas of the Living Resources of the Seas, Rome, 1972, by permission of the Food and Agricultural Organization of the Un

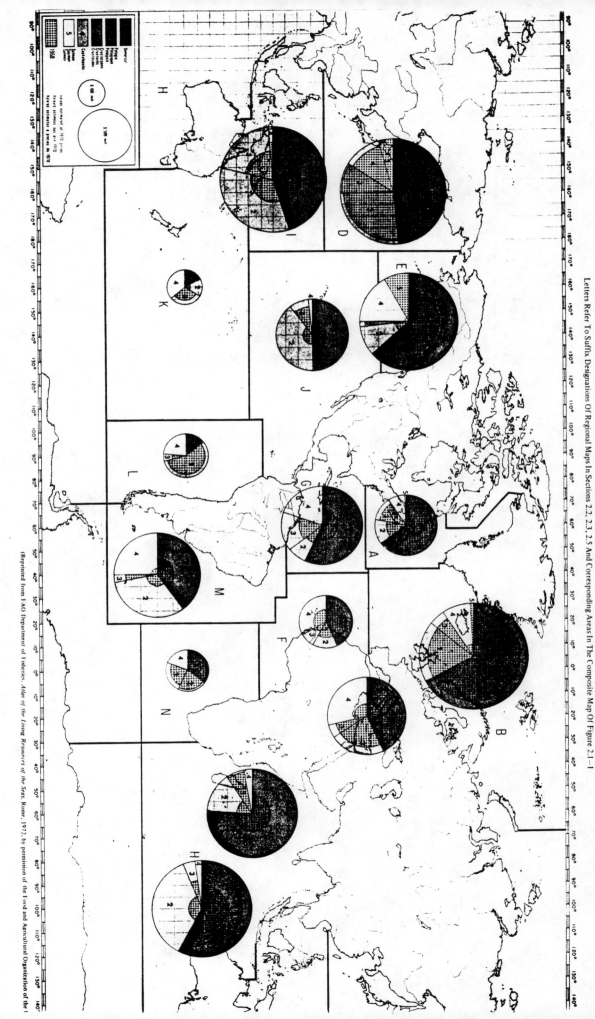

Figure 2.6-1   VALUE OF WORLD CATCH (1968) AND ESTIMATED POTENTIALS BY REGION

Letters Refer To Suffix Designations Of Regional Maps In Sections 2.2, 2.3, 2.5 And Corresponding Areas In The Composite Map Of Figure 2.1-1

(Reprinted from FAO Department of Fisheries, *Atlas of the Living Resources of the Seas*, Rome, 1972, by permission of the Food and Agricultural Organization of the U

Figure 3.1-1 DISTRIBUTION OF THE ABUNDANCE OF ZOOPLANKTON

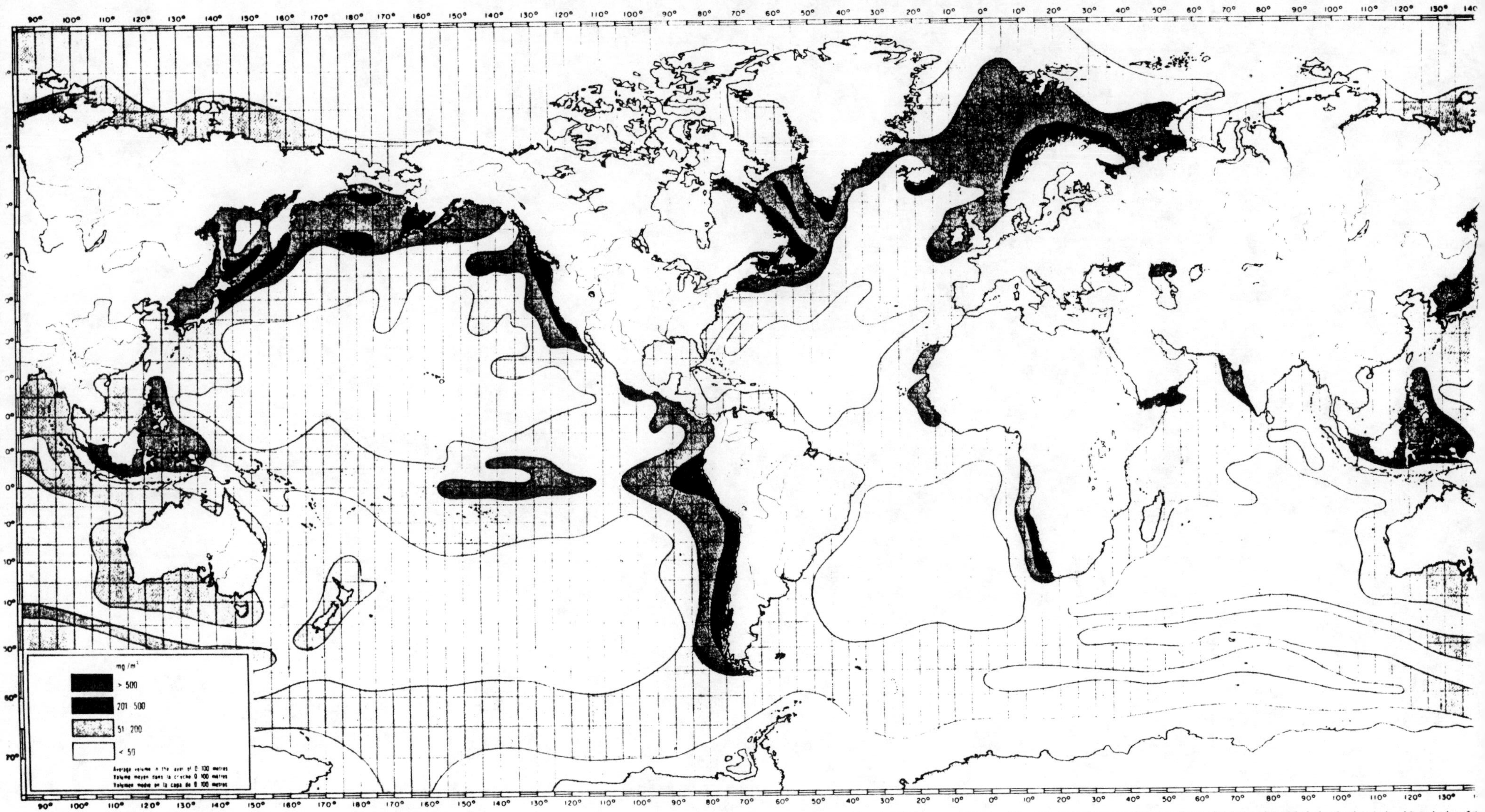

(Reprinted from FAO Department of Fisheries, *Atlas of the Living Resources of the Seas*, Rome, 1972, by permission of the Food and Agricultural Organization of the United Nations)

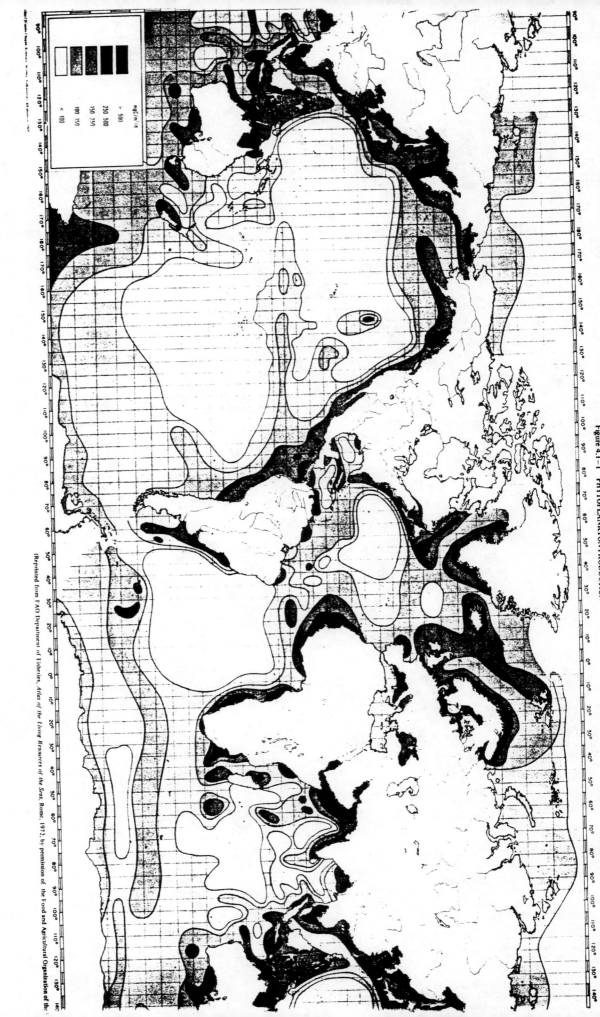

Figure 4.1–1 PHYTOPLANKTON PRODUCTION

(Reprinted from FAO Department of Fisheries, *Atlas of the Living Resources of the Seas*, Rome, 1972, by permission of the Food and Agricultural Organization of the United Nations)

mgC/m²/d

> 500

250–500

150–250

100–150

< 100

INDEX